SEMA SIMAI Springer Series

Volume 29

As of 2013, the SIMAI Springer Series opens to SEMA in order to publish a joint series aiming to publish advanced textbooks, research-level monographs and collected works that focus on applications of mathematics to social and industrial problems, including biology, medicine, engineering, environment and finance. Mathematical and numerical modeling is playing a crucial role in the solution of the complex and interrelated problems faced nowadays not only by researchers operating in the field of basic sciences, but also in more directly applied and industrial sectors. This series is meant to host selected contributions focusing on the relevance of mathematics in real life applications and to provide useful reference material to students, academic and industrial researchers at an international level. Interdisciplinary contributions, showing a fruitful collaboration of mathematicians with researchers of other fields to address complex applications, are welcomed in this series.

THE SERIES IS INDEXED IN SCOPUS

Domingo Barrera • Sara Remogna • Driss Sbibih
Editors

Mathematical and Computational Methods for Modelling, Approximation and Simulation

Springer

Editors
Domingo Barrera
Department of Applied Mathematics
University of Granada
Granada, Spain

Sara Remogna
Dipartimento di Matematica "Giuseppe Peano"
Università degli Studi di Torino
Torino, Italy

Driss Sbibih
Faculty of Sciences, LANO Laboratory
Mohammed First University
Oujda, Morocco

ISSN 2199-3041 ISSN 2199-305X (electronic)
SEMA SIMAI Springer Series
ISBN 978-3-030-94338-7 ISBN 978-3-030-94339-4 (eBook)
https://doi.org/10.1007/978-3-030-94339-4

This Springer imprint is published by the registered company Springer Nature Switzerland AG
The registered company address is: Gewerbestrasse 11, 6330 Cham, Switzerland

Preface

This volume contains three plenary lectures presented at the International Conference on Mathematical and Computational Modelling, Approximation and Simulation (MACMAS 2019) which was held from 9 to 11 September 2019 in Granada, Spain, as well as nine contributed papers from the different topics covered at the conference.

MACMAS 2019 was jointly organized by the University of Granada (Spain), the Mohammed I University (Morocco), and the University of Torino (Italy).

MACMAS is a new phase in the series of international conferences on approximation methods and numerical modelling in environment and natural resources held every 2 years from 2005 to 2017. The following are among its goals:

- Gather the researchers interested on the fields of approximation, numerical modelling, and their applications.
- Reinforce the scientific interchange between researches from different countries, with special emphasis on the Mediterranean basin and neighborhood.
- Encourage young researchers to present the results of their scientific works.
- Help new international research projects to arise by facilitating contacts between tentative partners.

Stefano De Marchi, Pablo Pedregal, and Alessandra Sestini gave plenary lectures, sharing their research results with the participants.

Stefano de Marchi is an associate professor at the University of Padova (Italy), with habilitation to Full Professor of Numerical Analysis. Since 2005, he has been the coordinator of the "CAA Research Group" (Constructive Approximation and Applications), between the Universities of Verona and Padova. In addition, he has been responsible of the Italian Mathematical Union Thematic Group on Approximation Theory and Applications. His current main research topics are mapped bases approximation, RBF variably scaled discontinuous kernels, RBF and meshless approximation, multivariate polynomial approximation, and cubature.

Stefano De Marchi's plenary lecture dealt with the solution of approximation problems by polynomials to reduce Runge and Gibbs phenomena. The main idea is to avoid resampling the function or data and to rely on the mapped polynomials or "fake" nodes approach. This technique turns out to be effective for stability by reducing the Runge effect and also in the presence of discontinuities by almost cancelling the Gibbs phenomenon. This technique is also applied to rational approximation and quadrature.

Pablo Pedregal obtained his bachelor's degree in mathematics in Madrid (Universidad Complutense de Madrid, 1986), and he moved to the United States to pursue his PhD, obtained at the University of Minnesota at the end of 1989, under the direction of David Kinderlehrer. He is currently full professor at the University of Castilla-La Mancha after a period at the Complutense University of Madrid as associate professor. His field of work is variational techniques applied to optimization problems in a very broad sense, covering the calculus of variations, non-convex vector problems, optimal design in continuous media, and optimal control.

Pablo Pedregal's plenary lecture focuses on two specific problems that show the enormous difficulties that arise when studying models that depend on the highly non-linear behavior of a system of partial differential equations. The first one is motivated by inverse problems in conductivity and the process to recover an unknown conductivity coefficient from measurements in the boundary. The second addresses an optimal control problem for soft robots. This paper has an outreach character to draw attention to these areas of research.

Alessandra Sestini is associate professor at the University of Firenze (Italy), with habilitation to Full Professor of Numerical Analysis. She has participated in several projects of the "Gruppo Nazionale per il Calcolo Scientifico," having been the responsible researcher in 2012 and 2019. In recent years, she has also conducted her research into the 5-year project "Splines for accUrate NumeRics: adaptIve models for Simulation Environments (SUNRISE)" of the INdAM *Istituto Nazionale di Alta Matematica ‹‹Francesco Severi››*. Her current research interests are numerical methods for approximation and for graphics, spline theory, numerical methods for ODEs, and isogeometric analysis.

Alessandra Sestini's plenary lecture is a collaborative work with Cesare Bracco, Carlotta Giannelli, David Großmann, Sofia Imperatore, and Dominik Mokriš, in which they present a two-stage scattered data fitting with truncated hierarchical B-splines (THB-splines) for the adaptive reconstruction of industrial models. The first stage deals with the computation of local least squares variational spline approximations. The second one involves hierarchical spline quasi-interpolation based on THB-splines to construct the adaptive spline surface approximating the whole scattered data set, and a suitable strategy to guide the adaptive refinement is introduced. A selection of examples on geometric models representing components of aircraft turbine blades highlights the performance of the scheme.

Following the end of MACMAS 2019, a call for papers was opened. The submissions received underwent a peer-review process and nine manuscripts were retained for publication.

The chapters by Abbadi and Ibáñez and Rahouti et al. deal with the construction of univariate quasi-interpolants, which somewhat complement the results in the literature and offer approximation schemes potentially useful in practice. Quasi-interpolation is a subject that has attracted and continues to attract the interest not only of those working in the field of approximation theory but also of those who require tools to be used in practice. An example of this is the chapter by Sestini. The chapters by Allouch et al., Bellour et al., and Barrera et al. use spline quasi-interpolation to numerically solve Fredholm and Hammerstein integral equations, which are relevant areas in terms of the application of numerical methods. The chapter by Mbarki and Oubrahim provides new results on the notion of convexity in probabilistic metric spaces, which are applied to the study of the existence and uniqueness of the solution of a Volterra equation. The chapter by Korikache and Paquet focuses on the completely discretized problem of the dual mixed formulation for the heat diffusion equation in a polygonal domain by the Crank-Nicolson scheme in time. The authors show the existence and the stability, and provide *a priori* error estimates for the solution. The chapter by Akhrif et al. proposes the application of smoothing splines to the study of decomposed statistical series. The general scheme is illustrated by applying it to economic data from Morocco and Venezuela. The chapter by Illescas-Manzano et al. deals with the analysis of poverty measures, which is a topic of increased interest to society. This chapter contributes to the literature by developing percentile ratio estimators when there are missing data.

We want to thank all the contributors who have coauthored the chapters contained in this volume, as well as the anonymous referees who have revised the manuscripts, improving them with their comments and suggestions.

We would also like to express our gratitude to Francesca Bonadei from Springer for her patience, attention, and support at every step of the editorial process.

Granada, Spain — Domingo Barrera
Torino, Italy — Sara Remogna
Oujda, Morocco — Driss Sbibih
April 2021

Contents

Editors and Contributors

About the Editors

Domingo Barrera is Full Professor of Applied Mathematics in the Department of Applied Mathematics at the University of Granada. His current main research line is spline quasi-interpolation and its application to the numerical solution of integral equations and in the use of advanced mathematical techniques for the extraction of parameters in the modelling of nanoelectronic devices and the study of the functional quality of digital models of terrain elevations in engineering.

Sara Remogna is Professor of Numerical Analysis at the University of Torino (Italy). Her research interests include univariate and multivariate spline approximation, numerical methods for computer-aided geometric design, and numerical methods for the solution of differential and integral problems based on spline approximation.

Driss Sbibih is Full Professor of Approximation and Numerical Analysis at the University Mohammed First in Morocco. His research interests are approximation theory by spline functions, computer graphics, and numerical analysis. He has a long experience in international cooperation, especially with universities in Spain, France, and Italy.

Contributors

Abdelaziz Abbadi University Mohammed I, Team ANAA, LANO Laboratory, Oujda, Morocco

Rim Akhrif University Abdelmalek Essaadi, F.S.J.E.S., de Tétouan, Morocco

Chafik Allouch The Multidisciplinary Faculty of Nador, Team of Modeling and Scientific Computing, Nador, Morocco

Antonio Arcos-Rueda Department of Statistics and O. R., University of Granada, Granada, Spain

Domingo Barrera Department of Applied Mathematics, University of Granada, Granada, Spain

Azzeddine Bellour Département de Mathématiques, Ecole Normale Supérieure de Constantine, Constantine, Algeria

Cesare Bracco Dipartimento di Matematica e Informatica "U. Dini", Università degli Studi di Firenze, Firenze, Italy

Stefano De Marchi Department of Mathematics "Tullio Levi-Civita", Padova, Italy

Elvira Delgado Márquez Department of Economics and Statistics, University of León, León, Spain

Carlotta Giannelli Dipartimento di Matematica e Informatica "U. Dini", Università degli Studi di Firenze, Firenze, Italy

David Großmann MTU Aero Engines AG, Munich, Germany

Ikram Hamzaoui The Multidisciplinary Faculty of Nador, Team of Modeling and Scientific Computing, Nador, Morocco

María José Ibáñez Department of Applied Mathematics, University of Granada, Granada, Spain

María Illescas-Manzano Department of Economy and Company, University of Almería, Almería, Spain

Sofia Imperatore Dipartimento di Matematica e Informatica "U. Dini", Università degli Studi di Firenze, Firenze, Italy

Reda Korikache LANO, ESTO, University Mohammed I, Oujda, Morocco

Abdelouahed Kouibia Department of Applied Mathematics, University of Granada, Granada, Spain

Sergio Martínez-Puertas Department of Economy and Company, University of Almería, Almería, Spain

Abderrahim Mbarki ANO Laboratory, National School of Applied Sciences, Mohammed First University, Oujda, Morocco

Dominik Mokriš MTU Aero Engines AG, Munich, Germany

Rachid Oubrahim ANO Laboratory, National School of Applied Sciences, Mohammed First University, Oujda, Morocco

Luc Paquet LAMAV, Polytechnic University Hauts-De-France, Valenciennes, France

Miguel Pasadas Department of Applied Mathematics, University of Granada, Granada, Spain

Pablo Pedregal ETSI Industriales, Ciudad Real, Spain

Afaf Rahouti FSO, Mohammed First University, Oujda, Morocco

María del Mar Rueda Department of Statistics and O. R., University of Granada, Granada, Spain

Abdelmonaim Saou ANAA Team, ANO Laboratory, Faculty of Sciences, Mohammed First University, Oujda, Morocco

Driss Sbibih Mohammed First University, LANO Laboratory, Oujda, Morocco

Abdelhafid Serghini ESTO, Mohammed First University, Oujda, Morocco

Alessandra Sestini Dipartimento di Matematica e Informatica "U. Dini", Università degli Studi di Firenze, Firenze, Italy

Mohamed Tahrichi ANAA Team, ANO Laboratory, Faculty of Sciences, Mohammed First University, Oujda, Morocco

Ahmed Tijini FSO, Mohammed First University, Oujda, Morocco

Ahmed Zidna LGIPM Laboratory, University of Lorraine, Metz, France

Contents

Editors and Contributors

About the Editors

Domingo Barrera is Full Professor of Applied Mathematics in the Department of Applied Mathematics at the University of Granada. His current main research line is spline quasi-interpolation and its application to the numerical solution of integral equations and in the use of advanced mathematical techniques for the extraction of parameters in the modelling of nanoelectronic devices and the study of the functional quality of digital models of terrain elevations in engineering.

Sara Remogna is Professor of Numerical Analysis at the University of Torino (Italy). Her research interests include univariate and multivariate spline approximation, numerical methods for computer-aided geometric design, and numerical methods for the solution of differential and integral problems based on spline approximation.

Driss Sbibih is Full Professor of Approximation and Numerical Analysis at the University Mohammed First in Morocco. His research interests are approximation theory by spline functions, computer graphics, and numerical analysis. He has a long experience in international cooperation, especially with universities in Spain, France, and Italy.

Contributors

Abdelaziz Abbadi University Mohammed I, Team ANAA, LANO Laboratory, Oujda, Morocco

Rim Akhrif University Abdelmalek Essaadi, F.S.J.E.S., de Tétouan, Morocco

Chafik Allouch The Multidisciplinary Faculty of Nador, Team of Modeling and Scientific Computing, Nador, Morocco

Antonio Arcos-Rueda Department of Statistics and O. R., University of Granada, Granada, Spain

Domingo Barrera Department of Applied Mathematics, University of Granada, Granada, Spain

Azzeddine Bellour Département de Mathématiques, Ecole Normale Supérieure de Constantine, Constantine, Algeria

Cesare Bracco Dipartimento di Matematica e Informatica "U. Dini", Università degli Studi di Firenze, Firenze, Italy

Stefano De Marchi Department of Mathematics "Tullio Levi-Civita", Padova, Italy

Elvira Delgado Márquez Department of Economics and Statistics, University of León, León, Spain

Carlotta Giannelli Dipartimento di Matematica e Informatica "U. Dini", Università degli Studi di Firenze, Firenze, Italy

David Großmann MTU Aero Engines AG, Munich, Germany

Ikram Hamzaoui The Multidisciplinary Faculty of Nador, Team of Modeling and Scientific Computing, Nador, Morocco

María José Ibáñez Department of Applied Mathematics, University of Granada, Granada, Spain

María Illescas-Manzano Department of Economy and Company, University of Almería, Almería, Spain

Sofia Imperatore Dipartimento di Matematica e Informatica "U. Dini", Università degli Studi di Firenze, Firenze, Italy

Reda Korikache LANO, ESTO, University Mohammed I, Oujda, Morocco

Abdelouahed Kouibia Department of Applied Mathematics, University of Granada, Granada, Spain

Sergio Martínez-Puertas Department of Economy and Company, University of Almería, Almería, Spain

Abderrahim Mbarki ANO Laboratory, National School of Applied Sciences, Mohammed First University, Oujda, Morocco

Dominik Mokriš MTU Aero Engines AG, Munich, Germany

Rachid Oubrahim ANO Laboratory, National School of Applied Sciences, Mohammed First University, Oujda, Morocco

Luc Paquet LAMAV, Polytechnic University Hauts-De-France, Valenciennes, France

Miguel Pasadas Department of Applied Mathematics, University of Granada, Granada, Spain

Pablo Pedregal ETSI Industriales, Ciudad Real, Spain

Afaf Rahouti FSO, Mohammed First University, Oujda, Morocco

María del Mar Rueda Department of Statistics and O. R., University of Granada, Granada, Spain

Abdelmonaim Saou ANAA Team, ANO Laboratory, Faculty of Sciences, Mohammed First University, Oujda, Morocco

Driss Sbibih Mohammed First University, LANO Laboratory, Oujda, Morocco

Abdelhafid Serghini ESTO, Mohammed First University, Oujda, Morocco

Alessandra Sestini Dipartimento di Matematica e Informatica "U. Dini", Università degli Studi di Firenze, Firenze, Italy

Mohamed Tahrichi ANAA Team, ANO Laboratory, Faculty of Sciences, Mohammed First University, Oujda, Morocco

Ahmed Tijini FSO, Mohammed First University, Oujda, Morocco

Ahmed Zidna LGIPM Laboratory, University of Lorraine, Metz, France

Part I
Plenary Lectures

Chapter 1
Mapped Polynomials and Discontinuous Kernels for Runge and Gibbs Phenomena

Stefano De Marchi

Abstract In this paper, we present recent solutions to the problem of approximating functions by polynomials for reducing in a substantial manner two well-known phenomena: Runge and Gibbs. The main idea is to avoid resampling the function or data and relies on the *mapped polynomials* or "fake" nodes approach. This technique turns out to be effective for stability by reducing the Runge effect and also in the presence of discontinuities by almost cancelling the Gibbs phenomenon. The technique is very general and can be applied to any approximant of the underlying function to be reconstructed: polynomials, rational functions or any other basis. A "natural" application is then quadrature, that we started recently to investigate and we propose here some results. In the case of jumps or discontinuities, where the Gibbs phenomenon appears, we propose a different approach inspired by approximating functions by kernels, in particular Radial Basis Functions (RBF). We use the so called *Variably Scaled Discontinuous Kernels (VSDK)* as an extension of the Variably Scaled Kernels (VSK) firstly introduced in Bozzini et al. (IMA J Numer Anal 35:199–219, 2015). VSDK show to be a very flexible tool suitable to substantially reducing the Gibbs phenomenon in reconstructing functions with jumps. As an interesting application we apply VSDK in *Magnetic Particle Imaging* which is a recent non-invasive tomographic technique that detects super-paramagnetic nanoparticle tracers and finds applications in diagnostic imaging and material science. In fact, the image generated by the MPI scanners are usually discontinuous and sampled at scattered data points, making the reconstruction problem affected by the Gibbs phenomenon. We show that VSDK are well suited in MPI image reconstruction also for identifying image discontinuities.

Keywords Runge and Gibbs phenomena · Mapped polynomial basis · "Fake" nodes · Variably scaled discontinous kernels

S. De Marchi (✉)
Department of Mathematics "Tullio Levi-Civita", Padova, Italy
e-mail: demarchi@math.unipd.it

D. Barrera et al. (eds.), *Mathematical and Computational Methods for Modelling, Approximation and Simulation*, SEMA SIMAI Springer Series 29,
https://doi.org/10.1007/978-3-030-94339-4_1

1.1 Introduction

Univariate approximation of functions and data is a well studied topic since the antiquity (Babylon and Greece), with many different developments by the Arabs and Persians in pre and late medieval period. The scientific revolution in astronomy, mainly due to Copernicus, Kepler and Galileo was the starting point for the investigations done later on by Newton, who gave strong impetus to the further advancement of mathematics, including what is now called "classical" interpolation theory (interested people on these historical aspects may read the nice survey by E. Meijering [50]).

Interpolation is essentially a way of estimating a given function $f : [a, b] \subset \mathbb{R} \to \mathbb{R}$ known only at a finite set $X_n = \{x_i,\ i = 0, \dots, n\}$ of $n+1$ (*distinct*) points, called *interpolation points* The corresponding set of values is denoted by $Y_n = \{y_i = f(x_i),\ i = 0, \dots, n\}$. Then, we are looking for a polynomial $p \in \mathbb{P}_n$, with $\mathbb{P}_n$ the linear space of polynomials of degree less and equal than n. The space $\mathbb{P}_n$ has finite dimension $n+1$ and spanned by the monomial basis $M = \{1, x, x^2, \dots, x^n\}$. Therefore, every $p \in \mathbb{P}_n$ is written as

$$p(x) = \sum_{j=0}^{n} a_j x^j . \tag{1.1}$$

The coefficients a_i are determined by solving the linear system $p(x_i) = y_i,\ i = 0, \dots, n$. Introducing the *Vandermonde* matrix $V_{ij} = (x_i^j)$, the coefficient vector $\boldsymbol{a} = (a_0, \dots, a_n)^t$ and the vector $\boldsymbol{y} = (y_0, \dots, y_n)^t$, the linear system can compactly be written as

$$V\boldsymbol{a} = \boldsymbol{y} . \tag{1.2}$$

As well-known, the solution of the system is guaranteed if the points are *distinct*, making V invertible [21, §2]. Moreover, the interpolating polynomial (1.1) can be expressed at any point $x \in [a, b]$ by the discrete inner product $p = \langle \boldsymbol{a}, \boldsymbol{x} \rangle$.

Instead of the monomial basis, we can use the cardinal basis of *elementary Lagrange polynomials*. That is $L = \{\ell_i,\ i = 0, \dots, n\}$ where $l_i(x) = \prod_{j=0, j\neq i} \dfrac{x - x_j}{x_i - x_j}$ or, alternatively by the ratio

$$\ell_i(x) = \frac{\det V_i(x)}{\det V} \tag{1.3}$$

where, $V_i(x)$ is the Vandermonde matrix in which we have substituted the i-th column with the vector $\boldsymbol{x} = (1, x, x^2, \dots, x^n)^T$. The formula (1.3) is essentially the Cramer's rule applied to the system

$$V\boldsymbol{\ell} = \boldsymbol{x}$$

showing immediately the main property of the elementary Lagrange polynomials: they form a set of cardinal polynomial functions, that is $\ell_i(x_j) = \delta_{ij}$. Using the Lagrange basis the interpolant becomes $p = \langle \boldsymbol{l}, \boldsymbol{y} \rangle$. Therefore, by unicity of interpolation we get

$$\langle \boldsymbol{a}, \boldsymbol{x} \rangle = \langle \boldsymbol{y}, \boldsymbol{\ell} \rangle .$$

Hence, in the Lagrange basis $\boldsymbol{\ell}$ the vector of coefficients $\boldsymbol{a}$ is at once $\boldsymbol{y}$, so that in (1.2) V is substituted by the identity matrix I of order $n + 1$.

There is another interesting formula that we can used to express the interpolant p. As pointed out in [20, §3 Prob. 14], the interpolant at every point x can be written in determinantal form as follows

$$p(x) = C \cdot \det \left[\begin{array}{c|cccc} 0 & 1 & x & \cdots & x^n \\ \hline y_0 & & & & \\ y_1 & & V & & \\ \vdots & & & & \\ y_n & & & & \end{array} \right] . \tag{1.4}$$

This expresses the interpolant as the determinant of an $(n + 2) \times (n + 2)$ matrix, obtained by bordering the Vandermonde matrix with two vectors $\boldsymbol{y}$, $\boldsymbol{x}^T$ and the scalar 0. The constant C appearing is (1.4) is a normalizing factor for expressing the interpolant in Lagrange form, that is $C = -1/\det(V)$.

This formula can be specialized for any set of linear independent functions, say $\{g_0, \ldots, g_n\}$ (cf. [20, §3, Prob. 15]) and in particular for the Lagrange basis L obtaining

$$p(x) = -\det \left[\begin{array}{c|cccc} 0 & \ell_0(x) & \ell_1(x) & \cdots & \ell_n(x) \\ \hline y_0 & & & & \\ y_1 & & I & & \\ \vdots & & & & \\ y_n & & & & \end{array} \right] , \tag{1.5}$$

with I the identity matrix of order $n + 1$.

Historically, but also nowadays in different frameworks and applications, the simplest way to take distinct samples x_i, is to consider equally spaced points (assuming for simplicity $x_0 = a$ and $x_n = b$). Two well-known phenomena are related to this choice of the interpolation points. The first one is the *Runge phenomenon*: when using polynomial interpolation with polynomials of high degree, the polynomial shows high oscillations at the edges of the interval. It was discovered by Carl

Runge when exploring the behavior of errors when using polynomial interpolation to approximate also analytic functions (cf. [56]). It is related to the *stability* of the interpolation process via the *Lebesgue constant*

$$\Lambda_n := \max_{x \in [a,b]} \sum_{i=0}^{n} |l_i(x)| \tag{1.6}$$

by means of the inequality

$$\|f - p\|_\infty \le (1 + \Lambda_n)\|f - p^*\|_\infty \tag{1.7}$$

where p^* represents the polynomial of best uniform approximation, that is $p^* := \inf_{p \in \mathbb{P}_n} \|f - p\|_\infty$, which uniquely exists. About the growth of the Lebesgue constant and its relevant properties we invite the readers to refer to the great survey by L. Brutman [19]. Here we simply recall the fundamental fact that Λ_n, which depends only on the choice of the node set X, grows exponentially (with n) when the interpolation points are equally spaced. Therefore it will be fundamental to look for a different choice than the equal distribution. As well-known, this is what do the *Chebyshev points* in $[a, b]$ or the *zeros of orthogonal polynomials* with respect to the interval $[a, b]$ and a given weight function (cf. e.g. [17, 19, 21]).

The second phenomenon and related to the Runge phenomenon is the *Gibbs phenomenon*, which tells us the peculiar manner in which the Fourier series of a piecewise continuously differentiable periodic function behaves at a jump discontinuity. This effect was originally discovered by Henry Wilbraham (1848) and successively rediscovered by J. Willard Gibbs (1899) (see [42]). The phenomenon observed that the n-th partial sum of the Fourier series has large oscillations near the jump, which might increase the maximum of the partial sum above that of the function itself. The overshoot does not die out as n increases, but approaches a finite limit. The Gibbs phenomenon is the cause of unpleasant artifacts in signal and image processing in presence of discontinuities.

The Gibbs phenomenon is also a well-known issue in higher dimensions and for other basis systems like wavelets or splines (cf. e.g. [44] for a general overview) and also in barycentric rational approximation [51]. Further, it appears also in the context of Radial Basis Function (RBF) interpolation [36] and in subdivision schemes (cf. [2, 3]). To soften the effects of this artifact, additional smoothing filters are usually applied to the interpolant. For radial basis function methods, one can for instance use linear RBFs in regions around discontinuities [25, 45]. Furthermore, post-processing techniques, such as Gegenbauer reconstruction procedure [40] or digital total variation [57], are available. This technique can be also applied in the non-polynomial setting by means of *discontinuous kernels*.

This survey paper consists of 6 main sections and various subsections as follows. In Sect. 1.2 we introduce our main idea for mitigating the Runge and Gibbs effects based on the mapped-basis approach or its equivalent form that we have termed "fake-nodes". In the successive Sect. 1.3 we present the algorithms for choosing a suitable map in the two instances studied in the paper. In Sect. 1.4 we show that the technique can be applied also using rational approximation instead of polynomial, while in Sect. 1.5 we discuss the application to quadrature. We continue then with Sect. 1.6 in which, for treating the Gibbs phenomenon we use a non-polynomial approach based on discontinuous kernels, in particular the so-called VSDK. Finally in Sect. 1.7 we discuss the application of these VSDK to the Magnetic Particle Imaging, a new quantitative imaging method for medical applications. We conclude with Sect. 1.8 by summarizing the main ideas and highlighting some further developments.

1.2 Mitigating the Runge and Gibbs Phenomena

Let $[a, b] \subset \mathbb{R}$ be an interval, X the set of distinct nodes (also called data sites) and $f : \Omega \longrightarrow \mathbb{R}$ a given function sampled at the nodes with $F_n = \{f_i = f(x_i)\}_{i=0,\ldots,n}$.

We now introduce a method that changes the interpolation problem (1.2) without resampling the function f. The idea rely on the so-called *mapped basis approach* where the map is used to mitigate the oscillations in the Runge and Gibbs phenomena. The idea of mapped polynomials is not new. Indeed, such kinds of methods have been used in the context of spectral schemes for PDEs.

For examples of well-known maps refer to e.g. [1, 41, 49]. However, for our purposes, that are devoted especially to applications when resampling cannot be performed, we consider a generic map $S : [a, b] \longrightarrow \mathbb{R}$. We investigate the conditions which this map S should satisfy in Sect. 1.2.1.

Let $\hat{x} \in \hat{S} := S([a, b])$ we can compute the polynomial $P_g : \hat{S} \longrightarrow \mathbb{R}$ interpolating the function values F_n at the *"fake" nodes* $S_X = \{S(x_i) = \hat{x}_i\}_{i=0,\ldots,n} \subset \hat{S}$ defined by

$$P_g(\hat{x}) = \sum_{i=0}^{n} c_i \, \hat{x}^i ,$$

for some smooth function $g : \hat{S} \longrightarrow \mathbb{R}$ such that

$$g_{|S_X} = f_{|X_n} .$$

Hence, for $x \in [a, b]$ we are interested in studying the function

$$R_f^S(x) := P_g(S(x)) = \sum_{i=0}^{n} c_i \, S(x)^i . \tag{1.8}$$

The function R_f^S in (1.8) can be considered as an interpolating function at the original set of nodes X_n and data values F_n, which is a linear combination of the basis functions $\mathcal{S}_n := \{S(x)^i,\ i = 0, \ldots, n\}$. As far as we know, a similar approach has been mentioned in [5], without being later worked out.

The analysis of this interpolation process can be performed in the following equivalent ways.

- The *mapped bases* approach on $[a, b]$: interpolate f on the set X_n via R^S_f in the function space $\mathcal{S}_n$.
- The *"fake" nodes* approach on $\hat{S}$: we interpolate g on the set S_X via P_g in the polynomial space Π_n.

1.2.1 The Mapped Bases Approach

The first question to answer is: how arbitrary is the choice of the map S?

Definition 1.1 We term S *admissible* if the resulting interpolation process admits a unique solution, that is the Vandermonde-like matrix $V^S := V(S_0, \ldots, S_n)$, is invertible.

Since the new point set in the mapped space is S_X, then

$$\det(V^S) = \prod_{0 \le i < j \le n} (S_j - S_i) \neq 0.$$

A consequence of this observation is the following proposition.

Proposition 1.1 *The map S is admissible if and only if for any $0 \le i, j \le n,\ i \neq j$ we have $S_i \neq S_j$. In other words, S is injective in $[a, b]$.*

In fact, $\det(V^S) \neq 0$ if and only if $S_j - S_i \neq 0$.

Remark 1.1 We point out that we can easily write

$$\det(V^S) = \sigma(S, X)\det(V),$$

where V is the classical Vandermonde matrix and

$$\sigma(S, X) := \prod_{0 \le i < j \le n} \frac{S_j - S_i}{x_j - x_i}.$$

This fact presents some similarities with the so-called *generalized Vandermonde determinants* that can be factored as the classical Vandermonde determinant and a Schur function, as outlined for instance in the paper [22].

We can associate to the interpolant R_f^S also its Lagrange form

$$R_f^S(x) = \sum_{i=0}^{n} f_i \ell_i^S(x),$$

with, in analogy with (1.3),

$$\ell_i^S(x) := \frac{\det(V_i^S(x))}{\det(V^S)} = \prod_{\substack{0 \le j \le n \\ j \ne i}} \frac{S(x) - S_j}{S_i - S_j}, \tag{1.9}$$

where $V_i^S(x) = V_i(S(x))$.

Consequently, we can consider the S-Lebesgue constant Λ_n^S associated to the mapped Lagrange basis $\mathcal{L}^S = \{\ell_0^S, \ldots, \ell_n^S\}$ and to the interpolation operator $\mathcal{L}_n^S : \Upsilon \longrightarrow S_n$ with $(\Upsilon, \|\cdot\|_\Omega)$ be a normed function space, which contains only real-valued functions on $\Omega = [a, b]$. Then, Λ_n^S is the operator norm of $\mathcal{L}_n^S$ with respect to $\|\cdot\|_\Omega$, that is

$$\Lambda_n^S = \sup_{\substack{v \in \Upsilon \\ v \ne 0}} \frac{\|\mathcal{L}_n^s(v)\|_\Omega}{\|v\|_\Omega},$$

and hence we can provide the sup-norm error estimator as follows:

$$\|f - R_f^S\|_\Omega \le (1 + \Lambda_n^s(\Omega)) E_n^{S,\star}(f),$$

where $E_n^{S,\star}(f)$ is the best polynomial approximation error in the sup-norm [53, Theorem I.1, p. 1].

We have just seen that ℓ_i and ℓ_i^S are defined as in (1.3) and (1.9), respectively. The next Proposition proposes a rough upper bound for the S-Lebesgue constant, useful for understanding the asymptotic behaviour of the error.

Proposition 1.2 *For all $x \in [a.b]$, $x \ne x_j$, we have*

$$\ell_i^S(x) = \underbrace{\frac{\beta_i(x)}{\alpha_i}}_{\gamma_i(x)} \ell_i(x), \tag{1.10}$$

with $\alpha_i := \prod_{\substack{0 \le j \le n \\ j \ne i}} \frac{S_i - S_j}{x_i - x_j}, \quad \beta_i(x) := \prod_{\substack{0 \le j \le n \\ j \ne i}} \frac{S(x) - S_j}{x - x_j}.$

Proof By construction $\ell_i^S(x_j) = \ell_i(x_j) = \delta_{i,j}$. Then, let $x \neq x_j$, from (1.3) and (1.9) we have that

$$\frac{\ell_i^S(x)}{\ell_i(x)} = \frac{\det(V_i^S(x))\det(V)}{\det(V_i(x))\det(V^s)} = \frac{\det(V_i^S(x))}{\sigma(S,X)\det(V_i(x))} := \gamma_i(x).$$

We can also write

$$\ell_i^S(x) = \prod_{\substack{0\le j\le n\\ j\ne i}} \frac{S(x)-S_j}{S_i-S_j} = \prod_{\substack{0\le j\le n\\ j\ne i}} \frac{x-x_j}{x_i-x_j}\cdot\frac{x_i-x_j}{x-x_j}\cdot\frac{S(x)-S_j}{S_i-S_j},$$

$$= \prod_{\substack{0\le j\le n\\ j\ne i}} \frac{x-x_j}{x_i-x_j} \prod_{\substack{0\le j\le n\\ j\ne i}} \frac{S(x)-S_j}{x-x_j} \prod_{\substack{0\le j\le n\\ j\ne i}} \frac{x_i-x_j}{S_i-S_j}.$$

By defining

$$\alpha_i := \prod_{\substack{0\le j\le n\\ j\ne i}} \frac{S_i-S_j}{x_i-x_j}, \quad \beta_i(x) := \prod_{\substack{0\le j\le n\\ j\ne i}} \frac{S(x)-S_j}{x-x_j},$$

we get formula (1.10) as claimed. □

As a consequence, we can bound Λ_n^S from above by Λ_n unless a constant $C(S, n)$ depending on the map S and n.

Theorem 1.1 *Letting*

$$L = \max_{0\le i\le n}\ \max_{x\in[a,b], i\ne j} \left|\frac{S(x)-S_j}{x-x_j}\right|, \quad D = \min_{0\le i\le n}\ \min_{j\ne i} \left|\frac{S_i-S_j}{x_i-x_j}\right|,$$

then

$$\Lambda_n^S \le C(S,n)\Lambda_n, \tag{1.11}$$

where $C = (L/D)^n$ *with* Λ_n *the classical Lebesgue constant in* (1.6).

Proof Using Proposition 1.2

$$|\ell_i^S(x)| = \left|\frac{\beta_i(x)}{\alpha_i}\right| |\ell_i(x)|.$$

An upper bound for $|\beta_i|$ is

$$|\beta_i(x)| = \prod_{\substack{0\le j\le n \\ j\ne i}} \left|\frac{S(x)-S_j}{x-x_j}\right| \le \prod_{\substack{0\le j\le n \\ j\ne i}} L_i,$$

where

$$L_i = \max_{x\in[a,b], i\ne j} \left|\frac{S(x)-S_j}{x-x_j}\right|.$$

Thus,

$$|\beta_i(x)| \le L_i^n .$$

A lower bound for $|\alpha_i|$ is

$$|\alpha_i| = \prod_{\substack{0\le j\le n \\ j\ne i}} \left|\frac{S_i-S_j}{x_i-x_j}\right| \ge \prod_{\substack{0\le j\le n \\ j\ne i}} D_i = D_i^n,$$

where

$$D_i := \min_{j\ne i} \left|\frac{S_i-S_j}{x_i-x_j}\right|.$$

We then have

$$|\ell_i^s(x)| \le \left(\frac{L_i}{D_i}\right)^n |\ell_i(x)|.$$

Therefore, letting $L := \max_{i\ne j} L_i$, $D := \min_i D_i$ and considering the sum of the Lagrange polynomials, we obtain

$$\Lambda_n^s \le \left(\frac{L}{D}\right)^n \Lambda_n.$$

We conclude by setting $C(S,n) = (L/D)^n$. □

1.2.2 The "Fake" Nodes Approach

The construction of the interpolating function R_f^S is equivalent to building a polynomial interpolant at the "fake" nodes, as defined in (1.8). Therefore, in what follows we concisely analyze the parallelism with the polynomial interpolation problem in $\hat{S}$.

If ℓ_i is the i-th Lagrange polynomial related to the set S_X, then for $\hat{x} \in \hat{S}$, we have

$$\ell_i(\hat{x}) = \prod_{\substack{0 \le j \le n \\ j \ne i}} \frac{\hat{x} - S(x_j)}{S(x_i) - S(x_j)},$$

and the Lebesgue constant is given by

$$\Lambda_n(S(\Omega)) = \max_{\hat{x} \in S(\Omega)} \sum_{i=0}^{n} |\ell_i(\hat{x})|. \tag{1.12}$$

For $x \in \Omega$, we observe that

$$\ell_i(\hat{x}) = \ell_i(S(x)) = \prod_{\substack{0 \le j \le n \\ j \ne i}} \frac{S(x) - S(x_j)}{S(x_i) - S(x_j)} = \ell_i^s(x).$$

As a consequence, we obtain

$$\Lambda_n^S(\Omega) = \Lambda_n(S(\Omega)),$$

and

$$\| f - R_f^S \|_\Omega = \| g - P_{n,g} \|_{S(\Omega)},$$

which in particular implies that

$$\| f - R_f^S \|_\Omega \le (1 + \Lambda_n(S(\Omega))) E_n^\star(g).$$

Since we can suppose without loss of generality that g is a regular function, for an appropriate choice of the map S, and hence of the nodes S_X, we might improve the results with respect to classical polynomial approximation in $[a, b]$. The main difficulties are in finding a *good* map. In the next section we thus propose two receipts for computing suitable maps that, as numerically shown later, enable us to naturally mitigate both Runge and Gibbs phenomena.

1.3 Choosing the Map S: Two Algorithms

In this Section, we describe how, given an *ordered set* of interpolation nodes $X_n = \{x_i \in [a,b] \mid x_0 = a,\ x_n = b,\ x_i < x_{i+1}\}$, we can effectively construct suitable maps S. We present two different ways of constructing the map S and in doing so we deal with the Runge and Gibbs phenomenon, respectively.

Treating the Runge Phenomenon In order to prevent the appearance of the Runge phenomenon, we construct a map S such that the resulting set of "fake" nodes S_X guarantees a stable interpolation process. The natural way is mapping X_n to the set of ordered Chebyshev-Lobatto (CL) nodes $C_n = \{c_0, \ldots, c_n\}$ on $[a,b]$.

Indeed, as well-known the Lebesgue constant of the CL nodes grows logarithmically with respect to n [19]. The map S on $[a,b]$ that does this transformation, for any $x \in [a,b]$, is the piecewise interpolant as obtained by the following algorithm that we term *S-Runge*.

Algorithm 1

Input: X_n.

1. Define the set of CL nodes C_n.
2. For $x \in [x_i, x_{i+1}]$, $i = 0, \ldots, n-1$, set $\beta_{1,i} = \frac{c_{i+1}-c_i}{x_{i+1}-x_i}$, $\beta_{2,i} = c_i$ so that $S(x) = \beta_{1,i}(x - x_i) + \beta_{2,i}$.

Outputs: the vectors $\boldsymbol{\beta}_1, \boldsymbol{\beta}_2$ of the coefficients that define S.

Since the CL nodes are distinct, the map is admissible by construction. For instance, if $X_n = \{x_k = a + k \cdot (b-a)/n, k = 0, \ldots, n\}$ is the ordered set of $n+1$ equispaced nodes in $[a,b]$, it can be analytically mapped to C_n by using the map

$$S(x) = \frac{a-b}{2} \cos\left(\pi \frac{x-a}{b-a}\right) + \frac{a+b}{2}. \tag{1.13}$$

This algorithm, is robust and does not require any additional hypothesis on X_n and it could work for scattered data or on random, nearly equispaced or *well-spaced* interpolation nodes (for the definition of well-spaced we refer to [15]). These algorithms are also quoted in Wikipedia at [61].

Treating the Gibbs Phenomenon Let us suppose that the underlying function f presents jump discontinuities, whose positions and magnitudes are encoded in the set

$$D_m := \{(\xi_i, d_i) | \xi_i \in (a,b),\ \xi_i < \xi_{i+1},\ i = 0, \ldots, m, \text{ and } d_i := |f(\xi_i^+) - f(\xi_i^-)|\}.$$

We assume to know the discontinuities and the related jumps. Such assumption is not restrictive. Indeed for the one dimensional case, but also in higher dimensions, algorithms for detecting the discontinuity points are available, see for instance the references [4, 54].

Near a discontinuity, the interpolant is forced to strongly vary making the Gibbs phenomenon more evident. The usual approach is to put more points nearby the jumps or use a so-called filter in order to get more accurate solution than rough Fourier expansions (cf e.g. [25]) or acceleration techniques of the Fourier expansions based on the ϵ-algorithm (see e.g. [6]).

Our strategy is very simple: we sufficiently increase the gap between the node right before and the one right after the discontinuity, so allowing the interpolant to become smoother. To accomplish this, we introduce a *shifting parameter* $k > 0$.

The next algorithm, that we call *S-Gibbs* implements this idea.

Algorithm 2

Inputs: D_m and k.

1. Letting $A_i = \sum_{j=0}^{i} kd_j, \; i = 0, \dots, m.$
2. Define S as

$$S(x) = \begin{cases} x, & \text{for } x \in [a, \xi_0[, \\ x + A_i, & \text{for } x \in [\xi_i, \xi_{i+1}[, \; 0 \le i < m, \; \text{or } x \in [\xi_m, b]. \end{cases}$$

Outputs: the values A_i.

As experimentally observed the choice of k is not critical. It suffices that in the mapped space the so-constructed function g has no steep gradients and this can be obtained by taking $k \gg 1$. Since the resulting "fake" nodes S_X are distinct, the so constructed map is admissible.

1.3.1 Numerical Tests

We show via the algorithms described in the previous Sect. 1.3 that we are able to substantially reduce the oscillating effects due to Runge and Gibbs phenomena. Our "fake" nodes approach is compared with the resampling at Chebyshev nodes. We note, that in many applications we unfortunately dispose of the data at equispaced samples. It is the reason why our approach becomes relevant for the underlying idea: *mapping without resampling*.

We consider the interval $[-5, 5]$ and both equispaced and randomly distributed interpolation nodes. Moreover, we evaluate the interpolants on a set of equispaced points $\Xi = \{\bar{x}_i, i = 0, \ldots, 330\}$ and compute the Relative Maximum Absolute Error (RMAE)

$$\text{RMAE} = \max_{i=0,\ldots,m} \frac{|R^s_{n,f}(\bar{x}_i) - f(\bar{x}_i)|}{|f(\bar{x}_i)|}. \tag{1.14}$$

The experiments have been carried out in PYTHON 3.6 using Numpy 1.15; see [30].

1.3.1.1 Application to Runge Phenomenon

For this test we consider the function $f_1(x) = \frac{1}{e^{-3x}+1}$ sampled at equispaced interpolation nodes E_n. We then compare f_1 evaluated at Ξ with respect to

i. the interpolating polynomial at equispaced points E_n, i.e. the original data set and function values;
ii. the interpolating polynomial at CL nodes C_n and resampled function values $f_1(C_n)$, i.e. we resample the function;
iii. the approximant built upon a polynomial interpolant at the fake nodes $S(E_n)$ corresponding to the CL, C_n obtained by the map (1.13), and function values $f_1(E_n)$.

In Fig. 1.1, we show different reconstructions of f_1 for a fixed number of nodes. In Fig. 1.2 we show the RMAE for the function f_1 varying the number of nodes while in Fig. 1.3 we plot the Lebesgue functions related to the proposed methods. As pointed out in the theoretical analysis, the behavior of the "fake" nodes in terms of Lebesgue constant is analogous to that of the classical polynomial interpolation at CL points.

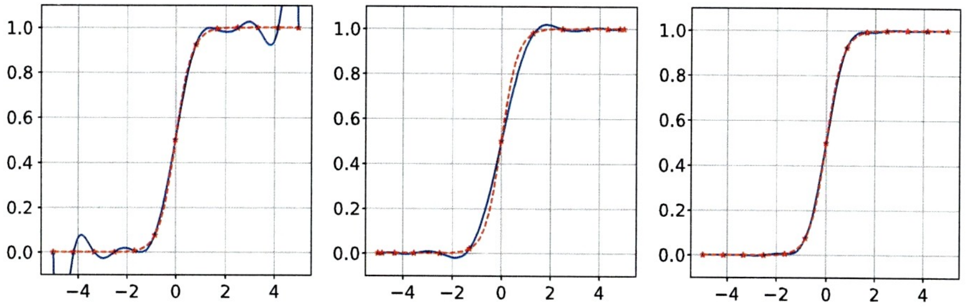

Fig. 1.1 Interpolation with 13 points of the function f_1 on $[-5, 5]$ using equispaced (left), CL (center) and "fake" nodes (right). The nodes are represented by stars, the original and reconstructed functions are plotted with continuous red and dotted blue lines, respectively

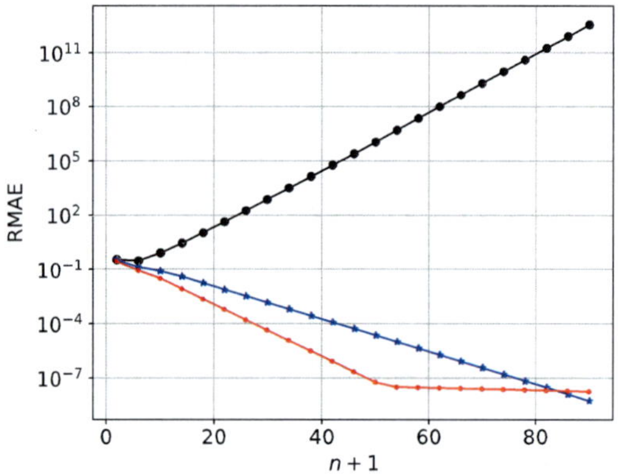

Fig. 1.2 The RMAE for the function f_1 varying the number of nodes. The results with equispaced, CL and fake nodes are represented by black circles, blue stars and red dots, respectively

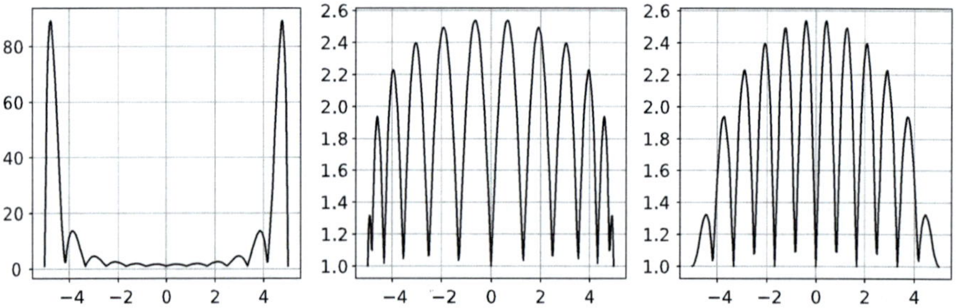

Fig. 1.3 Lebesgue functions of equispaced (left), CL (center) and fake CL (right) nodes

For more experiments and details the reader can refer to [27].

1.3.1.2 Applications to Gibbs Phenomenon

For the test, we consider the function with jump discontinuities at the origin

$$f_2(x) := \begin{cases} -\dfrac{x^2}{40} + 1, & x < 0, \\ \dfrac{x+4}{x+6}, & x \geq 0, \end{cases}$$

As done in the previous Sect. 1.3.1.1, we compare the interpolant in the three different situations indicated in the gray box.

In Fig. 1.4 we display the results obtained using 20 interpolation points. We observe that the Gibbs phenomenon affects the reconstruction obtained via resampling on CL nodes, while it is well mitigated if using the fake nodes. In Fig. 1.5,

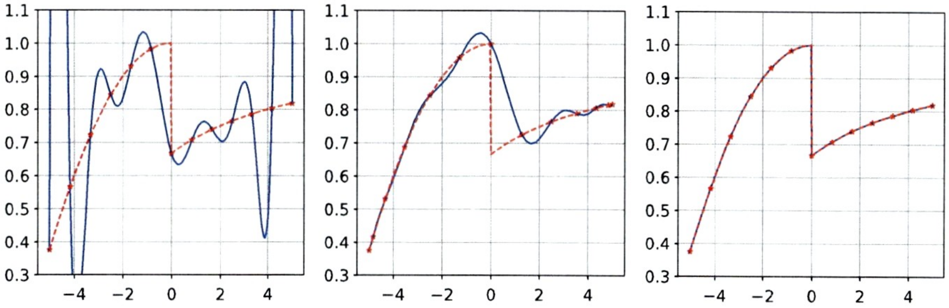

Fig. 1.4 Interpolation with 20 points of the function f_2 on $[-5, 5]$, using equispaced (left), CL nodes (center) and the discontinuous map (right). The nodes are represented by stars, the original and reconstructed functions are plotted with continuous red and dotted blue lines, respectively

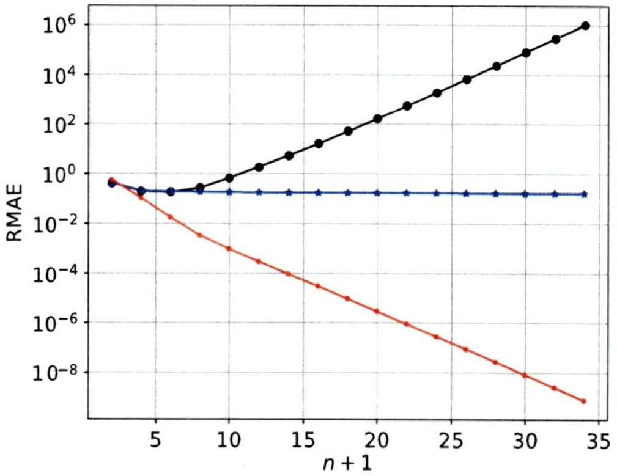

Fig. 1.5 The RMAE for the function f_2 varying the number of nodes. The results with equispaced, CL and fake nodes are represented by black circles, blue stars and red dots, respectively

we provide the RMAE behavior of these methods. The results are quite impressive, meaning that we are able to effectively reduce the Gibbs phenomenon by the S-Gibbs map of Algorithm 2.

Remarks In the presence of discontinuities, it is interesting noticing the behaviour of the elementary Lagrange functions under the mapping approach. In Fig. 1.6 we show this effect for the cubic case when there is no jump and jump, with and without scaling. What is interesting to see is that the cardinal functions become discontinuous in the presence of a discontinuities or jumps. This is the idea that we will develop in Sect. 1.6, about the use of discontinuous kernels when treating discontinuous functions.

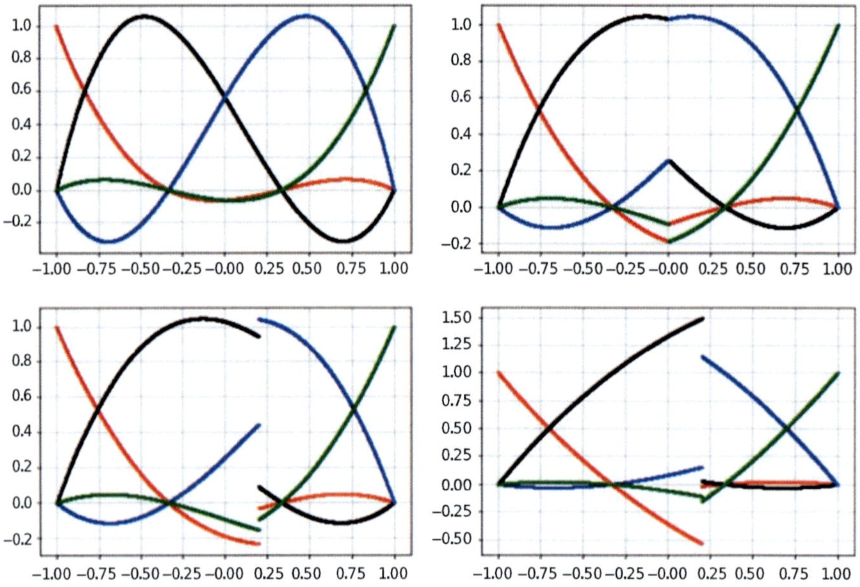

Fig. 1.6 Left-right, up-down: the original cardinals on 4 nodes, the cardinals around $\xi = 0$, $A = 1$ the cardinals around $\xi = 0.2$, $A = 1$,the cardinals around $\xi = 0.2$, $A = 5$

1.4 Application to Rational Approximation

We can extend the mapped bases or *"fake" nodes* approach to the barycentric rational interpolation. In particular we concentrate to the family of *Floater-Hormann (FH) rational approximants* [35] combined with *AAA-algorithm*. We consider the family of FH interpolants because they have shown good approximation properties for smooth functions, in particular using equidistant nodes (cf. [14]). The idea comes from the fact that our previous construction is a general purpose machinery, a kind of "black-box". Indeed, given a basis $\mathcal{B}$ for the approximation space, we can apply the S-map approach getting a new basis, say $\tilde{\mathcal{B}} = S \circ \mathcal{B}$. Considering that we are not resampling the values of the reconstructed function remains unchanged.

In [51], the *AAA-algorithm* has been discussed. This is a greedy algorithm to compute a barycentric rational approximant that is named by the authors *Adaptive Antoulas–Anderson algorithm*, reminding the names of the authors. This algorithm leads to impressively well-conditioned bases, which can be used in different fields, such as in computing conformal maps, or in rational minimax approximations (see e.g. [10]).

Unfortunately, both FH rational interpolants and those obtained by the AAA-algorithm, suffer of the *Gibbs phenomenon* when the underlying function presents jump discontinuities. For this reason we try to apply the *S-Gibbs* algorithm to mitigate or even better to reduce, this unpleasant effect.

We start by presenting a short overview of barycentric polynomial approximation and the FH family of rational approximants. Then we recall the main ideas about the AAA-algorithm and present the application of the S-Gibbs to both.

The numerical tests below will show an accurate interpolation of discontinuous functions.

1.4.1 Barycentric Polynomial Interpolation

As before let $X_n := \{x_i : i = 0, \dots, n\}$ the set of $n+1$ distinct nodes in $I = [a, b] \subset \mathbb{R}$, increasingly ordered and let $f : I \longrightarrow \mathbb{R}$ be known at the sample points. Again we set $F_n := \{f_i = f(x_i) : i = 0, \dots, n\}$.

It is well-known (see e.g. [7–9]) that it is possible to write the unique interpolating polynomial $P_{n;f}$ of degree at most n of f at X_n for any $x \in I$ in the second barycentric form

$$P_{n;f}(x) = \frac{\displaystyle\sum_{i=0}^{n} \frac{\lambda_i}{x - x_i} f_i}{\displaystyle\sum_{i=0}^{n} \frac{\lambda_i}{x - x_i}}, \tag{1.15}$$

where $\lambda_i = \displaystyle\prod_{j \neq i} \frac{1}{x_i - x_j}$ are called *weights* and this formula is one of the most stable way of evaluating $P_{n;f}$ (see [43]). If the weights λ_i are changed to other nonzero weights, say w_i, then the corresponding barycentric rational function

$$R_{n;f}(x) = \frac{\displaystyle\sum_{i=0}^{n} \frac{w_i}{x - x_i} f_i}{\displaystyle\sum_{i=0}^{n} \frac{w_i}{x - x_i}} \tag{1.16}$$

still satisfies the interpolation conditions $R_{n;f}(x_i) = f_i, \; i = 0, \dots, n$.

1.4.2 Floater-Hormann Rational Interpolation

Let $n \in \mathbb{N}$, $d \in \{0, \dots, n\}$. Let p_i, $i = 0, \dots, n-d$ denote the unique polynomial interpolant of degree at most d interpolating the $d+1$ points $(x_k, f_k), \; k =$

$i, \ldots,$ i+d. The Floater-Hormann rational interpolant is

$$R_{n,d;f}(x) = \frac{\displaystyle\sum_{i=0}^{n-d} \lambda_i(x) p_i(x)}{\displaystyle\sum_{i=0}^{n-d} \lambda_i(x)}, \text{ where } \lambda_i(x) = \frac{(-1)^i}{(x - x_i) \cdots (x - x_{i+d})},$$

which interpolates f at the set of nodes X_n. It has been proved in [35] that $R_{n,d;f}$ has no real poles and that it reduces to the unique interpolating polynomial of degree at most n when $d = n$.

One can derive the barycentric form of this family of interpolants as well. Indeed, with considering the sets $J_i = \{k \in \{0, 1, \ldots, n - d\} : i - d \leq k \leq i\}$, one has

$$R_{n,d;f}(x) = \frac{\displaystyle\sum_{i=0}^{n} \frac{w_i}{x - x_i} f_i}{\displaystyle\sum_{i=0}^{n} \frac{w_i}{x - x_i}}, \text{ where } w_i = (-1)^{i-d} \sum_{k \in J_i} \prod_{\substack{j=k \\ j \neq i}}^{j+d} \frac{1}{|x_i - x_j|}.$$

1.4.3 The AAA Algorithm

Let us consider a set of points X_N with a large value of N (which represents the discretization of our domain) and a function $f : I \longrightarrow \mathbb{R}$. The *AAA-algorithm* (cf. [51]) is a *greedy* technique that at the step $m \geq 0$ considers the set $X^{(m)} = X_N \setminus \{x_0, \ldots, x_m\}$ and constructs the interpolant

$$R_{m;f}(x) = \frac{\displaystyle\sum_{i=0}^{m} \frac{w_i}{x - x_i} f_i}{\displaystyle\sum_{j=0}^{m} \frac{w_j}{x - x_j}} = \frac{n(x)}{d(x)},$$

by solving the discrete least squares problem for the weight vector $\boldsymbol{w} = (w_0, \ldots, w_m)$

$$\min \|fd - n_{X^{(m)}}\| \qquad \|\boldsymbol{w}\|_2 = 1,$$

where $\|\cdot_{X^{(m)}\|}$ is the discrete 2-norm over $X^{(m)}$. The next data site is $x_{m+1} \in X^{(m)}$ that makes the residual $|f(x) - n(x)/d(x)|$ maximum with respect to $x \in X^{(m)}$. This choice confirms the greedy nature of the process.

1.4.4 Mapped Bases in Barycentric Rational Interpolation

Using the ideas of mapped based approach in [27], we apply it to the *Floater-Hormann* interpolants and to its approximants produced via the *AAA algorithm*.

First, since the interpolant $R_{n;f}$ defined in (1.16) can be written using the cardinal basis, that is $R_{n;f}(x) = \sum_{j=0}^{n} f_j b_j(x)$, where

$$b_j(x) = \frac{\dfrac{w_j}{x - x_j}}{\sum_{i=0}^{n} \dfrac{w_i}{x - x_i}}$$

is the j-th basis function, in the same spirit, we can write $R_{n;f}^S$ in the mapped cardinal basis form $R_{n;f}^S(x) = \sum_{i=0}^{n} f_i b_i^S(x)$, where $b_j^S(x) = \frac{\frac{w_j}{S(x)-S(x_j)}}{\sum_{i=0}^{n} \frac{w_i}{S(x)-S(x_i)}}$ is the j-th mapped basis function.

Using the S mapping approach, a more stable interpolant may arise. Indeed, the following result provides an upper bound for the S-Lebesgue constant that shows this improved stability.

Theorem 1.2 *Let* $\Lambda_n(X_n) = \max_{x \in I} \sum_{j=0}^{n} |b_j(x)|$ *and* $\Lambda_n^S(X_n) := \max_{x \in I} \sum_{j=0}^{n} |b_j^S(x)|$ *be the classical and the S-Lebesgue constants, respectively. We have*

$$\Lambda_n^S(X_n) \leq C \Lambda_n(X_n),$$

where $C = \dfrac{\max_k M^k}{\min_k m_k}$ *with*

$$M^k = \max_{x \in I} \prod_{\substack{l=0 \\ l \neq k}}^{n} \left| \frac{S(x) - S(x_l)}{x - x_l} \right|, \quad m_k = \min_{x \in I} \prod_{\substack{l=0 \\ l \neq k}}^{n} \left| \frac{S(x) - S(x_l)}{x - x_l} \right|.$$

Proof As done in the polynomial case, we should bound each basis function b_j^S in terms of b_j for all $x \in I$. For details see [10]. □

Equivalently to the above mapped basis instance, we may construct the interpolant $R_{n;f}^S$ via the "fake" nodes approach just following the same ideas developed in Sect. 1.4.

Let $\tilde{R}_{n;g}$ be the barycentric rational interpolant as in (1.16) that interpolates, at the set of "fake "nodes $S(X_n)$, where the function $g : S(I) \longrightarrow \mathbb{R}$ interpolates the values F_n

$$g_{|S(X_n)} = f_{|X_n}.$$

Observe that $R^S_{n;f}(x) = \tilde{R}_{n;g}(S(x))$ for every $x \in I$. Hence, we may also build $R^S_{n;f}]$ upon a standard barycentric interpolation process, thereby providing a more intuitive interpretation of the method.

As we already observed, the choice of the mapping S is crucial for the accuracy interpolation process. Here, we confine our attention to the case in which f presents jump discontinuities and by using the *S-Gibbs* Algorithm (SGA), presented above, we construct an effective mapping S.

1.4.5 A Numerical Example

In $I = [-5, 5]$ we consider the discontinuous functions

$$f_1(x) = \begin{cases} e^{\frac{1}{x.^2+1}}, & -5 \le x < -3 \\ \sin(3x), & -3 \le x < 2 \\ -\frac{x^3}{30} + 2, & 2 \le x \le 5. \end{cases} \qquad f_2(x) = \begin{cases} \cos(-\sin(x/2)), & -5 \le x < -2.5 \\ \tan(x/2), & -2.5 \le x < 2 \\ \log(e^{-\frac{1}{x-5.1}}), & 2 \le x \le 5. \end{cases}$$

We test the "fake" nodes approach with the S-Gibbs Algorithm in the framework of FH interpolants and the AAA algorithm for approximation. We fix $k = 10$ in the S-Gibbs. As observed in [27], also in this setting the choice of the shifting parameter is non-critical as long as it is taken "sufficiently large". We evaluate the constructed interpolants on a set of 5000 equispaced evaluation points $\Xi = \{\bar{x}_i = -5 + \frac{i}{1000} : i = 0, \dots, 5000\}$ and compute the Relative Maximum Absolute Error RMAE as defined in (1.14, both for $R_{n;f}$ and $R^S_{n;f}$.

1.4.5.1 The FH Interpolants

Here, we take various sets of equispaced nodes $\mathcal{X}_n = \{-5 + \frac{5i}{n} : i = 0, \dots, n\}$, varying the size of n. The results of the RMSE interpolation errors are displayed in Figs. 1.7 and 1.8, by doubling the number of the nodes for $n = 40$ up to 2560. We simply observe that the proposed reconstruction via the "fake" nodes approach outperforms the standard technique.

1.4.5.2 The AAA Algorithm

As the starting set for the AAA algorithm, we consider 10,000 nodes randomly uniformly distributed in I, which we denote by $\mathcal{X}_{rand}$.

Looking at Table 1.1, we observe that using the AAA algorithm with starting set $S(\mathcal{X}_{rand})$ (indicated in the Table as AAA^S), that is, constructing the approximants via the fake nodes approach, does not suffer from the effects of the Gibbs phenomenon. For both approximants we fix the maximum degree to 20 and to 40

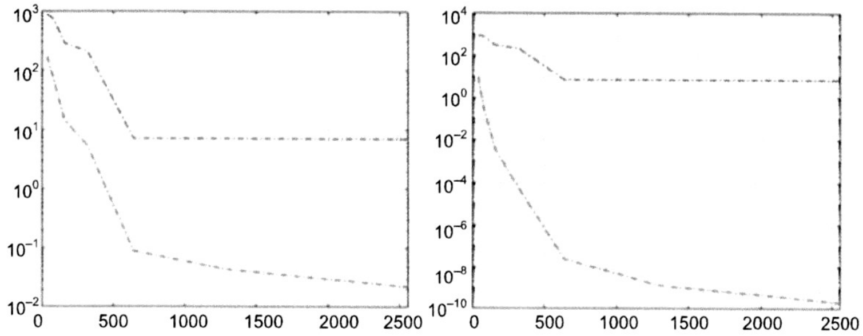

Fig. 1.7 RMSE interpolation error for f_1. Left with $d = 1$, right with $d = 4$. In blue, the standard interpolant $R_{n,d}$. In red, the proposed interpolant $R^s_{n,d}$

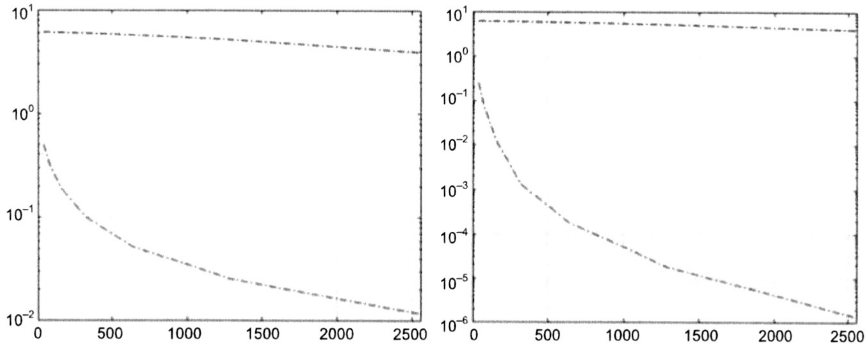

Fig. 1.8 RMSE interpolation error for f_2. Left with $d = 1$, right with $d = 4$. In blue, the standard interpolant $R_{n,d}$. In red, the proposed interpolant $R^s_{n,d}$

Table 1.1 RMAE for AAA and AAA^s approximants

f	m_{max}	AAA	AAA^s
f_1	20	5.8E+1	1.6E−4
	40	1.4E−1	2.5E−9
f_2	20	4.4E+1	7.4E−8
	40	8.0E−1	3.0E−12

(by default 100 in the algorithm) getting rational approximants of type (20, 20) and (40, 40), respectively.

Remarks This extension of the "fake" nodes approach to barycentric rational approximation, in particular to the family of FH interpolants, and its approximations by the AAA algorithm for the treatment of the Gibbs phenomenon via the S-Gibbs algorithm, shows that the proposed reconstructions outperform their classical versions, by erasing distortions and oscillations.

1.5 Application to Quadrature

We now drive our attention towards the use of mapped bases as an alternative to standard quadrature rules as started to discuss in the ongoing paper [29].

Given $f : I = [a, b] \to \mathbb{R}$, we are interested in approximating

$$\mathcal{I}(f) := \int_I f(x)\,\mathrm{d}x,$$

via quadrature formulae of interpolation type. To this aim we take a set of distinct (quadrature) nodes $X_n = \{x_i,\ i = 0, \ldots, n\}$ by assuming $x_0 = a$, $x_n = b$ and $n \geq 1$.

The classical quadrature of interpolation type substitutes f by its interpolant $P_{n,f}(\cdot) = \sum_{i=0}^{n} a_i\, b_i(\cdot)$, with b_i the i-th basis element of the polynomial space $\mathbb{P}_n$ (note that the most simple basis one can consider is the monomials one, that is $b_i = x^k$) so that

$$\mathcal{I}(f) \approx \mathcal{I}_n(f) := \langle \boldsymbol{w}, \boldsymbol{f} \rangle\,. \tag{1.17}$$

The vector $\boldsymbol{f}$ contains the function values $f_k = f(x_k)$ while the vector of the weights $\boldsymbol{w} = (w_0, \ldots, w_n)^\mathsf{T}$ is computed by solving the moments' system

$$V^\mathsf{T}\ \boldsymbol{w} = \boldsymbol{m} \tag{1.18}$$

where V is the Vandermonde matrix so that $V_{i,j} = b_i(x_j)$, the vector $\boldsymbol{m} = (m_0, \ldots, m_n)^\mathsf{T}$ consists of the moments $m_k = \mathcal{I}(b_k),\ \ k = 0, \ldots, n$.

Remark The quadrature (1.17) is a scalar product that can be computed as the determinant of formula (1.5) substituting the elementary Lagrange polynomials with the weights and changing sing

$$\langle \boldsymbol{w}, \boldsymbol{f} \rangle = \det \left[\begin{array}{c|cccc} 0 & -w_0 & -w_1 & \cdots & -w_n \\ -- & - & - & - & - \\ f_0 & & & & \\ f_1 & & I & & \\ \vdots & & & & \\ f_n & & & & \end{array}\right]. \tag{1.19}$$

1.5.1 Quadrature via "Fake" Nodes

In the mapped bases approach [27] instead of $P_{n;f}$ we consider the mapped interpolant $P^S_{n;f}$. It comes then "natural" to construct the quadrature formulae in the mapped basis, that is

$$\mathcal{I}(f) \approx \mathcal{I}^S_n(f) := \langle \boldsymbol{w}^S, \boldsymbol{f} \rangle .$$

The weights $\boldsymbol{w}^S = (w^S_0, \ldots, w^S_n)^\mathsf{T}$ are computed by solving the mapped-based system

$$(V^S)^\mathsf{T}\, \boldsymbol{w}^S = \boldsymbol{m}^S ,$$

with $\boldsymbol{m}^S$ the vector of S-mapped moments. For instance, taking the monomial basis, the i-th S-moment is $m^S_i = \mathcal{I}(S(x)^i)$.

Equivalently, for $x \in I$, the interpolant $P^S_{n;f}$ can be seen as a standard polynomial interpolant $P_{n;g}$ on the mapped set of nodes $S(I) = \hat{I}$. Moreover, if the map S is at least C^1 in I, letting $t = S(x)$, for $x \in I$, we get

$$\hat{S}(t) := \frac{\mathrm{d}S^{-1}(t)}{\mathrm{d}t} = \frac{1}{S'(S^{-1}(t))} . \tag{1.20}$$

Thus,

$$\mathcal{I}(P^S_{n;f}, I) = \mathcal{I}(P_{n;g}\hat{S}, \hat{I}) . \tag{1.21}$$

It then follows

Proposition 1.3 $w^S_i = \mathcal{I}(l_i \hat{S}, \hat{I})$.

Proof Considering the Lagrange basis of the polynomial interpolant so that

$$P_{n;g}(S(x)) = \sum_{i=0}^{n} f(x_i) l_i(S(x)) = \sum_{i=0}^{n} f(x_i) \prod_{\substack{0 \le j \le n \\ j \ne i}} \frac{S(x) - S(x_j)}{S(x_i) - S(x_j)},$$

and taken into account (1.20) we conclude □

The following then can be easily proven.

Proposition 1.4 *Let S be such that g is at least $C^{n+1}(\hat{I})$. Then,*

$$E^S_n(f) = E_n(g) = \frac{g^{(n+1)}(S(\xi))}{(n+1)!} \mathcal{I}(\omega_{n+1}\hat{S}, \hat{I}), \quad \xi \in I$$

with $\omega_{n+1}(t) = \prod_{i=0}^{n}(t - t_i)$.

Proof Indeed

$$E_n^S(f) = \mathcal{I}((g - P_{n;g})\hat{S}, \hat{I}) = \frac{g^{(n+1)}(S(\xi))}{(n+1)!}\mathcal{I}(\omega_{n+1}\hat{S}, \tilde{I}),$$

as claimed. □

Remark We observe that if S is such that g is a polynomial of degree n, then the quadrature has exactness n.

We now focus on the computation of the weights for the two maps introduced above for S-Gibbs and S-Runge, respectively.

First, we recall that considering $h(x) = 1/\sqrt{1-x^2}$, the Chebyshev weight function and C_n the CL nodes on $J = [-1, 1]$

$$\mathcal{I}(fh, J) \approx \sum_{i=0}^{n} f(c_i)w_i^c \tag{1.22}$$

where $w_i^c = \frac{\pi}{z_i n}$ with

$$z_i = \begin{cases} 2 & i = 0, n \\ 1 & \text{otherwise} \end{cases}$$

We are ready to prove the following.

Theorem 1.3 *Let X_n be the set of $n + 1$ equispaced points of $I = [a, b]$ and consider the S-Runge map*

$$S(x) = -\cos\left(\frac{x-a}{b-a}\pi\right). \tag{1.23}$$

Then,

$$w_k^S = \begin{cases} \frac{h}{2}, \; \textit{for}\; k \in \{0, n\}, \\ h, \; \textit{for}\; k = 1, \dots, n-1. \end{cases}$$

Proof Let us now take on $I = [a, b]$ the set of equispaced nodes X_n and the S-map (1.23). Thus, $S(X_n) = C_n$ and letting $t = S(x)$ then $l_i(S(x)) = l_i(t),\ i = 0, \dots, n$, where $l_i(t)$ is the Lagrange basis at the CL nodes of $J = [-1, 1]$. Therefore, by using (1.20) and (1.21)

$$dx = \hat{S}(t)\, dt = h(t)\frac{b-a}{\pi}\, dt,$$

we have

$$w_i^S = \frac{(b-a)}{\pi} \mathcal{I}(l_i\, h, J).$$

Finally, by observing that the quadrature rule (1.22) is exact for polynomials of degree n, we get

$$w_i^S = \frac{(b-a)}{\pi} \sum_{j=0}^{n} l_i(c_j) w_j^c = \frac{(b-a)}{\pi} \sum_{j=0}^{n} \delta_{i,j} \frac{\pi}{z_j n} = \frac{b-a}{z_i n}.$$

This concludes the proof. □

Remarks

1. This result turns out to be quite surprising but interesting: the weights of the "fake" nodes quadrature, coincide with those of the trapezoidal rule and up to the constant $\frac{\pi}{b-a}$ with those for the weighted quadrature based on the CL nodes in $[-1, 1]$.
2. In the case of composite quadrature rules, the weights for the "fake" nodes approach can be computed by applying the S-map on *each* subinterval.

1.5.1.1 Examples

We apply the "new" quadrature approach to the following test functions: a discontinuous one and a Runge-type one

$$f_1(x) = \begin{cases} \sin(x), & \text{for } x \leq 0, \\ \log(x^4+4)+7, & \text{for } x > 0, \end{cases} \quad \text{and} \quad f_2(x) = \frac{1}{4x^2+1}.$$

We compute their integrals over the interval $I = [-2, 2]$. In Fig. 1.9 the absolute error between the true value of the integral and its approximation is displayed. As approximants we compare the standard approach on a equispaced points, the classical quadrature on Chebyshev-Lobatto nodes and the integral computed with the "fake" nodes approach: we use the S-Gibbs and S-Runge for f_1 and f_2, respectively. We consider polynomials with degrees $n = 2k$ with $k = 1, \ldots, 20$. We observe that the "fake" nodes quadrature outperforms the computation of the integral on equispaced nodes while still competitive with the classical Gaussian quadrature based on CL nodes.

The experiments have been carried out in PYTHON 3.7 using NUMPY 1.15. The Python codes are available at https://github.com/pog87/FakeQuadrature.

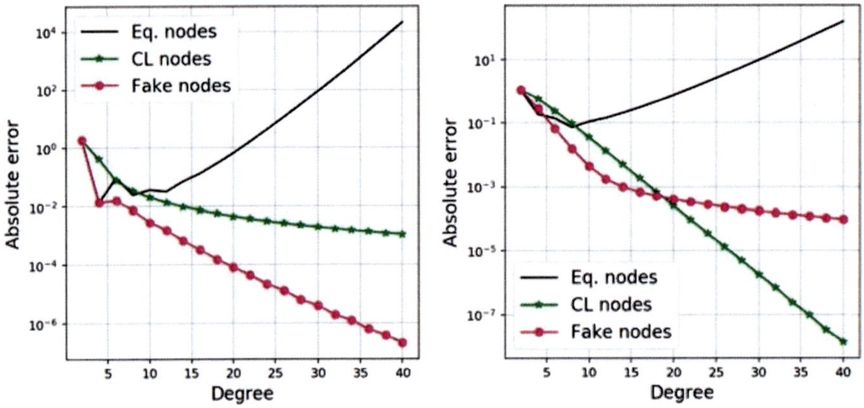

Fig. 1.9 Left: approximation of the integral over I of function f_1 and f_2, left and right respectively

1.6 Discontinuous Kernels

Interpolation by kernels, mainly by radial kernels known as *Radial Basis Functions* (RBF), are suitable tools for high-dimensional scattered data problems, solution of PDES, machine learning, image registration. For an overview of the topic we refer the reader to the monographs [34, 60] and references therein. Our interest is now confined to the approximation of data with discontinuities. Indeed, based on recent studies on *Variably Scaled Kernels (VSKs)* [18, 55] and their discontinuous extension [26], we use discontinuous kernel functions that reflect discontinuities in the data as a basis. These basis functions, referred to as Variably Scaled Discontinuous Kernels (VSDKs), enable us naturally to interpolate functions with given discontinuities.

1.6.1 A Brief Introduction to RBF Approximation

We start by introducing some basic notation and results about RBF interpolation.

Let $\mathcal{X}_N = \{\boldsymbol{x}_i,\ i = 1, \ldots, N\}$ be a set of distinct data points (data sites or nodes) arbitrarily distributed on a domain $\Omega \subseteq \mathbb{R}^d$ and let $\mathcal{F}_N = \{f_i = f(\boldsymbol{x}_i),\ i = 1, \ldots, N\}$ be an associated set of data values (measurements or function values) obtained by sampling some (unknown) function $f : \Omega \longrightarrow \mathbb{R}$ at the nodes $\boldsymbol{x}_i$. We can reconstruct f by interpolation, that is by finding a function $\mathcal{P} : \Omega \longrightarrow \mathbb{R}$ satisfying the interpolation conditions

$$\mathcal{P}(\boldsymbol{x}_i) = f_i, \quad i = 1, \ldots, N. \tag{1.24}$$

This problem (1.24) has unique solution provided $\mathcal{P} \in \text{span}\{\Phi_\varepsilon(\cdot, \boldsymbol{x}_i), \boldsymbol{x}_i \in \mathcal{X}\}$, where $\Phi_\varepsilon : \Omega \times \Omega \longrightarrow \mathbb{R}$ is a strictly positive definite and symmetric kernel. Notice that Φ depends on the so-called *shape parameter* $\varepsilon > 0$ which allows to change the shape of Φ, making it flatter (or wider) as $\varepsilon \to 0+$ or spiky and so more localized as $\varepsilon \to +\infty$. This has important consequences in error analysis and stability of interpolation by RBF (cf. e.g. [58]).

The resulting kernel-based interpolant, denoted by $\mathcal{P}_{\varepsilon,\mathcal{X}_N}$, can be written as

$$\mathcal{P}_{\varepsilon,\mathcal{X}_N}(\boldsymbol{x}) = \sum_{k=1}^{N} c_k \Phi_\varepsilon(\boldsymbol{x}, \boldsymbol{x}_k), \quad \boldsymbol{x} \in \Omega. \tag{1.25}$$

The interpolation problem (1.24) in matrix form becomes $A_\varepsilon \in \mathbb{R}^{N\times N}$ with $(A_\varepsilon)_{ik} = \Phi_\varepsilon(\boldsymbol{x}_i, \boldsymbol{x}_k),\ i, k = 1, \ldots, N$. Then, letting $\boldsymbol{f} = (f_1, \ldots, f_N)^\intercal$ the vector of data values, we can find the coefficients $\boldsymbol{c} = (c_1, \ldots, c_N)^\intercal$ by solving the linear system $A_\varepsilon \boldsymbol{c} = \boldsymbol{f}$. Since we consider strictly positive definite and symmetric kernels, the existence and uniqueness of the solution of the linear system is ensured. More precisely, the class of strictly positive definite and symmetric *radial kernels* Φ_ε can be defined as follows.

Definition 1.2 Φ_ε is called radial kernel if there exists a continuous function $\varphi_\varepsilon : [0, +\infty) \longrightarrow \mathbb{R}$, depending on the shape parameter $\varepsilon > 0$, such that

$$\Phi_\varepsilon(\boldsymbol{x}, \boldsymbol{y}) = \varphi_\varepsilon(\|\boldsymbol{x} - \boldsymbol{y}\|_2), \tag{1.26}$$

for all $\boldsymbol{x}, \boldsymbol{y} \in \Omega$.

Remark The notation (1.26) provides the “dimension-blindness” property of RBF. Hence, once we know the function φ and compose it with the Euclidean norm, we get a radial kernel.

In Table 1.2 we collect some of the most popular basis functions φ which are strictly positive definite. The Gaussian, Inverse Multiquadrics and Matérn (M0) are globally supported, while the Wendland (W2) and Buhmann (B2) are locally supported.

To Φ_ε we can associate a real pre-Hilbert space $H_{\Phi_\varepsilon}(\Omega)$

$$H_{\Phi_\varepsilon}(\Omega) = \text{span}\{\Phi_\varepsilon(\cdot, \boldsymbol{x}),\ \boldsymbol{x} \in \Omega\},$$

with Φ_ε as reproducing kernel. This space will be then equipped with the bilinear form $(\cdot, \cdot)_{H_{\Phi_\varepsilon}(\Omega)}$. Then we define the associate *native space* $\mathcal{N}_{\Phi_\varepsilon}(\Omega)$ of Φ_ε as the completion of $H_{\Phi_\varepsilon}(\Omega)$ with respect to the norm $||\cdot||_{H_{\Phi_\varepsilon}(\Omega)}$, that is $||f||_{H_{\Phi_\varepsilon}(\Omega)} =$

Table 1.2 Examples of strictly positive definite radial kernels depending on the shape parameter ε. The truncated power function is denoted by $(\cdot)_+$. W2 and B2 are compactly supported radial basis functions

Kernel	$\varphi(r)$
Gaussian C^∞ (GA)	$e^{-\varepsilon^2 r^2}$
Inverse Multiquadrics C^∞ (IM)	$(1+r^2/\varepsilon^2)^{-1/2}$
Matérn C^0 (M0)	$e^{-\varepsilon r}$
Wendland C^2 (W2)	$(1-\varepsilon r)_+^4\,(4\varepsilon r+1)$
Buhmann C^2 (B2)	$2r^4\log r - 7/2r^4 + 16/3r^3 - 2r^2 + 1/6$

$||f||_{\mathcal{N}_{\Phi_\varepsilon}(\Omega)}$ for all $f \in H_{\Phi_\varepsilon}(\Omega)$. For more details, as already quotes, see the monographs [34, 60]).

The accuracy of the interpolation process can be expressed, for instance, in terms of the *power function*. The power function is a positive function given as the ratio of two determinants (cf. [23])

$$P_{\Phi_\varepsilon,\mathcal{X}_N}(\boldsymbol{x}) := \sqrt{\frac{\det(A_\varepsilon(\mathcal{Y}_{N+1}))}{\det(A_\varepsilon(\mathcal{X}_N))}}\,. \tag{1.27}$$

where $A_\varepsilon(\mathcal{X}_N)$ is the interpolation matrix related to the set of nodes $\mathcal{X}_N$ and the kernel Φ_ε and $A_\varepsilon(\mathcal{Y}_{N+1})$ the matrix associated to the set $\mathcal{Y}_{N+1} := \{\boldsymbol{x}\}\cup\mathcal{X}_N,\ \boldsymbol{x}\in\Omega$.

The following pointwise error bound holds.

Theorem 1.4 *Let $\Phi_\varepsilon \in C(\Omega\times\Omega)$ be a strictly positive definite kernel and $\mathcal{X}_N \subseteq \Omega$ a set of N distinct points. For all $f \in \mathcal{N}_{\Phi_\varepsilon}(\Omega)$ we have*

$$|f(\boldsymbol{x}) - \mathcal{P}_{\varepsilon,\mathcal{X}_N}(\boldsymbol{x})| \le P_{\Phi_\varepsilon,\mathcal{X}_N}(\boldsymbol{x})\,\|f\|_{\mathcal{N}_{\Phi_\varepsilon}(\Omega)}, \quad \boldsymbol{x}\in\Omega.$$

Remarks We observe that this Theorem bounds the error in terms of the power function which depends on the kernel and the data points but is independent on the function values.

This theorem is a special instance of [60, Theorem 11.3, p. 182] where the fill-distance

$$h_{\mathcal{X}_N,\Omega} := \sup_{\boldsymbol{x}\in\Omega}\ \min_{\boldsymbol{x}_k\in\mathcal{X}_N} \|\boldsymbol{x}-\boldsymbol{x}_k\|\,,$$

is used instead of the power function.

1.6.2 From RBF to VSK Interpolation

The choice of the shape parameter ε is a crucial computational issue in RBF interpolation leading to instability effects without a clever choice of it. This is usually done by analyzing the behaviour of some kind of errors (like the Root Mean Square Error) versus the conditioning of the interpolation matrix (cf. e.g. [37, 52]) and so many techniques has been developed in order to overcome such problems. Many of these techniques allow to choose the *best* shape parameter based on a "trade-off" between conditioning and efficiency. There are approaches based on the choice of well-conditioned bases, like in the RBF-QR method for Gaussians [38] or in the more general setting discussed in [24].

In the seminal paper[18] the authors introduced the so called *Variably Scaled Kernels (or VSKs)* where the classical tuning strategy of finding the optimal shape parameter, is substituted by the choice of a *scale function* which plays the role of a density function. More precisely [18, Def. 2.1]

Definition 1.3 Letting $\mathcal{I} \subseteq (0, +\infty)$ and Φ_ε a positive definite radial kernel on $\Omega \times \mathcal{I}$ depending on the shape parameter $\varepsilon > 0$. Given a scale function $\psi : \Omega \longrightarrow \mathcal{I}$, a Variably Scaled Kernel Φ_ψ on Ω is

$$\Phi_\psi(\boldsymbol{x}, \boldsymbol{y}) := \Phi_1((\boldsymbol{x}, \psi(\boldsymbol{x})), (\boldsymbol{y}, \psi(\boldsymbol{y}))), \tag{1.28}$$

for $\boldsymbol{x}, \boldsymbol{y} \in \Omega$.

Defining then the map $\Psi(\boldsymbol{x}) = (\boldsymbol{x}, \psi(\boldsymbol{x}))$ on Ω, the interpolant on the set of nodes $\Psi(\mathcal{X}_N) := \{(\boldsymbol{x}_k, \psi(\boldsymbol{x}_k)), \ \boldsymbol{x}_k \in \mathcal{X}_N\}$ with fixed shape parameter $\varepsilon = 1$ (or any other constant value c) takes the form

$$\mathcal{P}_{1,\Psi(\mathcal{X}_N)}(\Psi(\boldsymbol{x})) = \sum_{k=1}^{N} c_k \Phi_1(\Psi(\boldsymbol{x}), \Psi(\boldsymbol{x}_k)), \tag{1.29}$$

with $\boldsymbol{x} \in \Omega, \ \boldsymbol{x}_k \in \mathcal{X}_N$.

By analogy with the interpolant in (1.25), the vector of coefficients $\boldsymbol{c} = (c_1, \ldots, c_N)^\mathsf{T}$ is determined by solving the linear system $A_\psi \boldsymbol{c} = \boldsymbol{f}$, where $\boldsymbol{f}$ is the vector of data values and $(A_\psi)_{ik} = \Phi_\psi(\boldsymbol{x}_i, \boldsymbol{x}_k)$ is strictly positive definite because of the strictly positiveness of Φ_ψ.

Once we have the interpolant $\mathcal{P}_{1,\Psi(\mathcal{X}_N)}$ on $\Omega \times \mathcal{I}$, we can project back on Ω the points $(\boldsymbol{x}, \psi(\boldsymbol{x})) \in \Omega \times \mathcal{I}$. In this way, we obtain the so-called VSK interpolant $\mathcal{V}_\psi$ on Ω. Indeed, by using (1.28), we get

$$\mathcal{V}_\psi(\boldsymbol{x}) := \sum_{k=1}^{N} c_k \Phi_\psi(\boldsymbol{x}, \boldsymbol{x}_k) = \sum_{k=1}^{N} c_k \Phi_1(\Psi(\boldsymbol{x}), \Psi(\boldsymbol{x}_k)) = \mathcal{P}_{1,\Psi(\mathcal{X}_N)}(\Psi(\boldsymbol{x})). \tag{1.30}$$

The error and stability analysis of this varying scale process on Ω coincides with the analysis of a fixed scale kernel on $\Omega \times \mathcal{I}$ (for details and analysis of these continuous scale functions see [18]).

1.6.3 *Variably Scaled Discontinuous Kernels (VSDK)*

To understand the construction of a VSDK let start from the one dimensional case. Let $\Omega = (a, b) \subset \mathbb{R}$ be an open interval, $\xi \in \Omega$ and the discontinuous function $f : \Omega \longrightarrow \mathbb{R}$

$$f(x) := \begin{cases} f_1(x), & a < x < \xi, \\ f_2(x), & \xi \le x < b, \end{cases}$$

where f_1, f_2 are real valued smooth functions for which exist finite the limits $\lim_{x\to a^+} f_1(x)$, $\lim_{x\to b^-} f_2(x)$ and $f_2(\xi) \neq \lim_{x\to\xi} f_1(x)$.

If we approximate f on some set of nodes, say $\mathcal{X} \subset \Omega$, the presence of a jump will result in unpleasant oscillations due to the *Gibbs phenomenon*. The idea is then to approximate f at $\mathcal{X}$ by interpolants of the form (1.30) with the main issue of considering discontinuous scale functions in the interpolation process. The strategy is that of approximating a function having jumps by using a scale function that has jumps discontinuities at the same positions of the considered function.

To this aim, take $\alpha, \beta \in \mathbb{R}$, $\alpha \neq \beta$, $\mathcal{S} = \{\alpha, \beta\}$ and the scale function $\psi : \Omega \longrightarrow \mathcal{S}$

$$\psi(x) := \begin{cases} \alpha, \; x < \xi, \\ \beta, \; x \ge \xi. \end{cases}$$

which is piecewise constant, having a discontinuity exactly at ξ as the function f.

Then we consider Φ_ε a positive definite radial kernel on $\Omega \times \mathcal{S}$, possibly depending on a shape parameter $\varepsilon > 0$ or alternatively a VSK Φ_ψ on Ω as in (1.28). For the sake of simplicity we start by taking the function φ_1 related to the kernel Φ_1 that is

$$\begin{aligned} \varphi_1(\|\Psi(x) - \Psi(y)\|_2) &= \varphi_1(\|(x, \psi(x)) - (y, \psi(y))\|_2) = \\ &= \varphi_1\left(\sqrt{(x-y)^2 + (\psi(x) - \psi(y))^2}\right), \end{aligned}$$

so that

$$\varphi_1(\|\Psi(x) - \Psi(y)\|_2) = \begin{cases} \varphi_1(|x - y|), & x, y < \xi \quad \text{or } x, y \ge \xi, \\ \varphi_1(\|(x, \alpha) - (y, \beta)\|_2), & x < \xi \le y \text{ or } y < \xi \le x, \end{cases}$$

noticing that $\varphi_1(\|(x,\alpha)-(y,\beta)\|_2) = \varphi_1(\|(x,\beta)-(y,\alpha)\|_2)$.

Then, consider the set $\mathcal{X}_N = \{x_k,\ k = 1,\dots,N\}$ of points of Ω and the interpolant $\mathcal{V}_\psi : \Omega \longrightarrow \mathbb{R}$ which is a linear combination of discontinuous functions $\Phi_\psi(\cdot, x_k)$ having a jump at ξ.

- if $a < x_k < \xi$

$$\Phi_\psi(x,x_k) = \begin{cases} \varphi_1(|x-x_k|), & x < \xi, \\ \varphi_1(\|(x,\alpha)-(x_k,\beta)\|_2), & x \geq \xi, \end{cases}$$

- if $\xi \leq x_k < b$

$$\Phi_\psi(x,x_k) = \begin{cases} \varphi_1(|x-x_k|), & x \geq \xi, \\ \varphi_1(\|(x,\alpha)-(x_k,\beta)\|_2), & x < \xi. \end{cases}$$

By this construction we can give the following definition that extends the idea when we have more than one jump.

Definition 1.4 Let $\Omega = (a,b) \subset \mathbb{R}$ be an open interval, $\mathcal{S} = \{\alpha,\beta\}$ with $\alpha,\beta \in \mathbb{R}_{>0}$, $\alpha \neq \beta$ and let $\mathcal{D} = \{\xi_j,\ j = 1,\dots,\ell\} \subset \Omega$ be a set of distinct points with $\xi_j < \xi_{j+1}$ for every j. Let $\psi : \Omega \longrightarrow \mathcal{S}$ the scale function defined as

$$\psi(x) := \begin{cases} \alpha, & x \in (a,\xi_1) \text{ or } x \in [\xi_j,\xi_{j+1}), \text{ where } j \text{ is even}, \\ \beta, & x \in [\xi_j,\xi_{j+1}), \qquad\qquad\qquad\quad \text{where } j \text{ is odd}, \end{cases}$$

and

$$\psi(x)_{|[\xi_\ell,b)} := \begin{cases} \alpha, & \ell \text{ is even}, \\ \beta, & \ell \text{ is odd}. \end{cases}$$

The kernel Φ_ψ is then called a Variably Scaled Discontinuous Kernel on Ω.

For the analysis of the VSDKs introduced in Definition 1.4 we cannot rely on some important and well-known results of RBF interpolation due to the discontinuous nature of these kernels. Therefore we may proceed as follows.

Let Ω and $\mathcal{D}$ be as in Definition 1.4 and $n \in \mathbb{N}$. We define $\psi_n : \Omega \longrightarrow \mathcal{I} \subseteq (0,+\infty)$ as

$$\psi_n(x) := \begin{cases} \alpha, & x \in (a,\xi_1 - 1/n) \text{ or } x \in [\xi_j + 1/n, \xi_{j+1} - 1/n) \ j \text{ is even}, \\ \beta, & x \in [\xi_j + 1/n, \xi_{j+1} - 1/n) \qquad\qquad\qquad\qquad\quad j \text{ is odd}, \\ \gamma_1(x), & x \in [\xi_j - 1/n, \xi_j + 1/n) \qquad\qquad\qquad\qquad\quad\ \ j \text{ is odd}, \\ \gamma_2(x), & x \in [\xi_j - 1/n, \xi_j + 1/n) \qquad\qquad\qquad\qquad\quad\ \ j \text{ is even}, \end{cases} \tag{1.31}$$

$$\psi_n(x)_{|[\xi_\ell+1/n,b)} := \begin{cases} \alpha, \ \ell \text{ is even}, \\ \beta, \ \ell \text{ is odd}, \end{cases}$$

where γ_1, γ_2 are continuous, strictly monotonic functions so that

$$\lim_{x \to \xi_{j+1}+1/n} \gamma_1(x) = \gamma_2(\xi_j - 1/n) = \beta, \qquad \lim_{x \to \xi_{j+1}+1/n} \gamma_2(x) = \gamma_1(\xi_j - 1/n) = \alpha.$$

From Definition 1.4, it is straightforward to verify that $\forall x \in \Omega$ the following pointwise convergence result holds

$$\lim_{n \to \infty} \psi_n(x) = \psi(x).$$

We point out that for every fixed $n \in \mathbb{N}$ the kernel Φ_{ψ_n} is a continuous VSK, hence it satisfies the error bound of Theorem 1.4. For VSDKs instead we have the following results whose proofs can be found in the paper [26].

Theorem 1.5 *For every $x, y \in \Omega$, we have*

$$\lim_{n \to \infty} \Phi_{\psi_n}(x, y) = \Phi_\psi(x, y),$$

where Φ_ψ is the kernel considered in Definition 1.4.

Corollary 1.1 *Let $H_{\Phi_{\psi_n}}(\Omega) = span\{\Phi_{\psi_n}(\cdot, x), \ x \in \Omega\}$ be equipped with the bilinear form $(\cdot, \cdot)_{H_{\Phi_{\psi_n}}(\Omega)}$ and let $\mathcal{N}_{\Phi_{\psi_n}}(\Omega)$ be the related native space. Then, taking the limit of the basis functions, we obtain the space $H_{\Phi_\psi}(\Omega) = span\{\Phi_\psi(\cdot, x), \ x \in \Omega\}$ equipped with the bilinear form $(\cdot, \cdot)_{H_{\Phi_\psi}(\Omega)}$ and the related native space $\mathcal{N}_{\Phi_\psi}(\Omega)$.*

We get an immediate consequence for the interpolant $\mathcal{V}_\psi$ too.

Corollary 1.2 *Let Ω, $\mathcal{S}$ and $\mathcal{D}$ be as in Definition 1.4. Let $f : \Omega \longrightarrow \mathbb{R}$ be a discontinuous function whose step discontinuities are located at the points belonging to $\mathcal{D}$. Moreover, let ψ_n and ψ be as in Theorem 1.5. Then, considering the interpolation problem with nodes $\mathcal{X}_N = \{x_k, \ k = 1, \ldots, N\}$ on Ω, we have*

$$\lim_{n \to \infty} \mathcal{V}_{\psi_n}(x) = \mathcal{V}_\psi(x),$$

for every $x \in \Omega$.

To provide error bounds in terms of the *power function*, we should first define the equivalent power function for a VSDK Φ_ψ on a set of nodes $\mathcal{X}_N$. From (1.27), we immediately have

$$P_{\Phi_\psi, \mathcal{X}}(x) = \sqrt{\frac{\det(A_\psi(\mathcal{Y}_{N+1}))}{\det(A_\psi(\mathcal{X}_N))}}.$$

From Theorem 1.5 and Corollary 1.1, we may define the power function for a discontinuous kernel as

$$P_{\Phi_\psi, \mathcal{X}_N}(x) = \lim_{n\to\infty} P_{\Phi_{\psi_n}, \mathcal{X}_N}(x), \; \forall x \in \Omega.$$

Finally, we can state the error bound for interpolation via VSDKs.

Theorem 1.6 *Let Φ_ψ be a VSDK on $\Omega = (a, b) \subset \mathbb{R}$, and $\mathcal{X}_N \subseteq \Omega$ consisting of N distinct points. For all $f \in \mathcal{N}_{\Phi_\psi}(\Omega)$ we have*

$$|f(x) - \mathcal{V}_\psi(x)| \leq P_{\Phi_\psi, \mathcal{X}_N}(x) \|f\|_{\mathcal{N}_{\Phi_\psi}(\Omega)}, \quad x \in \Omega.$$

Proof For all $n \in \mathbb{N}$ and $x \in \Omega$, since the VSK Φ_{ψ_n} is continuous, from Theorem 1.4, we get

$$|f(x) - \mathcal{V}_{\psi_n}(x)| \leq P_{\Phi_{\psi_n}, \mathcal{X}}(x) \|f\|_{\mathcal{N}_{\Phi_{\psi_n}}(\Omega)}.$$

Then, taking the limit $n \to \infty$ and recalling the results of this subsection, the thesis follows. □

1.6.4 VSDKs: Multidimensional Case

VSDKs rely upon the classical RBF bases, therefore as noticed are "dimension-blind" which make them a suitable and flexible tool to approximate data in any dimension. However, in higher dimensions, the geometry is more complex, so we must pay attention in defining the scale function ψ on a bounded domain $\Omega \subset \mathbb{R}^d$.

We consider the following setting.

(i) The bounded set $\Omega \subset \mathbb{R}^d$ is the union of n pairwise disjoint sets Ω_i and $\mathcal{P}_\Omega = \{\Omega_1, \ldots, \Omega_n\}$ forms a partition of Ω.
(ii) The subsets Ω_i satisfy an interior cone condition and have a Lipschitz boundary.
(iii) Let $\alpha_1, \ldots, \alpha_n \in \mathbb{R}$ and $\Sigma = \{\alpha_1, \ldots, \alpha_n\}$. The function $\psi : \Omega \to \Sigma$ is constant on the subsets Ω_i, i.e., $\psi(\boldsymbol{x}) = \alpha_i$ for all $\boldsymbol{x} \in \Omega_i$. In particular, ψ is piecewise constant on Ω and discontinuities appear only at the boundaries of the subsets Ω_i. We further assume that $\alpha_i \neq \alpha_j$ if Ω_i and Ω_j are neighboring sets.

A suitable scale function ψ for interpolating f via VSDKs on $\Omega \subset \mathbb{R}^d$ can be defined as follows.

Let $\Omega \subset \mathbb{R}^d$ satisfies the assumptions (i)–(iii) above. We define the scale function $\psi : \Omega \longrightarrow \mathcal{S}$ as

$$\psi_{|\Omega_i} := \alpha_i. \tag{1.32}$$

Definition 1.5 Given the scale function (1.32) the kernel Φ_ψ defined by (1.28) is a Variably Scaled Discontinuous Kernel on Ω.

Remarks In Definition 1.5 we choose a scale function which emulates the properties of the one-dimensional function of Definition 1.4. The difference is that the multidimensional ψ could be discontinuous not just at the same points as f, but also on other nodes. That is all the jumps of f lye along $(d-1)$-dimensional manifolds $\gamma_1, \ldots, \gamma_p$ which verify

$$\gamma_i \subseteq \bigcup_{i=1}^{n} \partial\Omega_i \setminus \partial\Omega, \quad \forall i = 1, \ldots, p.$$

More precisely, if we are able to choose $\mathcal{P}_\Omega$ so that

$$\bigcup_{i=1}^{p} \gamma_i = \bigcup_{i=1}^{n} \partial\Omega_i \setminus \partial\Omega,$$

then f and ψ have the same discontinuities. Otherwise, if

$$\bigcup_{i=1}^{p} \gamma_i \subset \bigcup_{i=1}^{n} \partial\Omega_i \setminus \partial\Omega,$$

then ψ is discontinuous along $\bigcup_{i=1}^{n} \partial\Omega_i \setminus \left(\partial\Omega \cup \bigcup_{i=1}^{p} \gamma_i\right)$, while f is not.

The theoretical analysis in the multidimensional case is done along the same path of the one-dimensional setting. The idea is to consider continuous scale functions $\psi_n : \Omega \longrightarrow \mathcal{I} \subseteq (0, +\infty)$ such that $\forall \boldsymbol{x} \in \Omega$ the limits hold

$$\lim_{n\to\infty} \psi_n(\boldsymbol{x}) = \psi(\boldsymbol{x}),$$

and

$$\lim_{n\to\infty} \mathcal{V}_{\psi_n}(\boldsymbol{x}) = \mathcal{V}_\psi(\boldsymbol{x}).$$

We omit this extension and all the corresponding considerations which are similar to those discussed above for the one-dimensional setting, while we state directly the error bound.

Theorem 1.7 *Let Φ_ψ be a VSDK as in Definition 1.5. Suppose that $\mathcal{X}_N = \{\boldsymbol{x}_i, i = 1, \dots, N\} \subseteq \Omega$ have distinct points. For all $f \in \mathcal{N}_{\Phi_\psi}(\Omega)$ we have*

$$|f(\boldsymbol{x}) - \mathcal{V}_\psi(\boldsymbol{x})| \leq P_{\Phi_\psi, \mathcal{X}_N}(\boldsymbol{x}) \|f\|_{\mathcal{N}_{\Phi_\psi}(\Omega)}, \quad \boldsymbol{x} \in \Omega.$$

Proof Just refer to Theorem 1.6 and considerations made above. □

For the characterization of the native spaces for the VSDKs (if the discontinuities are known) and Sobolev-type error estimates, based on the fill-distance, of the respective interpolation scheme the reader must refer to the very recent paper [28].

1.7 Application to MPI

As we already observed, interpolation is an essential tool in medical imaging. It is required for example in geometric alignment or registration of images, to improve the quality of images on display devices, or to reconstruct the image from a compressed amount of data. In Computerized Tomography (CT) and Magnetic Resonance Imaging (MRI), which are examples of medical inverse problems, interpolation is used in the reconstruction step in order to fit the discrete Radon data into the back projection step. Similarly in SPECT for regridding the projection data in order to improve the reconstruction quality while reducing the acquisition computational cost [59]. In Magnetic Particle Imaging (MPI), the number of calibration measurements can be reduced by using interpolation methods, as well (see the important paper [46]).

In the early 2000s, B. Gleich and J. Weizenecker [39], invented at Philips Research in Hamburg a new quantitative imaging method called Magnetic Particle Imaging (MPI). In this imaging technology, a tracer consisting of superparamagnetic iron oxide nanoparticles is injected and then detected through the superimposition of different magnetic fields. In common MPI scanners, the acquisition of the signal is performed following a generated field free point (FFP) along a chosen sampling trajectory. The determination of the particle distribution given the measured voltages in the receive coils is an ill-posed inverse problem that can be solved only with proper regularization techniques [47].

Commonly used trajectories in MPI are *Lissajous curves* [48], which are parametric space-filling curves of the square $Q_2 = [-1, 1]^2$. More precisely, by using the same notations in [31–33], let $\boldsymbol{n} = (n_1, n_2) \in \mathbb{N}^2$ be a vector of relatively prime integers and $\epsilon \in \{1, 2\}$, the two-dimensional Lissajous curve $\gamma_\epsilon^{\boldsymbol{n}} : [0, 2\pi] \to Q_2$ is defined as

$$\gamma_\epsilon^{\boldsymbol{n}}(t) := (\cos(n_2 t), \cos(n_1 t - (\epsilon - 1)\pi/(2n_2))) .$$

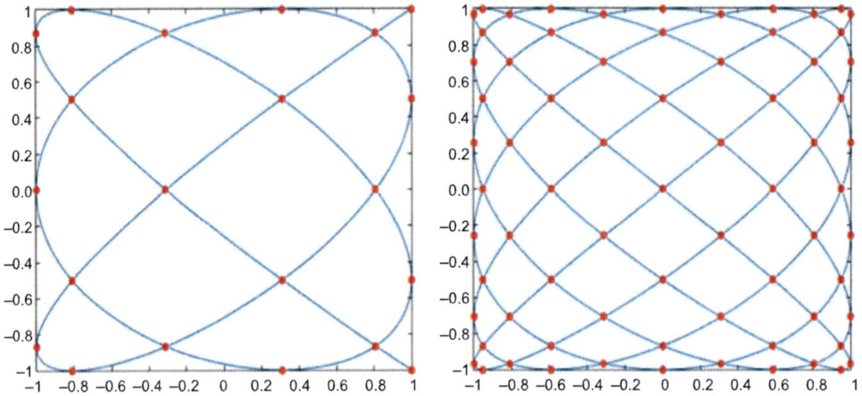

Fig. 1.10 Left: The degenerate curve $\gamma_1^{(5,6)}$. Right: the non-degenerate curve $\gamma_2^{(5,6)}$

The curve $\gamma_\epsilon^{\boldsymbol{n}}$ is called *degenerate* if $\epsilon = 1$, and *non-degenerate* if $\epsilon = 2$. The *Padua points* of degree n are a degenerate Lissajous curve which have generating curve $\gamma_{Pad}(t) = (-\cos((n+1)t), -\cos(nt)), \quad t \in [0, \pi]$ (see also [11, 16]). The set of Lissajous node points associated to the curve $\gamma_\epsilon^{\boldsymbol{n}}$ is given by

$$\mathrm{LS}_\epsilon^{\boldsymbol{n}} := \left\{\gamma_\epsilon^{\boldsymbol{n}}\left(\tfrac{\pi k}{\epsilon n_1 n_2}\right) : \qquad k = 0, \ldots, 2\epsilon n_1 n_2 - 1\right\}. \tag{1.33}$$

We define also the index set $\Gamma^{2\boldsymbol{n}} := \left\{(i, j) \in \mathbb{N}_0^2 : (i/2n_1) + (j/2n_2) < 1\right\} \cup \{(0, 2n_2)\}$ which give the cardinality of the set, that is

$$\#\mathrm{LS}_\epsilon^{\boldsymbol{n}} = \frac{(\epsilon n_1 + 1)(\epsilon n_2 + 1) - (\epsilon - 1)}{2}.$$

To reduce the amount of calibration measurements, it is shown in [46] that the reconstruction can be restricted to particular sampling points along the Lissajous curves, i.e., the Lissajous nodes $\mathbf{LS}_2^{(n)}$ introduced in (1.33). Furthermore, by using a polynomial interpolation method on the Lissajous nodes the entire density of the magnetic particles can then be restored (cf. [33]). As noticed, these sampling nodes and the corresponding polynomial interpolation are an extension of the theory of the *Padua points* (see [11, 16] and also https://en.wikipedia.org/wiki/Padua_points).

If the original particle density has sharp edges, the polynomial reconstruction scheme on the Lissajous nodes is affected by the Gibbs phenomenon. As shown in [25], post-processing filters can be used to reduce oscillations for polynomial reconstruction in MPI. In the following, we demonstrate that the usage of the VSDK interpolation method in combination with the presented edge estimator effectively avoids ringing artifacts in MPI and provides reconstructions with sharpened edges.

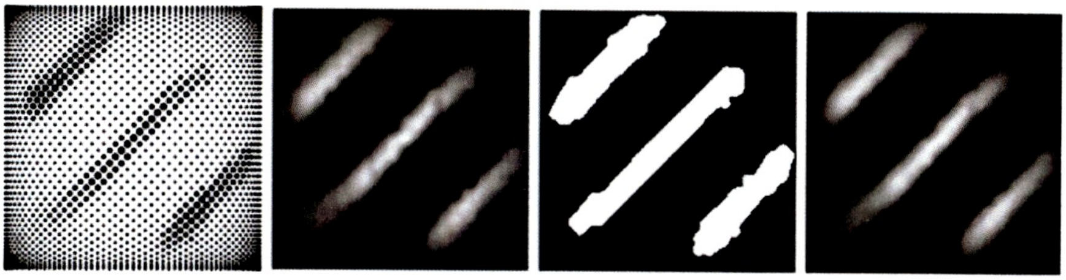

Fig. 1.11 Comparison of different interpolation methods in MPI. The reconstructed data on the Lissajous nodes $\mathbf{LS}_2^{(33,32)}$ (left) is first interpolated using the polynomial scheme derived in [33] (middle left). Using a mask constructed upon a threshold strategy described in [28] (middle right) the second interpolation is performed by the VSDK scheme (right)

1.7.1 An Example

As a test data set, we consider MPI measurements conducted in [46] on a phantom consisting of three tubes filled with Resovist, a contrast agent consisting of superparamagnetic iron oxide. By the proceeding described in [46] we then obtain a reconstruction of the particle density on the Lissajous nodes $\mathbf{LS}_2^{(33,32)}$ (due to the scanner available, as described in [46]). This reduced reconstruction on the Lissajous nodes is illustrated in Fig. 1.11 (left). A computed polynomial interpolant of this data is shown in Fig. 1.11 (middle, left). In this polynomial interpolant some ringing artifacts are visible The scaling function ψ for the VSDK scheme is then obtained by using the classification Algorithm [28] with a Gaussian function for the kernel machine. The resulting scaling function is visualized in Fig. 1.11 (middle, right). Using the C^0-Matérn (M0) kernel (see Table 1.2) for the VSDK interpolation, the final interpolant for the given MPI data is shown in in Fig. 1.11 (right).

1.8 Conclusion and Further Works

We have investigated the application of the polynomial mapped bases approach without resampling for reducing the Runge and Gibbs phenomena. The approach shows to be a kind of black-box that can be applied in many other frameworks. We indeed have applied it to barycentric rational approximation and quadrature. We have also studied the use of VSDK, a new family of variable scaled kernels, particularly effective in the presence of discontinuity in our data. A particular applications of VSDK is the image reconstruction from data coming from MPI scanners acquisitions.

Concerning the work in progress and the future works

- In the 2d case, we have results on discontinuous functions on the square, using polynomial approximation at the Padua points or tensor product meshes. In Fig. 1.12 we show the results of the interpolation of a discontinuous function

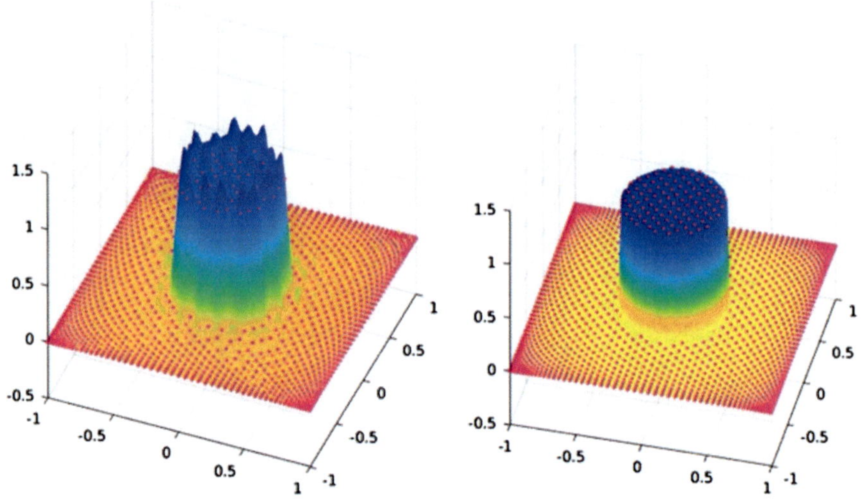

Fig. 1.12 Left: interpolation with Padua points of degree 60 of a function with a circular jump. Right: the same by mapping circularly the PD points, and using least-squares "fake-Padua"

along a disk of the square $[-1, 1]^2$, where the reconstruction has been done by interpolation on the Padua points of degree 60 on the left. On the right we show the same reconstruction where the points that do not fall inside the disk are mapped with a circular mapping. The mapping strategy indeed reduce the Gibbs oscillations, but outside the disk we cannot interpolate, we instead approximate by least-squares, because of the "fake Padua" points that are not anymore unisolvent.

- Again in 2d but also in 3d we can extract the so called *Approximate Fekete Points* of *Discrete Leja sequences* (cf. [12]) on various domains (disk, sphere, polygons, spherical caps, lunes and other domains). These points are numerically computed by numerical linear algebra methods and extracted from the so called *weakly admissible meshes* (WAM). For details about WAMs, refer to fundamental paper [13]

Finally we are working in improving the error analysis and finding more precise bounds for the Lebesgue constant(s). Among the applications of this approach we are interested to image registration in nuclear medicine and the reconstruction of periodic signals.

Acknowledgments I have to thanks especially who have collaborated with me in this project: Giacomo Elefante of the University of Fribourg and Wolfgang Erb, Francesco Marchetti, Emma Perracchione, Davide Poggiali of the University of Padova. We had many fruitful discussions that made the survey a nice overview of what can be done with "fake" nodes or discontinuous kernels for contrasting the Runge and Gibbs phenomena.

This research has been accomplished within Rete ITaliana di Approssimazione (RITA), partially funded by GNCS-INδAM, the NATIRESCO project BIRD181249 and through the European

Union's Horizon 2020 research and innovation programme ERA-PLANET, Grant Agreement No. 689443, via the GEOEssential project.

References

1. Adcock, B., Platte, R.B.: A mapped polynomial method for high-accuracy approximations on arbitrary grids. SIAM J. Numer. Anal. **54**, 2256–2281 (2016)
2. Amat, S., Dadourian, K., Liandrat, J.: On a nonlinear subdivision scheme avoiding Gibbs oscillations and converging towards C^s functions with $s > 1$. Math. Comput. **80**, 959–971 (2011)
3. Amat, S., Ruiz, J., Trillo, J.C., Yáñez, D.F.: Analysis of the Gibbs phenomenon in stationary subdivision schemes. Appl. Math. Lett. **76**, 157–163 (2018)
4. Archibald, R., Gelb, A., Yoon, J.: Fitting for edge detection in irregularly sampled signals and images. SIAM J. Numer. Anal. **43**, 259–279 (2005)
5. Bayliss, A., Turkel, E.: Mappings and accuracy for Chebyshev pseudo-spectral approximations. J. Comput. Phys. **101**, 349–359 (1992)
6. Beckermann, B., Matos, A.C., Wielonsky, F.: Reduction of the Gibbs phenomenon for smooth functions with jumps by the ϵ-algorithm. J. Comput. Appl. Math. **219**(2), 329–349 (2008)
7. Berrut, J.P., Elefante, G.: A periodic map for linear barycentric rational trigonometric interpolation. Appl. Math. Comput. **371** (2020). https://doi.org/10.1016/j.amc.2019.124924
8. Berrut, J.P., Mittelmann, H.D.: Lebesgue constant minimizing linear rational interpolation of continuous functions over the interval. Comput. Math. Appl. **33**, 77–86 (1997)
9. Berrut, J.P., Trefethen, L.N.: Barycentric Lagrange interpolation. SIAM Rev. **46**, 501–517 (2004)
10. Berrut, J.P., De Marchi, S., Elefante, G., Marchetti, F.: Treating the Gibbs phenomenon in barycentric rational interpolation via the S-Gibbs algorithm. Appl. Math. Lett. **103**, 106196 (2020)
11. Bos, L., Caliari, M., De Marchi, S., Vianello, M., Xu, Y.: Bivariate Lagrange interpolation at the Padua points: the generating curve approach. J. Approx. Theory **143**, 15–25 (2006)
12. Bos, L., De Marchi, S., Sommariva, A., Vianello, M.: Computing multivariate Fekete and Leja points by numerical linear algebra. SIAM J. Numer. Anal. **48**, 1984–1999 (2010)
13. Bos, L., Calvi, J.-P., Levenberg, N., Sommariva, A., Vianello, M.: Geometric weakly admissible meshes, discrete least squares approximation and approximate Fekete points. Math. Comput. **80**, 1601–1621 (2011)
14. Bos, L., De Marchi, D., Hormann, K.: On the Lebesgue constant of Berrut's rational interpolant at equidistant nodes. J. Comput. Appl. Math. **236**, 504–510 (2011)
15. Bos, L., De Marchi, S., Hormann, K., Sidon, J.: Bounding the Lebesgue constant of Berrut's rational interpolant at general nodes. J. Approx. Theory **169**, 7–22 (2013)
16. Bos, L., De Marchi, S., Vianello, M.: Polynomial approximation on Lissajous curves in the $d-$cube. Appl. Numer. Math. **116**, 47–56 (2017)
17. Boyd, J.P., Xu, F.: Divergence (Runge Phenomenon) for least-squares polynomial approximation on an equispaced grid and Mock-Chebyshev subset interpolation. Appl. Math. Comput. **210**, 158–168 (2009)
18. Bozzini, M., Lenarduzzi, L., Rossini, M., Schaback, R.: Interpolation with variably scaled kernels. IMA J. Numer. Anal. **35**, 199–219 (2015)
19. Brutman, L.: Lebesgue functions for polynomial interpolation - a survey. Ann. Numer. Math. **4**, 111–127 (1997)
20. Cheney, E.W.: Introduction to Approximation Theory, 2nd edn. AMS Chelsea Publishing, reprinted (1998)
21. Davis, P.J.: Interpolation and Approximation. Blaisdell, New York (1963)

22. De Marchi, S.: Polynomials arising in factoring generalized Vandermonde determinants: an algorithm for computing their coefficients. Math. Comput. Modelling **34**, 271–281 (2001)
23. De Marchi, S.: On optimal center locations for radial basis interpolation: computational aspects. Rend. Sem. Mat. Torino **61**(3), 343–358 (2003)
24. De Marchi, S., Santin, G.: A new stable basis for radial basis function interpolation. J. Comput. Appl. Math. **253**, 1–13 (2013)
25. De Marchi, S., Erb, W., Marchetti, F.: Spectral filtering for the reduction of the Gibbs phenomenon for polynomial approximation methods on Lissajous curves with applications in MPI. Dolomites Res. Notes Approx. **10**, 128–137 (2017)
26. De Marchi, S., Marchetti, M., Perracchione, E.: Jumping with variably scaled discontinuous kernels (VSDK). BIT Numer. Math. (2019). https://doi.org/10.1007/s10543-019-00786-z
27. De Marchi, S., Marchetti, M., Perracchione, E., Poggiali, D.: Polynomial interpolation via mapped bases without resampling. J. Comput. Appl. Math. **364**, 112347 (2020)
28. De Marchi, S., Erb, W., Marchetti, F., Perracchione, E., Rossini, M.: Shape-driven interpolation with discontinuous kernels: error analysis, edge extraction and application in MPI. SIAM J. Sci. Comput. **42**(2), B472–B491 (2020)
29. De Marchi, S., Elefante, G., Perracchione, E., Poggiali, D.: Quadrature at "fake" nodes. Dolomites Res. Notes Approx. **14**, 39–45 (2021)
30. De Marchi, S., Marchetti, F., Perracchione, E., Poggiali, D.: Python code for Fake Nodes interpolation approach. https://github.com/pog87/FakeNodes
31. Erb, W.: Bivariate Lagrange interpolation at the node points of Lissajous curves - the degenerate case. Appl. Math. Comput. **289**, 409–425 (2016)
32. Erb, W., Kaethner, C., Dencker, P., Ahlborg, M.: A survey on bivariate Lagrange interpolation on Lissajous nodes. Dolomites Res. Notes Approx. **8**(Special issue), 23–36 (2015)
33. Erb, W., Kaethner, C., Ahlborg, M., Buzug, T.M.: Bivariate Lagrange interpolation at the node points of non-degenerate Lissajous nodes. Numer. Math. **133**(1), 685–705 (2016)
34. Fasshauer, G.E., McCourt, M.J.: Kernel-Based Approximation Methods Using MATLAB. World Scientific, Singapore (2015)
35. Floater, M.S., Hormann, K.: Barycentric rational interpolation with no poles and high rates of approximation. Numer. Math. **107**(2), 315–331 (2007)
36. Fornberg, B., Flyer, N.: The Gibbs phenomenon for radial basis functions. In: The Gibbs Phenomenon in Various Representations and Applications. Sampling Publishing, Potsdam, NY (2008)
37. Fornberg, B., Wright, G.: Stable computation of multiquadrics interpolants for all values of the shape parameter. Comput. Math. Appl. **47**, 497–523 (2004)
38. Fornberg, B., Larsson, E., Flyer, N.: Stable computations with gaussian radial basis functions. SIAM J. Sci. Comput. **33**(2), 869–892 (2011)
39. Gleich, B., Weizenecker, J.: Tomographic imaging using the nonlinear response of magnetic particles. Nature **435**, 1214–1217 (2005)
40. Gottlieb, D., Shu, C.W.: On the Gibbs phenomenon and its resolution. SIAM Rev. **39**, 644–668 (1997)
41. Hale, N., Trefethen, L.N.: New quadrature formulas from conformal maps. SIAM J. Numer. Anal. **46**, 930–948 (2008)
42. Hewitt, E., Hewitt, R.E.: The Gibbs-Wilbraham phenomenon: an episode in Fourier analysis. Arch. Hist. Exact Sci. **21**(2), 129–160 (1979)
43. Higham, N.J.: The numerical stability of barycentric Lagrange interpolation. IMA J. Numer. Anal. **24**, 547–556 (2000)
44. Jerri, A.: The Gibbs Phenomenon in Fourier Analysis, Splines and Wavelet Approximations. Kluwer Academic Publishers, Dordrecht (1998)
45. Jung, J.-H.: A note on the Gibbs phenomenon with multiquadric radial basis functions. Appl. Numer. Math. **57**, 213–219 (2007)
46. Kaethner, C., Erb, W., Ahlborg, M., Szwargulski, P., Knopp, T., Buzug, T.M.: Non-equispaced system matrix acquisition for magnetic particle imaging based on Lissajous node points. IEEE Trans. Med. Imag. **35**(11), 2476–2485 (2016)

47. Knopp, T., Buzug, T.M.: Magnetic Particle Imaging. Springer, Berlin (2012)
48. Knopp, T., Biederer, S., Sattel, T., Weizenecker, J., Gleich, B., Borgert, J., Buzug, T.M.: Trajectory analysis for magnetic particle imaging. Phys. Med. Biol. **54**, 385–397 (2009)
49. Kosloff, D., Tal-Ezer, H.: A modified Chebyshev pseudospectral method with an $O(N^{-1})$ time step restriction. J. Comput. Phys. **104**, 457–469 (1993)
50. Meijering, E.: A chronology of interpolation: from ancient astronomy to modern signal and image processing. Proc. IEEE **90**(3), 319–342 (2002)
51. Nakatsukasa, Y., Sète, O., Trefethen, L.N.: The AAA algorithm for rational approximation. SIAM J. Sci. Comput. **40**(3), A1494–A1522 (2018)
52. Rippa, S.: An algorithm for selecting a good value for the parameter c in radial basis function interpolation. Adv. Comput. Math. **11**, 193–210 (1999)
53. Rivlin, T.J.: An Introduction to the Approximation of Functions. Dover, New York (1969)
54. Romani, L., Rossini, M., Schenone, D.: Edge detection methods based on RBF interpolation. J. Comput. Appl. Math. **349**, 532–547 (2019)
55. Rossini, M.: Interpolating functions with gradient discontinuities via variably scaled kernels. Dolomites Res. Notes Approx. **11**, 3–14 (2018)
56. Runge, C.: Über empirische Funktionen und die Interpolation zwischen äquidistanten Ordinaten. Z. Math. Phys. **46**, 224–243 (1901)
57. Sarra, S.A.: Digital total variation filtering as postprocessing for radial basis function approximation methods. Comput. Math. Appl. **52**, 1119–1130 (2006)
58. Schaback, R.: Error estimates and condition numbers for radial basis function interpolation. Adv. Comput. Math. **3**, 251–264 (1995)
59. Takaki, A., Soma, T., Kojima, A., Asao, K., Kamada, S., Matsumoto, M., Murase, K.: Improvement of image quality using interpolated projection data estimation method in SPECT. Ann. Nucl. Med. **23**(7), 617–626 (2009)
60. Wendland, H.: Scattered Data Approximation. Cambridge Monographs on Applied and Computational Mathematics. Cambridge University Press, Cambridge (2005)
61. https://en.wikipedia.org/wiki/Runge%27s_phenomenon#S-Runge_algorithm_without_resampling

Chapter 2
Steady Systems of PDEs. Two Examples from Applications

Pablo Pedregal

Abstract After some general comments on the tremendous difficulties associated with non-linear systems of PDEs, we will focus on two such illustrative situations where the underlying model depends on the highly nonlinear behavior coming from a system of PDEs. The first one is motivated by inverse problems in conductivity and the process to recover an unknown conductivity coefficient from measurements in the boundary; the second focuses on an optimal control problem for soft robots in which the underlying model comes from hyper-elasticity where the state system models the behavior of non-linear elastic materials capable of undergoing large deformations.

Keywords 35G60 · 49J20 · 49J45

2.1 Introduction: Systems of PDEs: Why Are They So Difficult?

It is not hard to find hundreds of excellent experts on (single) PDEs. The subject is mature and expanding rapidly in many fundamental and interesting directions. However, it is not so if one is interested in systems of PDEs. As a matter of fact, when one talks about systems of PDEs, some fundamental examples come to mind. Without pretending to be exhaustive, we can mention the following.

- Dynamical:
  - Navier-Stokes, and variants (Stokes, Euler, etc).
  - Systems of conservation laws, first-order hyperbolic systems.
  - Diffusion systems, non-linear waves.
  - Other.

P. Pedregal (✉)
ETSI Industriales, U. Castilla-La Mancha, Ciudad Real, Spain
e-mail: pablo.pedregal@uclm.es

D. Barrera et al. (eds.), *Mathematical and Computational Methods for Modelling, Approximation and Simulation*, SEMA SIMAI Springer Series 29,
https://doi.org/10.1007/978-3-030-94339-4_2

- Steady:
  - Reaction-diffusion systems.
  - Linear elasticity.
  - Non-linear elasticity, hyper-elasticity.
  - Other situations.

According to its complexity, a possible classification, as in any other area of Differential Equations, could be:

- Linear systems.
- Non-linear systems: non-linearity on lower order terms (most studied).
- Fully non-linear systems: non-linearity on highest order terms.

Though there are some other alternatives based on the implicit function theorem [10, 29], we would like to focus on the steady situation, where the main procedure to show existence of solutions is limited to equations or systems that are variational in nature [5, 10, 24]: there is an underlying energy functional whose Euler-Lagrange equation or system is precisely the one to be solved

$$E(\mathbf{u}) = \int_\Omega F(\mathbf{x}, \mathbf{u}(\mathbf{x}), \nabla\mathbf{u}(\mathbf{x}))\, d\mathbf{x},$$

$$\mathbf{u} : \Omega \subset \mathbb{R}^N \to \mathbb{R}^N, \quad \mathbf{u}, \text{ subjected to boundary conditions.}$$

The principal strategy for existence of solutions for the underlying Euler-Lagrnage system of PDEs is to try to minimize the functional E among those feasible fields $\mathbf{u}$'s. This method requires a fundamental structural property for the existence of minimizers: convexity of the integrand

$$F(\mathbf{x}, \cdot, \cdot) : \mathbb{R}^N \times \mathbb{R}^{N\times N} \to \mathbb{R}, \tag{2.1}$$

for a.e. $\mathbf{x} \in \Omega$. This convexity, in turn, amounts to ellipticity or monotonicity of the associated optimality system. As a matter of fact, among the systems of PDEs most examined in the literature, we can mention diffusion systems where the principal part of the system is the laplacian operator or some other linear, elliptic variant. This class of problems directly stems from the convexity of the underlying functional. They are non-linear systems where the non-linearity occurs on lower-order terms.

Why is convexity so important? Because existence is based on weak lower semicontinuity

$$\mathbf{u}_j \rightharpoonup \mathbf{u} \Rightarrow E(\mathbf{u}) \le \liminf_{j\to\infty} E(\mathbf{u}_j),$$

and this property rests in a crucial way on the convexity properties of the integrand F in (2.1). In the scalar case, when $\mathbf{u} = u$ is just a scalar function, weak lower semicontinuity is equivalent to usual convexity; but in the real vector case when $\mathbf{u}$ is a true field with several components, convexity is only sufficient, not necessary for weak lower semicontinuity [5, 11, 26]. The simplest example of this surprising and unexpected phenomenon is the determinant function

$$\det : \mathbf{M}^{2\times 2} \to \mathbb{R}.$$

Note that if

$$\mathbf{u}_j(\mathbf{x}) : \Omega \subset \mathbb{R}^2 \to \mathbb{R}^2, \quad \mathbf{u}_j = \mathbf{u}_0 \text{ on } \partial\Omega,$$
$$\mathbf{u}_j \rightharpoonup \mathbf{u} \text{ in } W^{1,p}(\Omega; \mathbb{R}^2) \Rightarrow \mathbf{u} = \mathbf{u}_0 \text{ on } \partial\Omega,$$

then for the area functional

$$\mathrm{A}(\mathbf{u}_0(\Omega)) = \mathrm{A}(\mathbf{u}_j(\Omega)) = \int_\Omega \det \mathbf{u}_j(\mathbf{x})\, d\mathbf{x} = \int_\Omega \det \mathbf{u}(\mathbf{x})\, d\mathbf{x}.$$

This means that the area functional is not only weakly lower semicontinuous but even weak continuous, and yet the function det is NOT convex. For vector problems (systems), there is a more general convexity condition, the "quasiconvexity", imposed on F to have the equivalence with the weak lower semicontinuity property.

To put this property in perspective, recall that usual convexity can be formulated in the form

$$\phi(\mathbf{F}) : \mathbb{R}^{m\times N} \to \mathbb{R},$$
$$\phi(\mathbf{F}) \le \int_\Omega \phi(\mathbf{F} + \mathbf{U}(\mathbf{x}))\, d\mathbf{x}, \quad \mathbf{U}(\mathbf{x}) : \Omega \subset \mathbb{R}^N \to \mathbb{R}^{m\times N}, \int_\Omega \mathbf{U}(\mathbf{x})\, d\mathbf{x} = \mathbf{0}.$$

From this viewpoint, quasiconvexity is then determined through

$$\phi(\mathbf{F}) : \mathbb{R}^{m\times N} \to \mathbb{R},$$
$$\phi(\mathbf{F}) \le \int_\Omega \phi(\mathbf{F} + \nabla\mathbf{u}(\mathbf{x}))\, d\mathbf{x}, \quad \mathbf{u}(\mathbf{x}) : \Omega \subset \mathbb{R}^N \to \mathbb{R}^m, \mathbf{u} = \mathbf{0} \text{ on } \partial\Omega,$$
$$\mathbf{U}(\mathbf{x}) = \nabla\mathbf{u}(\mathbf{x}) : \Omega \to \mathbb{R}^{m\times N}.$$

It can be understood by saying that quasiconvexity is convexity with respect to gradient fields. The main source of difficulties for systems of PDEs is that existence theorems require the quasiconvexity property, and this condition is difficult to manipulate. Indeed, it is still quite a mysterious concept.

One could think that if, after all, convexity is sufficient for weak lower semicontinuity one could be dispensed with dealing with quasiconvexity. The point is that usual convexity cannot be taken for granted in many of the situations of interest. A paradigmatic situation happens in hyper-elasticity, for a non-quadratic F, where convexity is incompatible with frame indifference [5, 10]. A typical situation is then to face a system of PDEs of interest that corresponds to the Euler-Lagrange system of a certain functional, but it is NOT convex. Two scenarios may demand attention:

1. either there are minimizers, in spite of this lack of convexity (though uniqueness might be compromised); or
2. there are no such minimizers, and then we do not have tools to decide whether the underlying system admits solutions.

We plan to examine two distinct situations in which one is forced to face a certain non-linear system of PDEs of interest. The first one deals with an alternative method to explore inverse problems in conductivity. The second one focuses on a control problem governed by a non-linear system of PDEs in the context of hyper-elasticity.

In the analysis of these two problems, several collaborators have contributed to its success: F. Maestre from U. de Sevilla (Spain), and J. Martínez-Frutos, R. Ortigosa and F. Periago from U. Politécnica de Cartagena (Spain).

Though we specifically focus on the two already-mentioned examples, there are a number of other fundamental problems where either the underlying topology of spaces changes, or the lack of convexity, though not fully and systematically treated yet, could be of interest. In particular, because it has been a dominant field of research in the last 30 years, we would like to bring reader's attention towards optimal transport problems [13, 27, 30]), among other possibilities.

2.2 Inverse Problems in Conductivity in the Plane

Suppose we are facing the problem of determining the conductivity coefficient $\gamma(\mathbf{x})$ of a media occupying a region $\Omega \subset \mathbb{R}^2$, by testing and measuring how it reacts to some electric stimulus. More precisely, we provide boundary datum $u_0(\mathbf{x})$ around $\partial\Omega$, and through the unknown solution $u(\mathbf{x})$ of the Dirichlet problem

$$\operatorname{div}[\gamma(\mathbf{x})\nabla u(\mathbf{x})] = 0 \text{ in } \Omega, \quad u(\mathbf{x}) = u_0(\mathbf{x}) \text{ on } \partial\Omega,$$

one can measure the response of the media by measuring the Neumann datum

$$v_0(\mathbf{x}) \equiv \gamma(\mathbf{x})\nabla u(\mathbf{x}) \cdot \mathbf{n}(\mathbf{x}) \text{ on } \partial\Omega.$$

The problem consists in recovering the coefficient $\gamma(\mathbf{x})$ from the pair of data (u_0, v_0). Typically, one has a broader set of measurement-pairs (u_i, v_i), $i = 1, 2, \ldots, m$. This is by now a classic and very well-studied problem, for which fundamental contributions have already been made. Starting from [8], from the mathematical point of view this kind of problems have led to deep and fundamental results (see [1, 4, 7, 9, 12, 16, 28], and references therein). The book [15] is a basic reference in this field. The main issues concerning inverse problems are *uniqueness* in various scenarios [4, 23], *stability* [2, 6] and *reconstruction* [3, 22].

From the conductivity equation

$$\operatorname{div}[\gamma(\mathbf{x})\nabla u(\mathbf{x})] = 0 \text{ in } \Omega \subset \mathbb{R}^2,$$

one can derive the algebraic identity

$$\gamma \nabla u + \mathbf{R}\nabla v = \mathbf{0}, \quad \mathbf{R} = \begin{pmatrix} 0 & 1 \\ -1 & 0 \end{pmatrix} \tag{2.2}$$

for a certain function v which is unique up to additive constants. The interplay between u and v is classically understood through the Beltrami equation, though we will stick to (2.2). The algebraic equation (2.2) furnishes three important pieces of information:

1. a conductivity equation for v:

$$\operatorname{div}\left[\frac{1}{\gamma(\mathbf{x})}\nabla v(\mathbf{x})\right] = 0 \text{ in } \Omega;$$

2. a direct expression for the conductivity coefficient

$$\gamma(\mathbf{x}) = \frac{|\nabla v(\mathbf{x})|}{|\nabla u(\mathbf{x})|};$$

3. Dirichlet boundary datum for v around $\partial\Omega$: if we multiply (2.2) by the outer normal vector $\mathbf{n}(\mathbf{x})$ to $\partial\Omega$, we see that

$$\gamma(\mathbf{x})\nabla u(\mathbf{x}) \cdot \mathbf{n}(\mathbf{x}) = \nabla v(\mathbf{x}) \cdot \mathbf{t}(\mathbf{x}) \text{ on } \partial\Omega, \tag{2.3}$$

where $\mathbf{t}(\mathbf{x})$ is the counter-clockwise, unit tangent vector to $\partial\Omega$. Note that (2.3) provides Dirichlet boundary condition v_0 for v upon integration around $\partial\Omega$ of the Neumann condition for u, and so it is a datum of the problem once a measurement pair (u_0, v_0)

Hence, if $\gamma(\mathbf{x})$ is unknown, we can envision to recover it through the pair (u, v) solution of the system of PDEs

$$\operatorname{div}\left(\frac{|\nabla v|}{|\nabla u|}\nabla u\right) = 0 \text{ in } \Omega, \quad u = u_0 \text{ on } \partial\Omega, \tag{2.4}$$

$$\operatorname{div}\left(\frac{|\nabla u|}{|\nabla v|}\nabla v\right) = 0 \text{ in } \Omega, \quad v = v_0 \text{ on } \partial\Omega. \tag{2.5}$$

In fact, we are using v_0 for two different things: for the second component of a measurement, and for the Dirichlet boundary condition of the second component of the non-linear systems of PDEs. One is the tangential derivative of the other around

$\partial\Omega$. We hope this will not create confusion, as it is clear that once a measurement pair (u_0, v_0) is known, the corresponding Dirichlet boundary condition for the second component v of our system of PDEs is easily calculated by integration around $\partial\Omega$ of v_0. We are also using the same notation v_0 for this Dirichlet datum.

We therefore focus on how to find and compute a solution pair (u, v) for our above system of PDEs. It is easily checked that, at least formally, it is the Euler-Lagrange system corresponding to the functional

$$I(\mathbf{u}) = \int_\Omega |\nabla u_1(\mathbf{x})|\,|\nabla u_2(\mathbf{x})|\,d\mathbf{x}, \quad \mathbf{u} = (u_1, u_2), u = u_1, v = u_2. \tag{2.6}$$

But the integrand for I

$$\phi(\mathbf{F}) : \mathbf{M}^{2\times 2} \to \mathbb{R}, \quad \phi(\mathbf{F}) = |\mathbf{F}_1|\,|\mathbf{F}_2|, \quad \mathbf{F} = \begin{pmatrix} \mathbf{F}_1 \\ \mathbf{F}_2 \end{pmatrix},$$

is not convex, not even quasiconvex. This means that a typical existence theorem for minimizers for I cannot be used to show existence of minimizers for I in appropriate function spaces, and so our unique possibility to prove existence of solutions of our system of PDEs, and to recover the conductivity coefficient γ, is absolutely blocked from the outset. Moreover, ϕ is not coercive either. We are left with no general alternative to show existence of solutions for our system of PDEs. As a matter of fact, existence of solutions depend on properties of boundary data (u_0, v_0).

2.2.1 A Situation in Which Existence Can be Achieved

We describe how to build pairs of data (u_0, v_0), closely connected to the underlying inverse problem, for which existence of solutions for our system of PDEs can be achieved. Generate pairs of Dirichlet data in the following way:

1. $\gamma(\mathbf{x}) \geq \gamma_0 > 0$ in Ω, and $u_0 \in H^{1/2}(\partial\Omega)$ freely.
2. Solve the problem
$$\operatorname{div}(\gamma\nabla u) = 0 \text{ in } \Omega, \quad u = u_0 \text{ on } \partial\Omega, \tag{2.7}$$
and compute the tangential derivative $w_0 = \nabla u \cdot \mathbf{t}$ around $\partial\Omega$.
3. Find a solution of the Neumann problem
$$\operatorname{div}\left(\frac{1}{\gamma}\nabla v\right) = 0 \text{ in } \Omega, \quad \frac{1}{\gamma}\nabla v \cdot \mathbf{n} = w_0 \text{ on } \partial\Omega. \tag{2.8}$$
4. Take $v_0 = v|_{\partial\Omega}$.

This mechanism can be used to generate synthetic pairs of compatible boundary measurements to be used in numerical experiments.

Proposition 2.1 *If data pairs* (u_0, v_0) *are built in the above way, the system*

$$\operatorname{div}\left(\frac{|\nabla v|}{|\nabla u|}\nabla u\right) = 0, \quad \operatorname{div}\left(\frac{|\nabla u|}{|\nabla v|}\nabla v\right) = 0 \quad in\ \Omega,$$

under $u = u_0$, $v = v_0$ *on* $\partial\Omega$, *admits, at least, one solution.*

Proof The proof is elementary. Equation (2.7) implies the existence of a function $w \in H^1(\Omega)$, unique up to an additive constant, such that

$$\gamma\nabla u + \mathbf{R}\nabla w = \mathbf{0}.$$

This vector equation can be rewritten in the form

$$\mathbf{R}\nabla u = \frac{1}{\gamma}\nabla w,$$

and hence we can conclude that

$$\operatorname{div}\left(\frac{1}{\gamma}\nabla w\right) = 0 \text{ in } \Omega, \quad \frac{1}{\gamma}\nabla w \cdot \mathbf{n} = \nabla u \cdot \mathbf{t} = w_0 \text{ on } \partial\Omega.$$

If we compare this information with (2.8), we can conclude that the auxiliary function w is, except for an additive constant, the function v. As a consequence, we see that the pair (u, v) is a solution of our non-linear system of PDEs. □

Much more information about this problem can be found in [18].

2.2.2 The Multi-Measurement Case

In practice, the determination of an unknown conductivity coefficient $\gamma(\mathbf{x})$ can be facilitated in one is allowed to make several measurements corresponding to pairs

$$(u^{(j)}_{1,0}, u^{(j)}_{2,0}), \quad j = 1, 2, \ldots, N.$$

In this case the system for an unknown field

$$\mathbf{u}(\mathbf{x}) : \Omega \subset \mathbb{R}^2 \to \mathbb{R}^{2N}$$

becomes

$$\operatorname{div}\left(\frac{|\nabla \mathbf{u}_2|}{|\nabla \mathbf{u}_1|}\nabla u_1^{(j)}\right) = 0 \text{ in } \Omega, \quad u_1^{(j)} = u_{1,0}^{(j)} \text{ on } \partial\Omega, \tag{2.9}$$

$$\operatorname{div}\left(\frac{|\nabla \mathbf{u}_1|}{|\nabla \mathbf{u}_2|}\nabla u_2^{(j)}\right) = 0 \text{ in } \Omega, \quad u_2^{(j)} = u_{2,0}^{(j)} \text{ on } \partial\Omega, \tag{2.10}$$

for $j = 1, 2, \ldots, N$, where

$$\mathbf{u} = (\mathbf{u}^{(j)})_{j=1,2,\ldots,N} = (u_1^{(j)}, u_2^{(j)})_{j=1,2,\ldots,N},$$

$$\mathbf{u}^{(j)} = (u_1^{(j)}, u_2^{(j)}), \quad \mathbf{u}_i = (u_i^{(j)})_{j=1,2,\ldots,N}, \quad i = 1, 2.$$

We also have an underlying functional whose Euler-Lagrange system is precisely (2.9)–(2.10), namely

$$I_N(\mathbf{u}) = \int_\Omega |\nabla \mathbf{u}_1|\,|\nabla \mathbf{u}_2|\,d\mathbf{x}. \tag{2.11}$$

This functional, as in the one-measurement case, is neither convex nor quasiconvex, and once again one runs into the same difficulties concerning existence of solutions. In those cases in which there is a solution $\mathbf{u} = (\mathbf{u}_1, \mathbf{u}_2)$ of our system, the recovered $\gamma(\mathbf{x})$ would be

$$\gamma(\mathbf{x}) = \frac{|\nabla \mathbf{u}_2(\mathbf{x})|}{|\nabla \mathbf{u}_1(\mathbf{x})|} \quad \text{a.e. } x \in \Omega,$$

and thus, we see that the information coming from every measurement for $j = 1, 2, \ldots, N$ is taken into account.

2.2.3 *Some Numerical Tests*

For those situations in which synthetic data sets have been determined through the technique of Sect. 2.2.1, one can proceed to approximate the solution of the corresponding non-linear system of PDEs, and consequently to find an approximation of the conductivity coefficient. We have explored three possible approximation procedures:

1. a typical Newton-Raphson method applied directly to the non-linear system either for the one-measurement or for the multi-measurement cases;
2. a descent algorithm based on minimizing the corresponding functional either (2.6) or (2.11);

3. a fix-point strategy consisting in iterating the action of the operation

$$(u, v) \mapsto (U, V) = \mathbf{T}(u, v)$$

where

$$\operatorname{div}\left(\frac{|\nabla v|}{|\nabla u|}\nabla U\right) = 0 \text{ in } \Omega, \quad U = u_0 \text{ on } \partial\Omega,$$

$$\operatorname{div}\left(\frac{|\nabla u|}{|\nabla v|}\nabla V\right) = 0 \text{ in } \Omega, \quad V = v_0 \text{ on } \partial\Omega,$$

for the one-measurement case, and a similar one for the multi-measurement situation.

All three methods worked fine for our experiments, though the descent method is the slowest in converging. See in Figs. 2.1, 2.2, 2.3, and 2.4.

It is worth-while to point out that we have a fine certificate of convergence for our numerical simulations. Indeed, the modified corresponding functional, adding

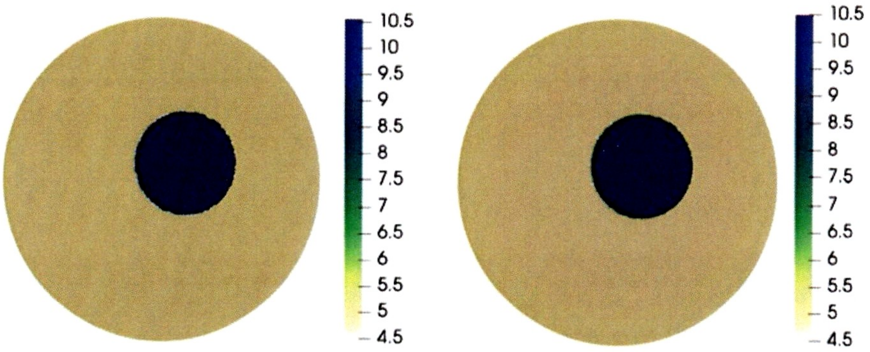

Fig. 2.1 Left. The target function γ with one inclusion. Right. The computed one

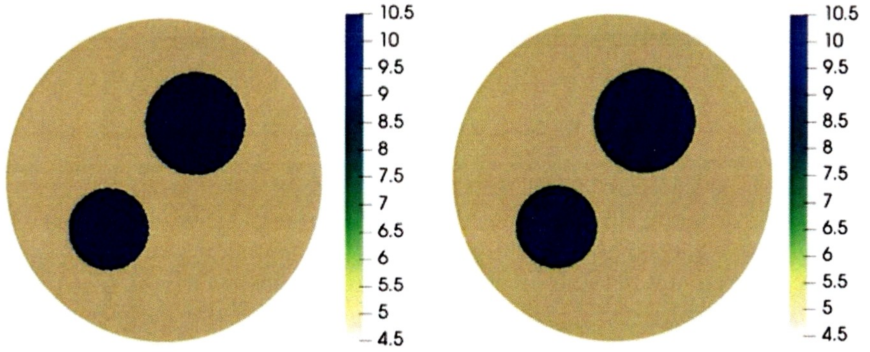

Fig. 2.2 Left. The target function γ with two disjoint inclusions. Right. The approximation

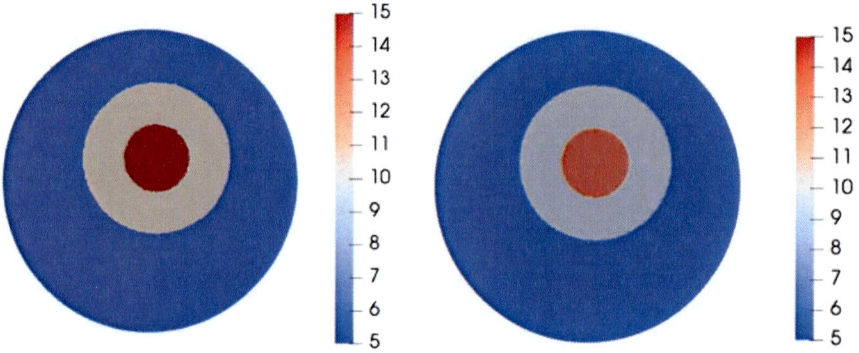

Fig. 2.3 Left. The target function γ with one inclusion inside another. Right. The approximated one

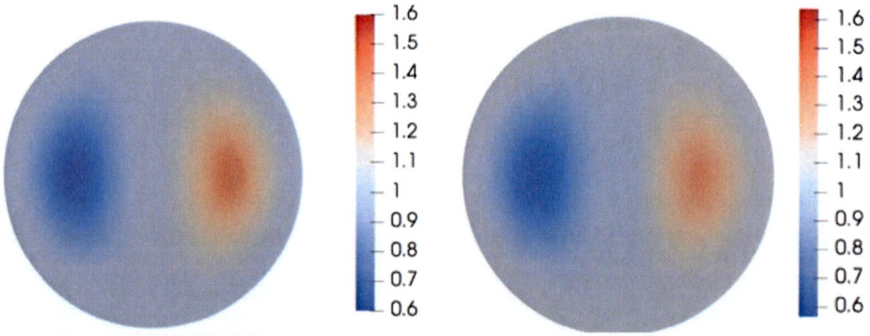

Fig. 2.4 Left. A continuous target γ. Right. The computed one

an additional term in (2.6) and in (2.11), become

$$I_1^*(\mathbf{u}) = \int_\Omega [|\nabla u_1(\mathbf{x})|\,|\nabla u_2(\mathbf{x})| - \det \nabla \mathbf{u}]\, d\mathbf{x}, \quad \mathbf{u} = (u_1, u_2),$$

$$I_N^*(\mathbf{u}) = \int_\Omega \left[|\nabla \mathbf{u}_1|\,|\nabla \mathbf{u}_2| - \sum_j \det \nabla \mathbf{u}^{(j)} \right] d\mathbf{x},$$

$$\mathbf{u} = (\mathbf{u}^{(j)})_{j=1,\dots,N},\, \mathbf{u}_i = (u_i^{(j)})_{j=1,\dots,N}.$$

Because we are adding a null-lagrangian in both cases, the underlying Euler-Lagrange systems of PDEs are exactly the same that the ones for the old functionals, namely (2.4)–(2.5) and (2.9)–(2.10), respectively. The advantage of these two new functionals over the old ones is that the infimum-minimum value for I_1^* and I_N^* should vanish for the solutions we are seeking. Hence, we can be sure that approximations $\mathbf{u}_k$ are fine provided

$$I_1^*(\mathbf{u}_k) \searrow 0, \quad I_N^*(\mathbf{u}_k) \searrow 0.$$

This fact was true in all of our above simulations.

More information is available in [18, 25].

2.3 Some Optimal Control Problems for Soft Robots

There is recently an ever-increasing interest in simulating and controlling the behavior of robots that may, eventually, undergo large deformations as opposed to the more traditional setting in which they are considered essentially as rigid structures, or allowed small deformations. For instance, one might be interested in simulating the heliotropic effect in plants where the stem under the flower aligns with the direction of the sun through an osmotic potential (inner pressure) inside the cells [14]. To pursue this objective, one should move from the linear-elasticity setting to the more complex non-linear or hyper-elastic framework. Another reference worth mentioning is [17] where existence of optimal boundary forces are shown to exist for a control problem arising in facial surgery models. A different viewpoint that is almost mandatory to examine at some point, but that we overlook here, is that of allowing uncertainties in the measurement of material parameters. Apparently, rather large variabilities have been reported in bulk and shear moduli measurements of hyper-elastic materials that are commonly used to simulate rubber-like or biological tissues. This possibility demands to address stochastic versions of corresponding optimal control problems (see [19]).

The following features are important for such models of material behavior:

- material properties: finite strain-elasticity with internal energy density of the form

$$W(\mathbf{F}) = a\|\mathbf{F}\|^2 + b\|\operatorname{cof}\mathbf{F}\|^2 + c(\det\mathbf{F})^2 - d\log(\det\mathbf{F}) + e \tag{2.12}$$

  where a, b, c, d, and e are material constants;
- reference configuration $\Omega \subset \mathbb{R}^3$;
- boundary conditions: displacement condition in a part of the boundary $\mathbf{v} = \mathbf{0}$ on $\Gamma_D \subset \partial\Omega$, traction-free on the complement $\Gamma_N = \partial\Omega \setminus \Gamma_D$;
- control: inner pressure $t(\mathbf{x})$ acting on every point in Ω;
- objective: given two fields $\mathbf{s}(\mathbf{x})$ and $\mathbf{s}_{des}(\mathbf{x})$ (identifying the direction of the stem and the direction indicating the sun, respectively), choose the inner pressure $t(\mathbf{x})$ so as to align, as closely as possible, $\mathbf{s}(\mathbf{x})$ with $\mathbf{s}_{des}(\mathbf{x})$.

Under such conditions, equilibrium equations for feasible solution $\mathbf{u}$ for the direct problem become:

$$\int_\Omega W_\mathbf{F}(\nabla\mathbf{u}) : \nabla\mathbf{v}\, d\mathbf{x} = 0$$

for every test $\mathbf{v}$ with $\mathbf{v} = \mathbf{0}$ on Γ_D. This is indeed the weak form of

$$\text{div}[W_{\mathbf{F}}(\nabla \mathbf{u})] = \mathbf{0} \text{ in } \Omega, \quad \mathbf{u} = \mathbf{0} \text{ on } \Gamma_D, \, W_{\mathbf{F}}(\nabla \mathbf{u})\mathbf{n} = \mathbf{0} \text{ on } \Gamma_N.$$

But such a energy density W is not quadratic nor convex. The corresponding system of PDEs is non-linear, and non-elliptic in the classical sense. Existence of solutions is achieved through polyconvexity [5], and the remarkable properties of minors. Uniqueness does not hold in general. More specifically, one can show the existence of minimizers for the direct variational problem

$$\text{Minimize in } \mathbf{u} \in \mathcal{A}: \quad E(\mathbf{u}) = \int_\Omega W(\nabla \mathbf{u}(\mathbf{x}))\, d\mathbf{x},$$
$$\mathcal{A} = \{\mathbf{u} \in H^1(\Omega; \mathbb{R}^3) : \mathbf{u} = \mathbf{0} \text{ in } \Gamma_D\}.$$

Such minimizer will be a solution of the above system of PDEs.

2.3.1 *The Control Problem*

The control problem we would like to explore can be formulated in the following terms

$$\text{Minimize in } t(\mathbf{x}) : \quad \frac{1}{2}\int_\Omega \|(\mathbf{id} + \nabla \mathbf{u})\mathbf{s} - \mathbf{s}_{des}\|^2 \, d\mathbf{x} + \frac{M}{2}\int_\Omega t^2 \, d\mathbf{x} + \frac{\epsilon}{2}\int_\Omega |\nabla t|^2 \, d\mathbf{x},$$

subject to

$$\int_\Omega \nabla_{\mathbf{F}} W(\mathbf{id} + \nabla \mathbf{u}(\mathbf{x})) : \nabla \mathbf{v}(\mathbf{x})\, d\mathbf{x} = \int_\Omega t(\mathbf{x}) \,\text{cof}(\mathbf{id} + \nabla \mathbf{u}(\mathbf{x})) : \nabla \mathbf{v}(\mathbf{x})\, d\mathbf{x} \tag{2.13}$$

for all $\mathbf{v}$ in

$$\mathbb{V}_D := \left\{\mathbf{v} \in H^1\left(\Omega; \mathbb{R}^N\right) : \mathbf{v} = 0 \text{ on } \Gamma_D\right\}.$$

The internal energy density W is assumed of the form (2.12). The state system is the non-linear system of PDEs

$$\text{div}[W_{\mathbf{F}}(\mathbf{id} + \nabla \mathbf{u}) - t \,\text{cof}(\mathbf{id} + \nabla \mathbf{u})] = \mathbf{0},$$

that corresponds to a minimizer of the associated integral functional with density

$$W(\mathbf{F}) - t\det\mathbf{F}, \quad W(\mathbf{F}) = a\|\mathbf{F}\|^2 + b\|\operatorname{cof}\mathbf{F}\|^2 + c(\det\mathbf{F})^2 - d\log(\det\mathbf{F}) + e,$$

whose weak formulation is precisely (2.13).

Theorem 2.1 *Under the conditions explained above, there is an optimal pressure for our control system.*

To understand the main difficulty in the proof of such a result, consider a minimizing sequence $\{t_j\}$ with corresponding states $\{\mathbf{u}_j\}$, i.e.

$$\int_\Omega \nabla_{\mathbf{F}} W(\mathbf{id} + \nabla\mathbf{u}_j(\mathbf{x})) : \nabla\mathbf{v}(\mathbf{x})\, d\mathbf{x} = \int_\Omega t_j(\mathbf{x}) \operatorname{cof}(\mathbf{id} + \nabla\mathbf{u}_j(\mathbf{x})) : \nabla\mathbf{v}(\mathbf{x})\, d\mathbf{x} \tag{2.14}$$

for $\mathbf{v} \in \mathbb{V}_D$. Under these circumstances, it is easy to find that $t_j \to t$, $\mathbf{u}_j \to \mathbf{u}$ (weak convergences), but the whole point is to check whether limits t and $\mathbf{u}$ are such that

$$\int_\Omega \nabla_{\mathbf{F}} W(\mathbf{id} + \nabla\mathbf{u}(\mathbf{x})) : \nabla\mathbf{v}(\mathbf{x})\, d\mathbf{x} = \int_\Omega t(\mathbf{x}) \operatorname{cof}(\mathbf{id} + \nabla\mathbf{u}(\mathbf{x})) : \nabla\mathbf{v}(\mathbf{x})\, d\mathbf{x}$$

for every $\mathbf{v} \in \mathbb{V}_D$. That is to say, whether t and $\mathbf{u}$ are related through the state system.

If we take limits in j in (2.14), we realize that we need to have

$$\nabla_{\mathbf{F}} W(\mathbf{id} + \nabla\mathbf{u}_j) \to \nabla_{\mathbf{F}} W(\mathbf{id} + \nabla\mathbf{u}), \quad \operatorname{cof}(\mathbf{id} + \nabla\mathbf{u}_j) \to \operatorname{cof}(\mathbf{id} + \nabla\mathbf{u}),$$
$$t_j \operatorname{cof}(\mathbf{id} + \nabla\mathbf{u}_j) \to t \operatorname{cof}(\mathbf{id} + \nabla\mathbf{u}).$$

Surprisingly, these three convergences are correct if $t_j \to t$, (strong) and $\mathbf{u}_j \rightharpoonup \mathbf{u}$ (weak), due essentially to the remarkable convergence properties of the subdeterminants

$$\mathbf{u}_j \rightharpoonup \mathbf{u} \Rightarrow M(\nabla\mathbf{u}_j) \rightharpoonup M(\nabla\mathbf{u})$$

for every subdeterminant M, and the presence of the quadratic term in the derivatives of $t(\mathbf{x})$ in the cost functional. The proof of the existence result is finished by standard weak lower semicontinuity through convexity.

2.3.2 The Fiber Tension Case

This is another fundamental control problem in the field of soft robots, in which given a field

$$\mathbf{a} = \mathbf{a}(\mathbf{x}) : \Omega \to \mathbb{R}^N,$$

a control

$$m = m(\mathbf{x}) : \Omega \to \mathbb{R}$$

is sought to minimize the cost functional

$$\frac{1}{2}\int_\Omega \|\mathbf{u}(\mathbf{x}) - \mathbf{u}_{\text{des}}(\mathbf{x})\|^2 \, d\mathbf{x} + \frac{M}{2}\int_\Omega m(\mathbf{x})^2 \, d\mathbf{x} + \frac{\epsilon}{2}\int_\Omega |\nabla m(\mathbf{x})|^2 \, d\mathbf{x}$$

where

$$\int_\Omega \nabla_{\mathbf{F}} W\left(\mathbf{id} + \nabla\mathbf{u}\right) : \nabla\mathbf{v} \, d\mathbf{x} = \int_\Omega m\mathbf{a}^T \left(\mathbf{id} + \nabla\mathbf{u}\right)^T \nabla\mathbf{v}\,\mathbf{a} \, d\mathbf{x}$$

for all $\mathbf{v}$ in

$$\mathbb{V}_D := \left\{\mathbf{v} \in H^1\left(\Omega; \mathbb{R}^N\right) : \mathbf{v} = 0 \text{ on } \Gamma_D\right\}.$$

The underlying state system is the weak formulation of the non-linear system of PDEs which is the Euler-Lagrange system corresponding to the functional

$$\int_\Omega \left[W\left(\mathbf{id} + \nabla\mathbf{u}\right) - \frac{1}{2}m|\left(\mathbf{id} + \nabla\mathbf{u}\right)\mathbf{a}|^2\right] d\mathbf{x},$$

for an inner density $W(\mathbf{F})$ as before. A similar existence result can be shown in this context as well.

2.3.3 Some Simulations

We present some simulations on both control problems described in the previous section. More information can be found in [20] and [21].

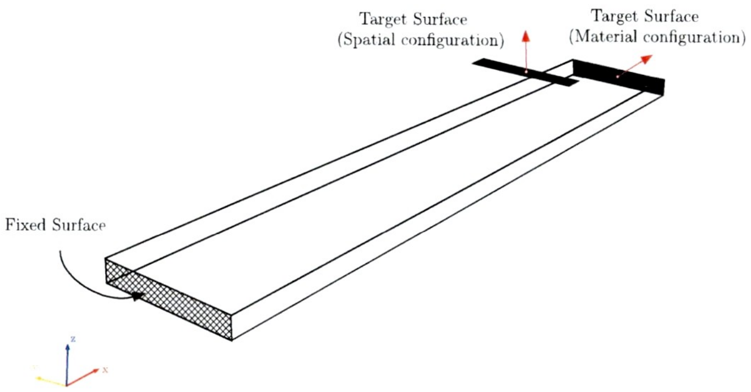

Fig. 2.5 Objective and boundary conditions for the control problem

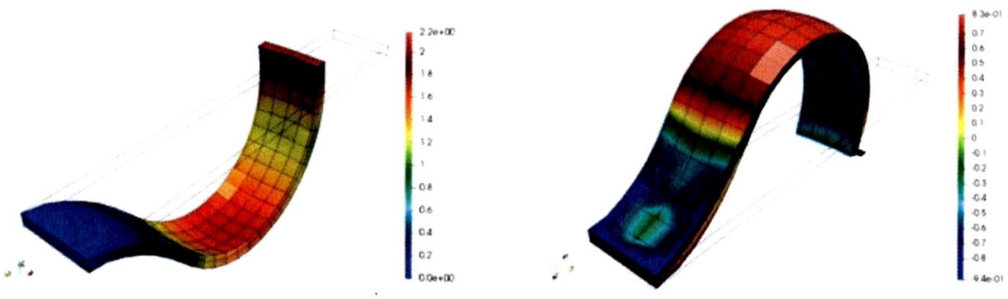

Fig. 2.6 Actuated control and pressure distribution

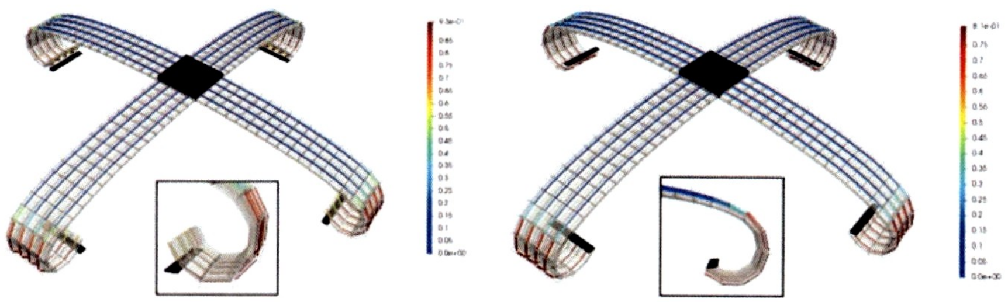

Fig. 2.7 A control situation for the fiber case. Left: desired effect. Right: mechanism activated

For the first one, we consider the situation depicted in Fig. 2.5. It is a clamped elastic plate whose free end surface is desired to be aligned vertically through an optimal use of an inner pressure field. Figure 2.6 shows the optimal result, and the optimal distribution of pressure.

Concerning the fiber problem, one desires to activate optimally a griping mechanism as in Fig. 2.7.

2.4 Conclusions

When facing a steady, non-linear systems of PDEs, one needs to care about the structure of the problem. If it is not variational, there is no much one can do.

Assuming it is variational, i.e. it corresponds to the Euler-Lagrange system for a certain integral functional with an integrand $W(\mathbf{F})$, the success of the analysis depends on the properties of $W : \mathbb{R}^{N\times n} \to \mathbb{R}$.

- If W is convex in the usual sense, under additional technical hypotheses like coercivity, there should be no problem in proving existence of solutions.
- If convexity is not correct, or it cannot be easily shown, two situations may happen:
  - Quasiconvexity holds: existence without convexity (optimal control for soft robots).
  - Quasiconvexity does not hold: results heavily depend on ingredients of problem (inverse problems in conductivity).

The major trouble is that we do not fully understand this quasiconvexity condition as it may be very difficult to decide if a given W enjoys such property, unless we have designed it so that it does. Much remains to be understood about non-linear systems of PDEs.

Acknowledgments Research supported by grant MTM2017-83740-P, and of 2019-GRIN-26890 of U. de Castilla-La Mancha.

References

1. Alessandrini, G.: Stable determination of conductivity by boundary measurements. Appl. Anal. **27**, 153–172 (1988)
2. Alessandrini, G., de Hoop, M.V., Gaburro, R., Sincich E.: Lipschitz stability for the electrostatic inverse boundary value problem with piecewise linear conductivities. J. Math. Pures Appl. (9) **107**(5), 638–664 (2017)
3. Ammari, H., Kang, H.: Reconstruction of Small Inhomogeneities from Boundary Measurements. Lecture Notes in Mathematics. Springer, Berlin (2004)
4. Astala, K., Päivärinta, L.: Calderón's inverse conductivity problem in the plane. Ann. Math. (2) **163**(1), 265–299 (2006)
5. Ball, J.M.: Convexity conditions and existence theorems in nonlinear elasticity. Arch. Rational Mech. Anal. **63**(4), 337–403 (1976–1977)
6. Barcel, T., Faraco, D., Ruiz, A.: Stability of Calderón inverse conductivity problem in the plane. J. Math. Pures Appl. **88**(6), 522–556 (2007)
7. Borcea, L.: Electrical impedance tomography. Inverse Problems **18**(6) (2002) R99–R136
8. Calderón, A.P.: On an inverse boundary value problem. In: Seminar on Numerical Analysis and its Applications to Continuum Physics (Rio de Janeiro, 1980), pp. 65–73. Bulletin of the Brazilian Mathematical Society, Rio de Janeiro (1980)
9. Cheney, M., Isaacson, D., Newell, J.C.: Electrical impedance tomography. SIAM Rev. **41**, 85–101 (1999)

10. Ciarlet, P.G.: Mathematical Elasticity. Vol. I. Three-Dimensional Elasticity. Studies in Mathematics and its Applications, vol. 20. North-Holland, Amsterdam (1988)
11. Dacorogna, B.: Direct Methods in the Calculus of Variations. Applied Mathematical Sciences, 2nd edn., vol. 78. Springer, New York (2008)
12. Faraco, D., Kurylev, Y., Ruiz, A.: G-convergence, Dirichlet to Neumann maps and invisibility. J. Funct. Anal. **267**(7), 2478–2506 (2014)
13. Gigli, N.: Introduction to optimal transport: theory and applications. Publicações Matemáticas do IMPA. [IMPA Mathematical Publications] 28o Colóquio Brasileiro de Matemática. [28th Brazilian Mathematics Colloquium] Instituto Nacional de Matemática Pura e Aplicada (IMPA), Rio de Janeiro (2011)
14. Günnel, A., Herzog, R.: Optimal control problems in finite strain elasticity by inner pressure and fiber Tension. Front. Appl. Math. Stat. **2**(4), (2016)
15. Isakov, V.: Inverse problems for partial differential equations. Applied Mathematical Sciences, 2nd edn., vol. 127. Springer, New York (2006)
16. Kohn, R.V., Vogelius, M.: Determining conductivity by boundary measurements. II. Interior results. Commun. Pure Appl. Math. **38**, 643–667 (1985)
17. Lubkoll, L., Schuela, A., Weiser, M.: An optimal control problem in polyconvex hyperelasticity. SIAM J. Control Optim. **52**(3), 1403–1422 (2014)
18. Maestre, F., Pedregal, P.: Some non-linear systems of PDEs related to inverse problems in conductivity. Calc. Var. Partial Differ. Equ. **60**(3), Paper No. 110, 26p. (2021)
19. Martínez-Frutos, J., Periago, F.: Optimal Control of PDEs Under Uncertainty. An Introduction with Application to Optimal Shape Design of Structures. Springer Briefs in Mathematics. BCAM SpringerBriefs. Springer, Berlin (2018)
20. Martínez-Frutos, J., Ortigosa, R., Pedregal, P., Periago, F.: Robust optimal control of stochastic hyper-elastic materials. Appl. Math. Model. **88**, 888–904 (2020)
21. Ortigosa, R., Martínez-Frutos, J., Mora-Corral, C., Pedregal, P., Periago, F.: Optimal control of soft materials using a Hausdorff distance functional. SIAM J. Control Optim. **59**(1), 393–416 (2021)
22. Nachman, A.: Reconstructions from boundary measurements. Ann. Math. **128**, 531–576 (1988)
23. Nachman, A.: Global uniqueness for a two-dimensional inverse boundary value problem. Ann. Math. **143**, 71–96 (1996)
24. Pedregal, P.: Variational Methods in Nonlinear Elasticity. SIAM, Philadelphia (2000)
25. Pedregal, P.: Inverse quasiconvexification. Milan J. Math. **89**(1), 123–145 (2021)
26. Rindler, F.: Calculus of Variations. Universitext. Springer, Cham (2018)
27. Santambrogio, F.: Optimal Transport for Applied Mathematicians. Calculus of Variations, PDEs, and Modeling. Progress in Nonlinear Differential Equations and their Applications, vol. 87. Birkhäuser/Springer, Cham (2015)
28. Uhlmann, G.: Electrical impedance tomography and Calderón's problem. Inverse Problems **25**(12), 123011, 39 pp. (2009)
29. Valent, T.: Boundary Value Problems of Finite Elasticity. Springer Tracts in Natural Philosophy, vol. 31. Springer, Berlin (1988)
30. Villani, C.: Optimal Transport. Old and New. Grundlehren der Mathematischen Wissenschaften [Fundamental Principles of Mathematical Sciences], vol. 338. Springer, Berlin (2009)

Chapter 3
THB-Spline Approximations for Turbine Blade Design with Local B-Spline Approximations

Cesare Bracco, Carlotta Giannelli, David Großmann, Sofia Imperatore, Dominik Mokriš, and Alessandra Sestini

Abstract We consider two-stage scattered data fitting with truncated hierarchical B-splines (THB-splines) for the adaptive reconstruction of industrial models. The first stage of the scheme is devoted to the computation of local least squares variational spline approximations, exploiting a simple fairness functional to handle data distributions with a locally varying density of points. Hierarchical spline quasi-interpolation based on THB-splines is considered in the second stage of the method to construct the adaptive spline surface approximating the whole scattered data set and a suitable strategy to guide the adaptive refinement is introduced. A selection of examples on geometric models representing components of aircraft turbine blades highlights the performances of the scheme. The tests include a scattered data set with voids and the adaptive reconstruction of a cylinder-like surface.

Keywords Scattered data · Quasi-interpolation · THB-splines · Smoothing splines · Turbine blade design

3.1 Introduction

Scattered data fitting is nowadays of fundamental importance in a variety of fields, ranging from geographic applications to medical imaging and geometric modeling. The topic can be addressed in different approximation spaces, either by using spline

C. Bracco · C. Giannelli · S. Imperatore · A. Sestini (✉)
Dipartimento di Matematica e Informatica "U. Dini", Università degli Studi di Firenze, Firenze, Italy
e-mail: cesare.bracco@unifi.it; carlotta.giannelli@unifi.it; sofia.imperatore@unifi.it; alessandra.sestini@unifi.it

D. Großmann · D. Mokriš
MTU Aero Engines AG, Munich, Germany
e-mail: David.GROSSMANN@mtu.de; Dominik.Mokris@mtu.de

D. Barrera et al. (eds.), *Mathematical and Computational Methods for Modelling, Approximation and Simulation*, SEMA SIMAI Springer Series 29,
https://doi.org/10.1007/978-3-030-94339-4_3

spaces or radial basis functions. In particular, thanks to their low computational cost and also to their better control of conditioning, schemes relying on local approximations have received a lot of attention, being formulated either as partition of unity interpolation or as two-stage approximation methods, see for example [4, 6–8] and references therein.

In this work we are interested in two-stage surface reconstruction of industrial models starting from scattered data obtained by optical scanning acquisitions. In order to reduce the noise, the available data are always preprocessed because the interest is in highly accurate reconstructions. Anyway, since the available data can not be considered exact, interpolation is not required and a less costly approximation scheme can be used to compute any local fitting. Furthermore, considering that splines represented in B-spline form are the standard choice for industrial computer-aided design applications (see e.g. [22]), we are interested in determining the final approximation in a space spanned by suitable extensions of tensor-product B-splines.

As well known, an approximating spline can be obtained by using different approximation approaches and also operating in different kinds of spline spaces, with the related schemes usually divided into two main classes. The first is the class of *global* schemes which use simultaneously all the given information and thus require the solution of a linear system whose size is (substantially) equal to the cardinality of the considered spline space. The second class collects all kinds of *local* schemes which avoid the solution of a global linear system. Intermediate alternatives are also possible, see e.g., [17], where spline spaces on triangulations are considered together with a domain decomposition technique. Many local methods can be collocated in the field of Quasi-Interpolation (QI) which is a fundamental methodology in the context of spline approximation, see e.g. [18, 19] for an introduction. In some cases, e.g. when dealing with scattered data, QI can require the solution of small linear systems whose size does not depend on the cardinality of the entire data set or on the size of the global spline space. These local systems descend from the application of *local* approximation schemes, each one considering a small number of data whose associated parameter values belong to a modest portion of the parametric domain intersecting the support of a compactly supported basis function. The computational advantage of the QI approach is evident, since these local linear systems are of small size and independent of each other. However, the development of an effective scattered data fitting approach based on local methods is never trivial, since the quality of the final spline approximation depends in this case not only on the considered global spline space but also on the approximation power of the local scheme and on the choice of the corresponding local data set. Here we are interested in considering a method of this kind but having also the adaptivity feature, in particular relying on a QI two stage methodology and working in adaptive spline spaces, since they allow local refinement and generalize the standard tensor-product model. Since scattered data can be characterized by a highly varying distribution, by also including voids, a flexible and reliable approach which automatically (re-) constructs the geometric model may strongly improve the efficiency of the overall scheme by suitably adapting the solution to the shape and configuration of the given point clouds.

In recent years the need for local refinement both in computer aided geometric design and isogeometric analysis has motivated a lot of research on adaptive spline spaces, leading to the introduction of several adaptive spline constructions, such as spline spaces over T-meshes [9], T-splines [23], LR B-splines [10] and hierarchical B-splines [16]. Spline spaces over T-meshes is the most straightforward approach: given a T-mesh, that is a rectangular mesh allowing T-junctions (vertices with valence 3), a space of functions which are polynomial over each cell can be easily defined. While this is very natural, it has been shown that the dimension of the corresponding space is stable (namely, it depends only on the topology of the mesh and not on the geometry) only under certain conditions, see [20] and [3]. Consequently, an efficient construction of a global basis is, in general, still an open question. T-splines [23] and LR B-splines [10] are both formally B-splines, but defined on local knot vectors depending on the topology of the mesh, and therefore allowing local refinement. They are very flexible, but their linear independence is guaranteed only under certain conditions on the mesh (different for the two types of spline), which any refinement algorithm must then preserve, see e.g. [1] and [21]. The construction of hierarchical B-splines [16] is based on a multi-level construction, where the refinement is obtained by replacing coarser elements with finer ones in the mesh hierarchy. As a consequence, the corresponding basis is composed of B-splines constructed on meshes of different levels. The multilevel nature of these splines allows the design of efficient local refinement schemes and their suitable integration into existing computer aided design software [15]. Moreover, the truncated version of hierarchical B-splines (THB-splines) [13] can be used for an easy extension to hierarchical spline spaces of any QI scheme formulated in a standard spline space [24, 25]. These features make THB-splines a natural choice to solve our scattered data fitting problem and are then the solution here considered.

The first proposal based on adaptive THB-spline fitting of scattered data for the reconstruction of industrial models was introduced in [15], where a global adaptive smoothed least-squares scheme was developed. Successively, in order to increase its locality and reduce the computational cost, the same problem has been addressed in different papers by combining a two-stage approach with hierarchical spline approximations. The first contribution where this kind of schemes was used by some of the authors was presented in [2] to deal with gridded data of Hermite type. In [4] these kinds of approximants were extended for the first time to scattered data, by using in the first stage of the scheme local polynomial least squares approximations of variable degree. A preliminary application of this scheme to industrial data reconstruction was given in [5], where its theoretical analysis was also presented.

In this paper a variant of the approach considered in [4, 5] is presented, to further decrease the number of degrees of freedom necessary to reach a certain accuracy and also to reduce the artifacts in the resulting surface. Since the local polynomials used in [4, 5] need to be converted into B-spline form for being usable by the quasi-interpolation approach considered in the second stage of scheme, we now work directly in local spline spaces. In this way, the algorithm for the first stage of the scheme is simplified. In order to avoid rank-deficiency problems by simultaneously controlling the smoothness of local approximants, a smoothing term is added to the least squares objective function. To improve the stability of the proposed method for general data configuration, we also inserted an automatic control on the choice of the local data sets.

The structure of the paper is as follows. Section 3.2 presents the model problem, while the first stage of the new scheme, devoted to the computation of local smoothed least squares B-spline approximations, is described in Sect. 3.3. Section 3.4 introduces the construction of the adaptive THB-spline surface approximating the scattered data set. A selection of examples on geometric models representing components of aircraft turbine blades is presented in Sect. 3.5, and compared with the results obtained in [5]. The numerical experiments include a new scattered data set with voids and the adaptive reconstruction of a surface closed in one parametric direction.

3.2 The Problem

The industrial models here considered are components of aircraft turbine blades which can be suitably represented in parametric form by using just one map, with the possibility of being periodic in one parametric direction. The problem can be mathematically described as follows. Let

$$F := \left\{ \mathbf{f}_i \in \mathbb{R}^3, i = 1, \ldots, n \right\}$$

be a scattered set of distinct points in the 3D space which can be reasonably associated by a one-to-one map to a set $X := \left\{ \mathbf{x}_i := (x_i, y_i) \in \Omega \subset \mathbb{R}^2, i = 1, \ldots, n \right\}$ of distinct parameter values belonging to a closed planar parametric domain Ω. Since the choice of a suitable parametrization method for the definition of the set X, which can naturally influence the quality of the final approximation, is not our focus, in this paper we relate to classical parametrization methods based on a preliminary triangulation of the scattered data set F, see e. g., [11, 12]. Consequently, both F and X are considered input data for the approximation problem.

Focusing on two-stage spline approximation schemes, we can introduce the general idea referring for simplicity to their formulation in a standard space V of tensor product splines of bi-degree $\mathbf{d} = (d_1, d_2)$, where it is assumed $d_i \geq 1, i = 1, 2$, in order to deal with at least continuous functions. With this kind of methods, a quasi-interpolation operator Q is defined so that $Q(F, X) = \mathbf{s}$, with $\mathbf{s}$ denoting a vector function, possibly periodic in one parametric direction, with components in the spline space V. Using a suitable spline basis $\mathcal{B} := \{B_J\}_{J \in \Gamma}$ of V, such vector spline $\mathbf{s}$ can be expressed as follows,

$$\mathbf{s}(\mathbf{x}) := \sum_{J \in \Gamma} \boldsymbol{\lambda}_J(F_J, X_J) B_J(\mathbf{x}), \qquad \mathbf{x} \in \Omega, \tag{3.1}$$

where each coefficient vector $\boldsymbol{\lambda}_J(F_J, X_J)$, $J \in \Gamma$, is computed in the first stage of the scheme by using a certain local subset $F_J \subset F$ of data and the corresponding set of parameter values $X_J \subset X$, so that $\mathbf{s}(\mathbf{x}_i) \approx \mathbf{f_i}$, $i = 1, \ldots, n$.

When dealing with discrete data, measuring the accuracy of the spline approximation with the maximum of the errors $||\mathbf{s}(\mathbf{x}_i) - \mathbf{f_i}||_2$ at each parameter site can appear reasonable at the first sight. However, the quality of the approximant is also strictly related to the lack of unwanted artifacts, a feature of fundamental importance for industrial applications of any approximation scheme. In this context, it is then a common practice to require the error under a prescribed tolerance only at a certain percentage of sites in X.

3.3 First Stage: Local B-Spline Approximations

For computing each vector coefficient $\boldsymbol{\lambda}_J$, $J \in \Gamma$, necessary in (3.1) to define the approximation $\mathbf{s}$, we consider a local data subset $F_J \subset F$,

$$F_J := \left\{\mathbf{f}_i : i \in I_J\right\} \quad \text{with} \quad I_J := \{i : \mathbf{x}_i \in X \cap \Omega_J\},$$

associated to the set

$$X_J := \{\mathbf{x}_i : i \in I_J\}$$

of parameter values in a local subdomain Ω_J of Ω which has non empty intersection with the support of the basis function B_J, namely $\Omega_J \cap \text{supp}(B_J) \neq \emptyset$. By denoting with $\mathcal{B}_J := \{B_I : I \in \Lambda_J \subset \Gamma\}$ the set of B-splines in $\mathcal{B}$ not vanishing in Ω_J (which necessarily includes B_J), the value of $\boldsymbol{\lambda}_J$ is defined as the vector coefficient associated with B_J in a local spline approximation $\mathbf{s}_J \in \mathcal{S}_J$, with $\mathbf{s}_J : \Omega_J \to \mathbb{R}^3$ and

$$\mathcal{S}_J := \text{span}\{B_I, I \in \Lambda_J\}.$$

Concerning this local spline space, the following proposition is proved, since it shows that $\mathcal{S}_J$ has reasonable approximation power and in particular that it includes the restriction to Ω_J of any linear polynomial,

Proposition *The following space inclusion holds true*

$$\Pi^2_{\mathbf{d}}|_{\Omega_J} \subseteq \mathcal{S}_J := \text{span}\{B_I : I \in \Lambda_J\},$$

where $\Pi^2_{\mathbf{d}}|_{\Omega_J}$ *denotes the restriction to* Ω_J *of the tensor product space of bivariate polynomials of bi-degree* $\mathbf{d}$.

Proof Let c_K be a cell of $\mathcal{G}$ such that $c_K \cap \Omega_J \neq \emptyset$. Then the definition of Λ_J implies that $c_K \subseteq \text{supp}(B_I) \Rightarrow I \in \Lambda_J$. Thus we can say that span$\{B_I$:

$c_K \subseteq \text{supp}(B_I)\} \subseteq \mathcal{S}_J$. Since span$\{B_I : c_K \subseteq \text{supp}(B_I)\} = \Pi^2_{\mathbf{d}}|c_K$, the proof is completed considering that c_K is any cell of $\mathcal{G}$ with non vanishing intersection with Ω_J. □

The variational fitting method adopted to determine $\mathbf{s}_J$ in $\mathcal{S}_J$ consists in minimizing the following objective function

$$\sum_{i \in I_J} \|\mathbf{s}_J(\mathbf{x}_i) - \mathbf{f}_i\|_2^2 + \mu\, E(\mathbf{s}_J)\,, \tag{3.2}$$

where $\mu > 0$ is a smoothing coefficient and $E(s_J)$ the thin-plate energy,

$$E(\mathbf{s}_J) := \int_{\Omega_J} \left\| \frac{\partial^2 \mathbf{s}_J}{\partial x^2} \right\|_2^2 + 2 \left\| \frac{\partial^2 \mathbf{s}_J}{\partial x \partial y} \right\|_2^2 + \left\| \frac{\partial^2 \mathbf{s}_J}{\partial y^2} \right\|_2^2 \mathrm{d}x\mathrm{d}y\,.$$

As recalled in the Appendix, the assumption of a positive μ ensures that this local approximation problem admits always a unique solution, provided that the sites belonging to X_J are not collinear.

Since the scheme is locally applied, an automatic (data-dependent) selection of the parameter μ could be considered. For example, the choice may take into account the cardinality $|X_J|$ of the local sample or the area of Ω_J, which influence the value of the first and of the second addend in (3.2), respectively. In view of this influence however, we may observe that a constant value of μ implies that the balancing between the fitting and the smoothing term in the objective function usually increases when $|X_J|$ or the area of Ω_J increases, being this true in the second case because second derivatives are involved in the smoothing term. Both these behaviors seem reasonable and are confirmed by the quality of the results obtained in our experiments, where a constant value for μ is suitably chosen.

Differently from [4, 5], in order to better avoid overfitting, a lower bound $n_{\min}$ for the cardinality of X_J is now required, being $n_{\min} \geq 3$ the only additional input parameter required by the algorithm, besides μ. To fulfill this condition, Ω_J is initialized as supp(B_J) and enlarged until $|X_J| \geq n_{\min}$. Note that the refinement strategy presented in Sect. 3.4 automatically guarantees that the inequality $|X_J| \geq n_{\min}$ becomes fulfilled without requiring an excessive enlargement of the set Ω_J, which would compromise the locality of the approximation. For this reason, it is not necessary to set a maximum value for controlling the enlargement of the local data set. Considering that the smaller $n_{\min}$ is, the higher is the obtainable level of detail but also the probability of overfitting, in our experiments (which always adopt $d_1 = d_2 = d = 2, 3$), a good low range for its selection has always been $d^2 \leq n_{\min} \leq (d+1)^2$.

Summarizing, the computation of each $\boldsymbol{\lambda}_J$, $J \in \Gamma$, is done according to the following algorithm.

Note that exploiting a regularized least square approximation and, as a consequence, being able to directly employ the local spline space has significantly

Algorithm 1 Local smoothing spline approximant

Inputs

- $F \subset \mathbb{R}^3$: scattered data set;
- $X \subset \Omega \subset \mathbb{R}^2$: set of parameter values corresponding to the data in F;
- $\mathcal{G}$: uniform tensor-product mesh in Ω (possibly with auxiliary cells);
- V: tensor-product spline space of bi-degree $\mathbf{d}$ associated with $\mathcal{G}$ with B-spline basis $\mathcal{B}$;
- $J \in \Gamma$: index of the considered basis function $B_J \in \mathcal{B}$;
- μ: smoothing spline parameter ($\mu > 0$);
- $n_{\min}$: minimum required number of local data ($3 \leq n_{\min} \ll |F|$);

1. Initialization
   a. initialize $\Omega_J = \text{supp}(B_J)$;
   b. initialize $I_J = \{i : \mathbf{x}_i \in X \cap \Omega_J\}$, $F_J = \left\{\mathbf{f}_i : i \in I_J\right\}$ and $X_J = \{\mathbf{x}_i : i \in I_J\}$;
2. while $|F_J| < n_{\min}$
   a. enlarge Ω_J with the first ring of cells in $\mathcal{G}$ surrounding Ω_J;
   b. update $I_J = \{i : \mathbf{x}_i \in X \cap \Omega_J\}$, $F_J = \left\{\mathbf{f}_i : i \in I_J\right\}$ and $X_J = \{\mathbf{x}_i : i \in I_J\}$;
3. if the sites in X_J are not collinear, then:
   a. compute the local approximation $\mathbf{s}_J = \sum_{I \in \Lambda_J} \mathbf{c}_I^{(J)} B_I$ minimizing the objective function in (3.2) for the data F_J and X_J;
   b. set $\boldsymbol{\lambda}_J = \mathbf{c}_J^{(J)}$;

   else set $\boldsymbol{\lambda}_J = \frac{1}{|F_J|} \sum_{I \in I_J} \mathbf{f}_I$.

Output

- vector coefficient $\boldsymbol{\lambda}_J$.

simplified the algorithm originally proposed for the first stage in [4, 5], where a variable-degree local polynomial approximation was considered. In particular, the new scheme does not require the selection of a suitable degree for the computation of any coefficient $\boldsymbol{\lambda}_J$ and eliminates the conversion of the computed approximant from the polynomial to the B-spline basis.

In the following section, after introducing the THB-spline basis, the operator Q is easily extended to hierarchical spline spaces, by also introducing the automatic refinement algorithm here considered. Note that this extension rule ensures that the coefficient associated to a THB-spline basis function remains unchanged on a refined hierarchical mesh if this function remains active on the updated hierarchical configuration.

3.4 Second Stage: THB-Spline Approximation

Let us consider a sequence $V^0 \subset \ldots \subset V^{M-1}$ of M spaces of tensor-product splines of bi-degree $\mathbf{d} := (d_1, d_2)$, $d_i \geq 1, i = 1, 2$, defined on the closed domain Ω, and each one associated with the tensor-product grid $\mathcal{G}^\ell$ and the basis $\mathcal{B}^\ell_{\mathbf{d}} := \left\{ B^\ell_J \right\}_{J \in \Gamma^\ell_{\mathbf{d}}}$. Let $\Omega^0 \supseteq \ldots \supseteq \Omega^M$ be a sequence of closed domains, with $\Omega^0 := \Omega$ and $\Omega^M := \emptyset$. Each Ω^ℓ, for $\ell = 1, \ldots, M-1$ is the union of cells of the tensor-product grid $\mathcal{G}^{\ell-1}$. Let $\mathcal{G}_{\mathcal{H}}$ be the hierarchical mesh defined by

$$\mathcal{G}_{\mathcal{H}} := \{ Q \in \mathcal{G}^\ell,\ 0 \leq \ell \leq M-1 \} \quad \text{with} \quad \mathcal{G}^\ell := \{ Q \in G^\ell :\ Q \subset \Omega^\ell \backslash \Omega^{\ell+1} \},$$

where each $\mathcal{G}^\ell$ is called the set of *active cells* of level ℓ. The hierarchical B-spline (HB-spline) basis $\mathcal{H}(\mathcal{G}_{\mathcal{H}})$ with respect to the mesh $\mathcal{G}_{\mathcal{H}}$ is defined as

$$\mathcal{H}(\mathcal{G}_{\mathcal{H}}) := \{ B^\ell_J :\ J \in A^\ell, \ell = 0, \ldots, M-1 \},$$

where

$$A^\ell := \left\{ J \in \Gamma^\ell : \operatorname{supp}(B^\ell_J) \subseteq \Omega^\ell \wedge \operatorname{supp}(B^\ell_J) \nsubseteq \Omega^{\ell+1} \right\},$$

is the set of *active multi-indices* of level ℓ, and $\operatorname{supp}(B^\ell_J)$ denotes the intersection of the support of B^ℓ_J with Ω^0. The corresponding hierarchical space is defined as $\mathcal{S}_{\mathcal{H}} := \operatorname{span} \mathcal{H}(\mathcal{G}_{\mathcal{H}})$.

For any $s \in V^\ell$, $\ell = 0, \ldots, M-2$, let

$$s = \sum_{J \in \Gamma^{\ell+1}} \sigma^{\ell+1}_J B^{\ell+1}_J$$

be its representation in terms of B-splines of the refined space $V^{\ell+1}$. We define the truncation of s with respect to level $\ell+1$ and its (cumulative) truncation with respect to all finer levels as

$$\operatorname{trunc}^{\ell+1}(s) := \sum_{J \in \Gamma^{\ell+1} :\, \operatorname{supp}(B^{\ell+1}_J) \nsubseteq \Omega^{\ell+1}} \sigma^{\ell+1}_J B^{\ell+1}_J,$$

and

$$\operatorname{Trunc}^{\ell+1}(s) := \operatorname{trunc}^{M-1}(\operatorname{trunc}^{M-2}(\ldots(\operatorname{trunc}^{\ell+1}(s))\ldots)),$$

respectively. For convenience, we also define $\mathrm{Trunc}^M(s) := s$ for $s \in V^{M-1}$. The THB-spline basis $\mathcal{T}(\mathcal{G}_\mathcal{H})$ of the hierarchical space $S_\mathcal{H}$ was introduced in [13] and can be defined as

$$\mathcal{T}(\mathcal{G}_\mathcal{H}) := \left\{ T_J^\ell := \mathrm{Trunc}^{\ell+1}(B_J^\ell) : J \in A^\ell, \ell = 0, \ldots, M-1 \right\}.$$

The B-spline B_J^ℓ is called the *mother* B-spline of the truncated basis function T_J^ℓ.

We recall that THB-splines are linearly independent, non-negative, preserve the coefficients of the underlying sequence of B-splines, and form a partition of unity [13, 14]. Besides that, following the general approach introduced in [25], using such basis we can easily construct the vector THB-spline approximation of the whole scattered data set in terms of the hierarchical quasi-interpolant $\mathbf{s} = Q(F, X)$ as follows,

$$\mathbf{s}(\mathbf{x}) := \sum_{\ell=0}^{M-1} \sum_{I \in A^\ell} \boldsymbol{\lambda}_I^\ell(F_I, X_I) T_I^\ell(\mathbf{x}), \tag{3.3}$$

where each vector coefficient $\boldsymbol{\lambda}_I^\ell$ is the one of the mother function B_I^ℓ and is obtained by computing the local regularized B-spline approximation $\mathbf{s}_I^\ell$ on the data set F_I^ℓ associated to B_I^ℓ as described in Sect. 3.3.

In order to define an adaptive approximation scheme, the following algorithm is used to iteratively construct the hierarchical mesh $\mathcal{G}_\mathcal{H}$, the corresponding spline space $S_\mathcal{H}$ and the final approximating spline $\mathbf{s}$. As any automatic refinement strategy, the algorithm requires in input some parameters which drive the refinement process. One of them is the error tolerance $\epsilon > 0$ whose value has to be chosen not only considering the accuracy desired for the reconstruction but also taking into account the level of noise affecting the given points cloud (here assumed without significant outliers). The percentage bound parameter η specifies the number of data points for which the error is required to be within the given tolerance and a value strictly less than 100% is suggested to reduce the influence of outliers with moderate size on the approximation. Another required parameter is the maximum number of levels $M_{\max}$ which has to be chosen considering the maximal level of detail desired for the reconstruction. The choice of the other input parameters, n_{loc}, n_1 and n_2, is discussed after the algorithm.

The refinement criterion has been motivated by the observation that, when the parameter values corresponding to the local data set F_I^ℓ are concentrated in a small part of the support of B_I^ℓ, the quality of the approximation may be affected. For this reason, we consider a splitting of the two sides of $\mathrm{supp}(B_J^{M-1})$ in n_1 and n_2 uniform segments, respectively, and subdivide the support of B_J^{M-1} in the resulting $n_1 n_2$ subregions, where we then check the presence of at least $\lceil n_{\mathrm{loc}}/(n_1 n_2) \rceil$ data points. To simplify the usage of the algorithm by default we set $n_1 = n_2 = 1$ but different values can be chosen if suitable, see e.g. the data set with voids considered in Example 3.3 of Sect. 3.5.

Algorithm 2 Adaptive hierarchical spline fitting

Inputs

- $F \subset \mathbb{R}^3$: scattered data set with $n = |F|$;
- $X \subset \Omega \subset \mathbb{R}^2$: set of parameter values corresponding to the data in F;
- $\mathcal{G}^0$: initial uniform tensor-product mesh in Ω (possibly with auxiliary cells);
- V^0: tensor-product spline space of bi-degree $\mathbf{d}$ associated with $\mathcal{G}^0$;
- ϵ : error tolerance;
- η: percentage bound of data points for which the error is required to be within the tolerance ϵ (default: $\eta = 95\%$);
- $M_{\max}$: integer specifying the maximum number of levels;
- n_{loc}: minimum number of local data required for refinement ($3 \le n_{\text{loc}} \ll |F|$);
- n_1 , n_2: positive integers specifying the number of uniform horizontal or vertical splittings for the support of a tensor-product B-spline (default: $n_1 = n_2 = 1$).

1. Initialization
   a. set $\mathcal{G}_{\mathcal{H}} = \mathcal{G}^0$and $S_{\mathcal{H}} = V^0$;
   b. set the current number of level $M = 1$;
   c. use Algorithm 1 to compute the coefficients of the hierarchical QI vector spline $\mathbf{s} \in S_{\mathcal{H}}$ introduced (3.3).
   d. evaluate the errors at the data sites $\mathbf{x}_i, \in X, i = 1, \ldots, n$,

$$e(\mathbf{x}_i) := \|\mathbf{s}(\mathbf{x}_i) - \mathbf{f}_i\|_2, \qquad i = 1, \ldots, n.$$

2. While $|\{i : e(\mathbf{x}_i) > \epsilon\}|/n > \eta$ and $M \le M_{\max}$, repeat the following steps:
   a. (*marking*) for each $\ell = 0, \ldots, M-1$, mark the cells of level ℓ in $\mathcal{G}_{\mathcal{H}}$ which are included in the support of a B_J^ℓ , $J \in A^\ell$ such that:
      - there exists at least one data site $\mathbf{x}_i \in X \cap \text{supp}(B_J^\ell)$ such that $e(\mathbf{x}_i) > \epsilon$;
      - there are at least $\lceil n_{\text{loc}}/(n_1 n_2) \rceil$ data sites in any of the $n_1 n_2$ subrectangles which uniformly split $\text{supp}(B_J^\ell)$ (note that such splitting is just temporarily considered to check whether this refinement requirement is satisfied);
   b. (*update the hierarchical mesh*) update $\mathcal{G}_{\mathcal{H}}$ by dyadically refining (in the two parametric directions) all the marked cells;
   c. (*update the number of levels*) set M equal to the current number of levels;
   d. (*update the hierarchical space*) update the THB-spline basis of $S_{\mathcal{H}}$ and the sets A^ℓ, $\ell = 0, \ldots, M-1$;
   e. (*spline update*) use Algorithm 2 to compute the coefficients of the hierarchical QI spline $\mathbf{s}$ defined in (3.3)—only coefficients associated with THB-splines having new mother B-splines have to be computed;
   f. (*error update*) evaluate the new errors $e(\mathbf{x}_i)$, $i = 1, \ldots, n$ at the data sites;

Outputs

- $\mathcal{G}_{\mathcal{H}}$: hierarchical mesh;
- $\mathcal{T}(\mathcal{G}_{\mathcal{H}})$: THB-spline basis of $S_{\mathcal{H}}$;
- coefficients of the hierarchical spline $\mathbf{s}$ defined in (3.3).

Concerning the parameter n_{loc}, we may observe that the requirement $n_{loc} \geq n_{min}$ guarantees that the points needed to compute the coefficients associated with the new functions in the first stage of the next iteration can be found not too far from the support of the functions themselves. Indeed in the algorithm introduced in the previous section, after a few enlargements, Ω_J will surely include the support of a refined function of the previous level intersecting $\text{supp}(B_J)$. As a consequence, analogously to n_{min}, a high value of n_{loc} contributes to the reduction of oscillations deriving from overfitting, but this value should also be low enough to guarantee that the refinement strategy can generate a hierarchical spline space with enough degrees of freedom for satisfying the given tolerance ϵ. For this reason, some tuning is necessary for a good selection of n_{loc}.

The proposed adaptive approximation method also extends to the case, not addressed in the previous works, of surfaces closed in one (or even two) parametric directions. Note that the local nature of the considered approximation approach makes the implementation especially easy, since coefficients associated with a THB present at successive steps of the adaptive refinement procedure (even if possibly further truncated) do not depend on such steps.

3.5 Examples

We present a selection of tests for the approximation of industrial data obtained by an optical scanning process of four different aircraft engine parts. For each of these surfaces, as a characterizing dimension, we report the length R of the diagonal of the minimal axis-aligned bounding box associated to the given point cloud. The parameter values are computed in all examples with standard parametrization methods based on a triangulation of the scattered data sets, see e. g., [11, 12]. The bi-degree $\mathbf{d}$ is set equal to $(2, 2)$ in the first considered example and always equal to 3 in the other examples. The results highlight the effects of considering a minimum number of local data points (also) in the first stage of method, as well as the improvements obtained by introducing a regularized B-spline approximation for each local fitting with respect to the scattered data fitting scheme considered in [5]. By combining these two changes, the two-stage approximation algorithm is more stable and unwanted oscillations are further reduced.

Concerning the parameters in input to Algorithm 1, we have always set $\mu = 10^{-6}$, except for Example 3.2 where it was chosen even smaller. In order to try to produce a very detailed reconstruction, a quite small value has been chosen for n_{min}, always selecting it between d^2 and $(d+1)^2$. Concerning the parameter selection for Algorithm 2, in all the presented experiments we have set the maximum number of levels M_{max} equal to 8, fixing the percentage bound parameter η equal to its default value ($\eta = 95\%$). The error tolerance ϵ has been always set to $5 \cdot 10^{-5}$ m, except for Example 3.2 where we used the about halved value chosen for the same experiment in [5] (as a reference value, for each example consider the dimension R characterizing the related point cloud). The integer parameters n_1 and n_2 also in input to Algorithm 2 have been always set to their unit default values, except for Example 3.3 which required a different selection because of the voids present in the considered data set. The only parameter which has required a finer tuning for the reported experiments has been n_{loc} which is anyway always assumed greater than n_{min}.

Example 3.1 (Tensile) In this example, we consider THB-spline approximations to reconstruct a part of a tensile from the set of 9281 scattered data shown in Fig. 3.1 (top) which has reference dimension $R = 2.5 \cdot 10^{-2}$ m. We compare the new local scheme based on local B-spline approximations with the algorithm based on local polynomial approximations of variable degree presented in [5], where this test was originally considered. Note that for this example we have never dealt with local sets of collinear points in our experiments.

As the first test, we ran both algorithms with the same setting considered in [5], namely, by starting with a 4×4 tensor-product mesh with $\mathbf{d} = (2, 2)$, tolerance $\epsilon = 5 \cdot 10^{-5}$ m, percentage bound η equal to the default 95% and $n_{loc} = 20$. The algorithm with local polynomial approximations with the parameter choice considered in [5] ($\sigma = 10^8$) led to an approximation with 2501 degrees of freedom, 96.25% of points below the tolerance and a maximum error of $1.22745 \cdot 10^{-4}$ m. The new scheme based on local B-spline approximations with $n_{min} = 6$ and $\mu = 10^{-6}$ generated a THB-spline surface with 1855 degrees of freedom that satisfies the required tolerance in 98.88% of points with a maximum error of $8.06007 \cdot 10^{-5}$ m. The number of levels used is $M = 5$, but all the cells of the first two levels are refined.

As the second test, we ran both algorithms by starting with a 16×4 tensor-product mesh with $\mathbf{d} = (2, 2)$, percentage η equal to the default, tolerance $\epsilon = 5 \cdot 10^{-5}$ m, and $n_{loc} = 15$. The algorithm with local polynomial approximations led to an approximation with 5922 degrees of freedom, 98.18% of points below the tolerance and a maximum error of $1.44222 \cdot 10^{-4}$ m. The surface and the corresponding hierarchical mesh are shown in Fig. 3.1 (center). This approximation clearly shows strong oscillations on the boundary of the reconstructed surface, due to a lack of available data points for the local fitting in correspondence of high refinement levels. The scheme based on local B-spline approximations, with $n_{min} = 7$ and $\mu = 10^{-6}$ produced a THB-spline surface with 1960 degrees of freedom that satisfies the required tolerance in the 99.36% of the data points with a maximum error of $8.10814 \cdot 10^{-5}$ m. The surface, free of unwanted oscillations along the boundary, and the corresponding hierarchical mesh are shown in Fig. 3.1 (bottom). The number of levels used is $M = 4$, with all the cells of level 0 refined.

Example 3.2 (Blade) In this example, we test the second example considered in [5] on the set of 27191 scattered data representing a scanned part of a blade shown in Fig. 3.2 (top) whose reference dimension R is equal to $5 \cdot 10^{-2}$ m. Again, to compare the new local scheme with the algorithm based on local polynomial approximations there considered, we ran both algorithms with the same setting of [5], namely, by starting with a 4×4 tensor-product mesh with $\mathbf{d} = (3, 3)$, tolerance $\epsilon = 2 \cdot 10^{-5}$ m, percentage bound η equal to the default 95% and $n_{loc} = 60$. The algorithm

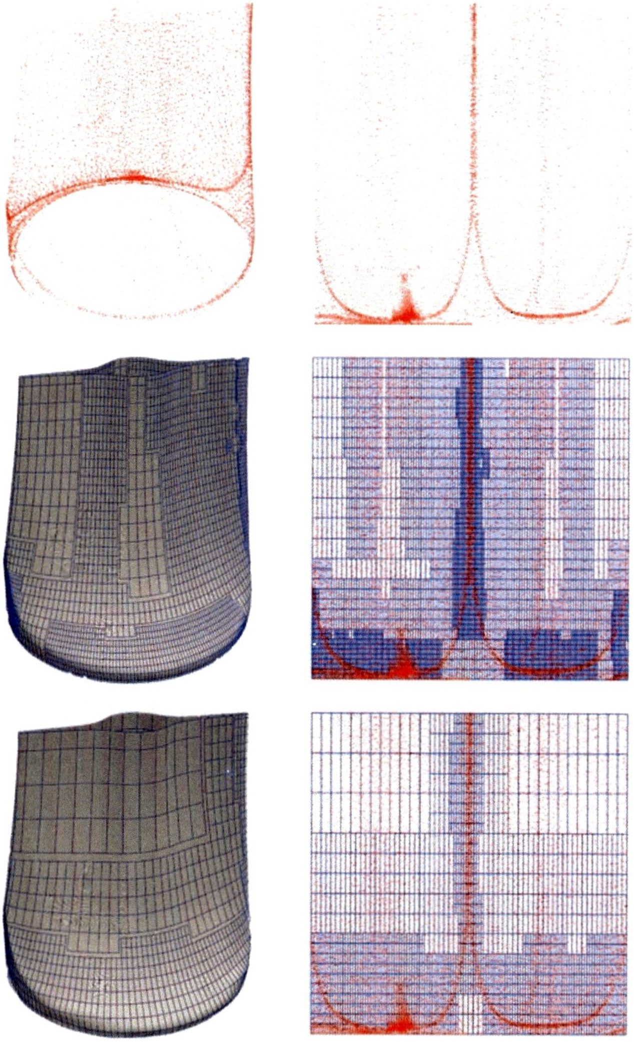

Fig. 3.1 Example 3.1: scattered data set corresponding to a critical part of a tensile (top), the reconstructed surfaces and the corresponding hierarchical meshes obtained with the algorithm presented in [5] (center) and the new scheme (bottom)

with local polynomial approximations with the parameter choice considered in [5] ($\sigma = 10^8$) led to an approximation with 12721 degrees of freedom, 97.06% of points below the tolerance and a maximum error of $1.08043 \cdot 10^{-4}$ m. The new scheme based on local B-spline approximations with $n_{\min} = 6$ and $\mu = 10^{-8}$ generated a THB-spline surface with 8314 degrees of freedom that satisfies the required tolerance in 99.94% of points with a maximum error of $1.32976 \cdot 10^{-4}$ m. The surface and the corresponding hierarchical mesh are shown in Fig. 3.2 (bottom).

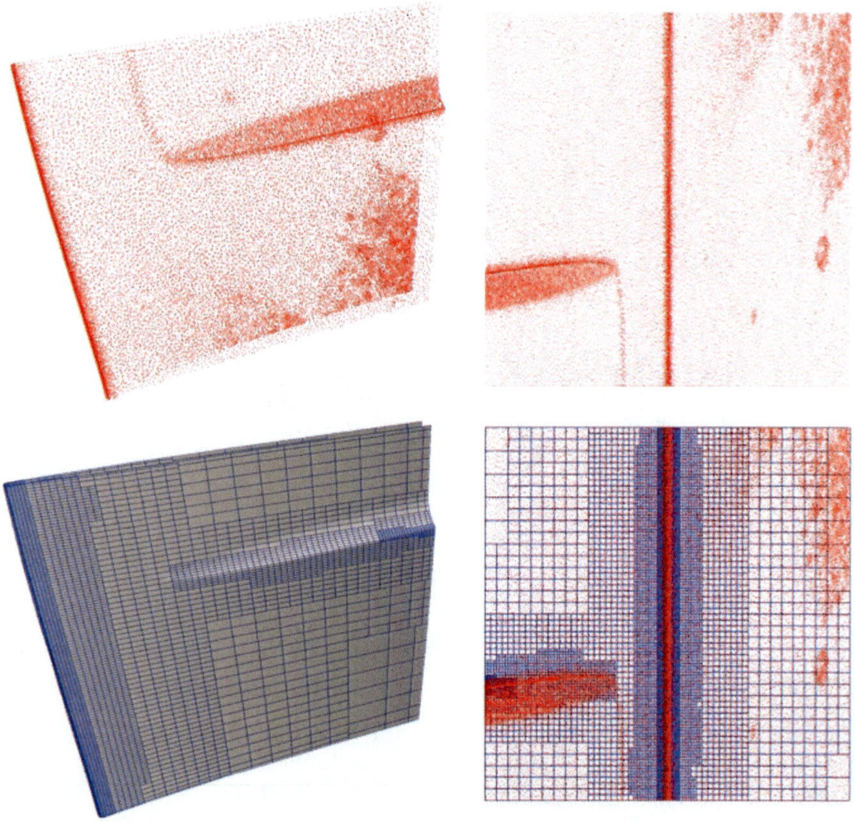

Fig. 3.2 Example 3.2: scattered data set corresponding to a critical part of a blade (top), the reconstructed surface and the corresponding hierarchical mesh obtained with the new scheme (bottom)

The number of levels used with the new scheme is 7 (the first two are not visible in the mesh because their cells are fully refined), one less than with the old approach. Finally for completeness we precise that the local collinearity check in Algorithm 1 is active only for 5 coefficients.

Example 3.3 (Endwall) In this example we illustrate the behavior of the adaptive algorithm on data sets with voids by considering the reconstruction of an endwall part from the scattered data set of 43,869 points shown in Fig. 3.3 (top) ($R = 5 \cdot 10^{-1}$ m). The figure shows that in this case the data set represents a model with three different holes, where no input data are available. The aim of this reconstruction is to avoid artifacts due to lack of points and obtain a sufficiently regular surface (e.g. by avoiding self-intersections), that can be post-processed with standard geometric software tools to obtain a suitably trimmed model. Consequently, not only the number of points in the local data sets is important to reach this aim, but also their

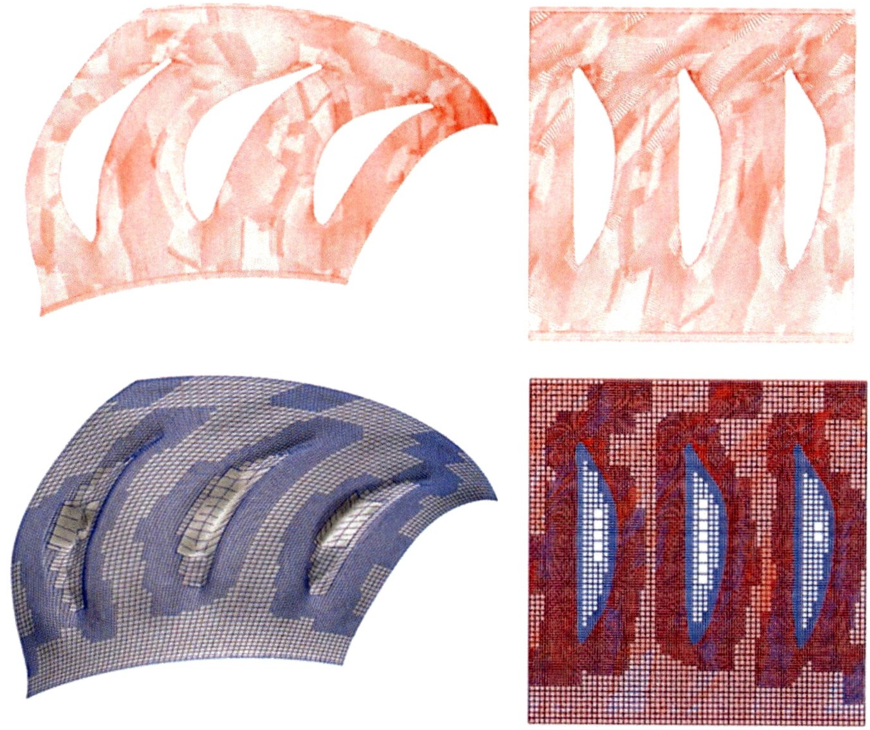

Fig. 3.3 Example 3.3: scattered data set corresponding to a critical part of an endwall (top), the reconstructed surface and the corresponding hierarchical mesh obtained with the new scheme (bottom)

distribution. To properly address this issue, we consider a real density parameter with value between 0 and 1 which determines whether the distribution of the points in the local set is reliable or not for the fitting. The distribution of the local data points is computed as the number of mesh cells of level ℓ inside the support of B^ℓ or its enlargement, which contain at least one point, over the total number of mesh cells, either in the support of B^ℓ or its enlargement. If this ratio is below a density threshold, then more data points are required and the function support is enlarged for the computation of the local approximation in the first stage of the method. The approximation is developed by starting from a 32×32 tensor-product mesh, with $\mathbf{d} = (3, 3)$, $n_{\text{loc}} = 15$, $n_{\min} = 11$, $\mu = 10^{-6}$ and $n_1 = n_2 = 2$. A choice of the density parameter δ equal to 0.3 permits to take care of the difficult distribution of data points in the construction of the approximation. By considering a tolerance $\epsilon = 5 \cdot 10^{-5}$ m and a percentage bound η equal to the default 95%, the refinement generated a THB-spline approximation with $M = 3$ and 11211 degrees of freedom, 98.70% of points below the threshold and a maximum error of $5.68999 \cdot 10^{-4}$ m. The surface and the corresponding hierarchical mesh are shown in Fig. 3.3 (bottom). In this case there are 18 coefficients of the last level and 15 of the last but one (all

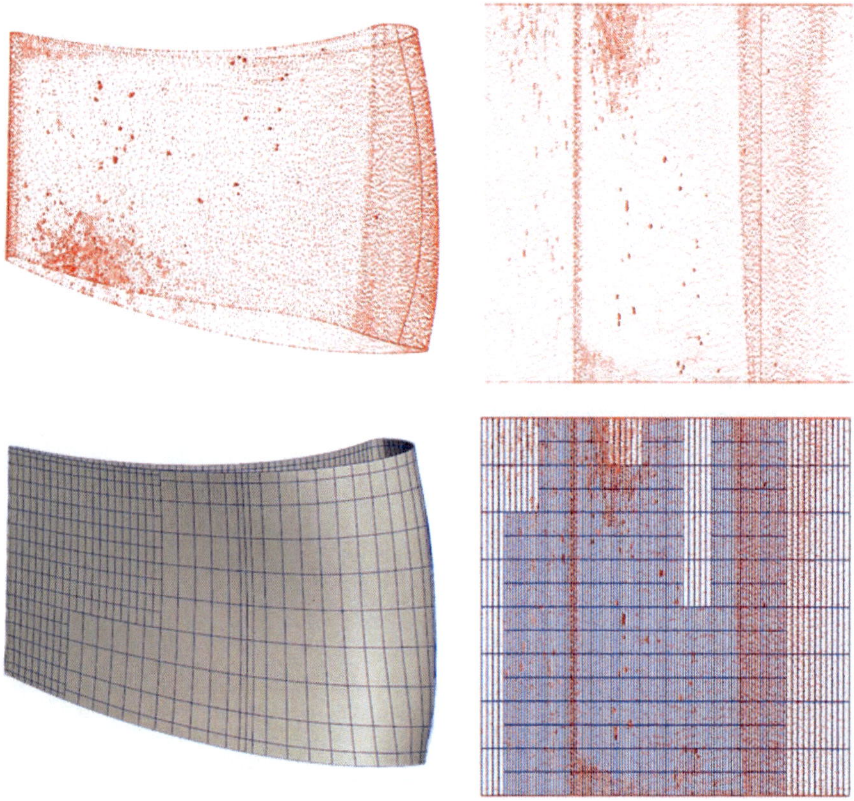

Fig. 3.4 Example 3.4: scattered data set corresponding to a critical part of an airfoil (top), the reconstructed surface and the corresponding hierarchical mesh obtained with the new scheme (bottom)

associated with B-splines whose support intersects a void) such that the related X_J is made up of all collinear points.

Example 3.4 (Airfoil) This example illustrates the behavior of the new adaptive fitting algorithm with local B-spline approximations for surfaces closed in one parametric direction. We test the scheme to reconstruct a blade airfoil from the set of 19669 scattered data shown in Fig. 3.4 (top) ($R = 10^{-1}$ m). We ran the method by starting with a 32×4 tensor-product mesh with $\mathbf{d} = (3, 3)$, setting $\eta = 95\%$, $\epsilon = 5 \cdot 10^{-5}$ m, and $n_{\mathrm{loc}} = 30$ in Algorithm 2 and using Algorithm 1 with $n_{\min} = 12$ and $\mu = 10^{-6}$. The refinement strategy produced an approximation with $M = 3$ and 1856 degrees of freedom distributed in the last two levels, that satisfies the required tolerance in 95.06% of the data points with maximum error $1.87742 \cdot 10^{-4}$ m (observe also that, as well as for Example 3.1, at the local stage X_J is never made up of all collinear points). The surface and the corresponding hierarchical mesh are shown in Fig. 3.4 (bottom). By trying to force additional refinement, some oscillations appear. In this case, they are consistent with the data distribution since there are clusters of high density points, due to scan noise. Consequently, they do not represent artifacts caused by regions with very low density of data and cannot be prevented by exploiting the bound for cardinality of the local data sample governed by n_{loc}.

Acknowledgments We thank the anonymous reviewer for his/her useful suggestions. Cesare Bracco, Carlotta Giannelli and Alessandra Sestini are members of the INdAM Research group GNCS. The INdAM support through GNCS and Finanziamenti Premiali SUNRISE is also gratefully acknowledged.

Appendix

In this appendix we give the proof that, assuming $\mathbf{d} = (d_1, d_2)$ with $d_i \geq 1, i = 1, 2$, the local vector spline $\mathbf{s}_J$ defined in Sect. 3.3 exists and is unique, provided that the sites in X_J are not collinear. First of all we observe that the objective function in (3.2) can be split in the sum of three analogous objective functions, one for each component $s_J^{(k)}$, $k = 1, 2, 3$, of $\mathbf{s}_J$,

$$\sum_{i \in I_J} \|\mathbf{s}_J(\mathbf{x}_i) - \mathbf{f}_i\|_2^2 + \mu\, E(\mathbf{s}_J) = \sum_{k=1}^{3} \Big(\sum_{i \in I_j} \big(s_J^{(k)}(\mathbf{x}_i) - (\mathbf{f}_i)_k\big)^2 + \mu\, \rho(s_J^{(k)}) \Big),$$

where

$$\rho\big(s_J^{(k)}\big) := \int_{\Omega_J} \left(\frac{\partial^2 s_J^{(k)}}{\partial x^2}\right)^2 + 2\left(\frac{\partial^2 s_J^{(k)}}{\partial x \partial y}\right)^2 + \left(\frac{\partial^2 s_J^{(k)}}{\partial y^2}\right)^2 \mathrm{d}x\mathrm{d}y\,.$$

Thus the study can be developed in the scalar case and for brevity we remove the subscript or superscript k ranging between 1 and 3. The analysis is developed in the following theorem, where $s_J : \Omega_J \to \mathbb{R}$ denotes the local spline in $\mathcal{S}_J$ associated to the coefficient vector $\mathbf{c} \in \mathbb{R}^{\ell_J}$, with $\ell_J := |\Lambda_J|$,

$$s_J(\mathbf{x}) = \sum_{I \in \Lambda_J} c_I B_I(\mathbf{x})\,.$$

Theorem *Let the considered spline bi-degree* $\mathbf{d} = (d_1, d_2)$ *be such that* $d_i \geq 1, i = 1, 2$. *When the points* $\mathbf{x}_i \in \Omega_J, i \in I_J$ *are not collinear, there exists a unique local spline* $s_J \in \mathcal{S}_J$ *minimizing the following objective function,*

$$\sum_{i \in I_J} \big(s_J(\mathbf{x}_i) - f_i\big)^2 + \mu\, \rho(s_J)\,, \tag{3.4}$$

where $\mu > 0$. If such points in Ω_J are collinear, then such minimizer does not exist or is not unique.

Proof Let us observe that $\rho(s_J) = \mathbf{c}^T M\mathbf{c}$, where $M \in \mathbb{R}^{\ell_J \times \ell_J}$ is such that

$$M_{i,r} := \int_{\Omega_J} \left(\frac{\partial^2 B_I}{\partial x^2}\right)\left(\frac{\partial^2 B_R}{\partial x^2}\right) + 2\left(\frac{\partial^2 B_I}{\partial x \partial y}\right)\left(\frac{\partial^2 B_R}{\partial x \partial y}\right) + \left(\frac{\partial^2 B_I}{\partial y^2}\right)\left(\frac{\partial^2 B_R}{\partial y^2}\right) \mathrm{d}x\mathrm{d}y\,,$$

where we are assuming that, in the adopted ordering of the B-spline basis elements of $\mathcal{S}_J$, B_R and B_I are the r–th and the i–th ones. On the other hand it is

$$\sum_{i \in I_J} \left(s_J(\mathbf{x}_i) - f_i\right)^2 = \|V\mathbf{c} - \mathbf{F}\|_2^2 = \mathbf{c}^T A^T A\mathbf{c} - 2\mathbf{F}^T A\mathbf{c} + \mathbf{F}^T\mathbf{F}\,,$$

where $\mathbf{F} \in \mathbb{R}^{|I_J|}$ denotes the vector collecting all the $f_i, i \in I_J$ and A is the $|I_J| \times \ell_J$ collocation matrix of the tensor-product B-spline basis generating $\mathcal{S}_J$. Thus the objective function in (3.4) can be written also as the following quadratic function,

$$\mathbf{c}^T(A^T A + \mu M)\mathbf{c} - 2\mathbf{F}^T A\mathbf{c} + \mathbf{F}^T\mathbf{F}\,.$$

As well known a quadratic function admits a global unique minimizer if and only if the symmetric matrix defining its homogeneous quadratic terms is positive definite and in such case the minimizer is given by its unique stationary point. In our case such matrix is $A^T A + \mu M$ and the stationary points are the solutions of the following linear system of ℓ_J equations in as many unknowns,

$$(A^T A + \mu M)\mathbf{c} = A^T\mathbf{F}\,.$$

Now, for all positive μ the matrix $A^T A + \mu M$ is symmetric and positive semidefinite since, for any vector $\boldsymbol{\zeta} \in \mathbb{R}^{\ell_J}, \boldsymbol{\zeta} \neq \mathbf{0}$ it is $\boldsymbol{\zeta}^T A^T A\boldsymbol{\zeta} \geq 0$ and $\boldsymbol{\zeta}^T M\boldsymbol{\zeta} \geq 0$, the last inequality descending from the fact that $\boldsymbol{\zeta}^T M\boldsymbol{\zeta} = \rho(s_{\zeta})$, where $s_{\zeta}(\mathbf{x}) = \sum_{I \in \Lambda_J} \zeta_I B_I(\mathbf{x})$. Now if the points $\mathbf{x}_i, i \in I_j$ are distributed in Ω_J along a straight line $ax + by + c = 0$, the proposition proved in Sect. 3.3 implies that it is possible to find $\boldsymbol{\zeta} \in \mathbb{R}^{\ell_J}, \boldsymbol{\zeta} \neq \mathbf{0}$ such that $s_{\zeta}(\mathbf{x}) \equiv ax + by + c, \forall \mathbf{x} \in \Omega_J$. This implies that $s_{\zeta}(\mathbf{x}_i) = 0, \forall i \in I_J$, that is the vector $A\boldsymbol{\zeta} \in \mathbb{R}^{|I_J|}$ vanishes. On the other hand, clearly it is also $0 = \rho(s_{\zeta}) = \boldsymbol{\zeta}^T M\boldsymbol{\zeta}$, since $s_{\zeta}|\Omega_J$ is a linear polynomial. This proves that the symmetric positive semidefinite matrix $(A^T A + \mu M)$ is not positive definite when all the $\mathbf{x}_i, i \in I_J$ are collinear. This is the only possible data distribution associated to a non positive definite matrix. Indeed if the points $\mathbf{x}_i, i \in I_J$ are not collinear, if $\boldsymbol{\zeta} \in \mathbb{R}^{\ell_J}, \boldsymbol{\zeta} \neq \mathbf{0}$ is associated to a non vanishing linear polynomial, it is $\boldsymbol{\zeta}^T M\boldsymbol{\zeta} = \rho(s_{\zeta}) = 0$ but $A\boldsymbol{\zeta} \neq \mathbf{0}$ and so $\boldsymbol{\zeta}^T A^T A\boldsymbol{\zeta} > 0$; on the other hand if $\boldsymbol{\zeta} \in \mathbb{R}^{\ell_J}, \boldsymbol{\zeta} \neq \mathbf{0}$ is not associated to a linear polynomial, then $\boldsymbol{\zeta}^T M\boldsymbol{\zeta} = \rho(s_{\zeta}) > 0$. □

References

1. Beirão da Veiga, L., Buffa, A., Sangalli, G., Vazquez, R.: Analysis-suitable T-splines of arbitrary degree: definition and properties.Math. Mod. Meth. Appl. Sci. **23**, 1979–2003 (2013)
2. Bracco, C., Giannelli, C., Mazzia, F., Sestini, A.: Bivariate hierarchical Hermite spline quasi-interpolation. BIT Numer. Math. **56**, 1165–1188 (2016)
3. Bracco, C., Lyche, T., Manni, C., Roman, F., Speleers, H.: On the dimension of Tchebycheffian spline spaces over planar T-meshes. Comput. Aided Geom. Des. **45**, 151–173 (2016)
4. Bracco, C., Giannelli, C., Sestini, A.: Adaptive scattered data fitting by extension of local approximations to hierarchical splines. Comput Aided Geom. Des. **52–53**, 90–105 (2017)
5. Bracco, C., Giannelli, C., Großmann, D., Sestini, A.: Adaptive fitting with THB-splines: error analysis and industrial applications. Comput. Aided Geom. Des. **62**, 239–252 (2018)
6. Cavoretto, R., De Rossi, A., Perracchione, E.: Optimal selection of local approximants in RBF-PU interpolation. J. Sci. Comput. **74**, 1–22 (2018)
7. Davydov, O., Schumaker, L.: Interpolation and scattered data fitting on manifolds using projected Powell–Sabin splines. IMA J. Numer. Anal. **28**, 785–805 (2008)
8. Davydov, O., Morandi, R., Sestini, A.: Local hybrid approximations for scattered data fitting with bivariate splines. Comput. Aided Geom. Des. **23**, 703–721 (2006)
9. Deng, J., Chen, F., Feng, Y.: Dimensions of spline spaces over T-meshes. J. Comput. Appl. Math. **194**, 267–283 (2006)
10. Dokken, T., Lyche, T., Pettersen, K.F.: Polynomial splines over locally refined box-partitions. Comput. Aided Geom. Des. **30**, 331–356 (2013)
11. Floater, M.S.: Parametrization and smooth approximation of surface triangulations. Comput. Aided Geom. Des. **14**, 231–250 (1997)
12. Floater, M.S., Hormann, K.: Surface parameterization: a tutorial and survey. In: Dodgson N.A., Floater M.S., Sabin M.A. (eds.) Advances in Multiresolution for Geometric Modelling. Mathematics and Visualization, pp. 157–186. Springer, Berlin (2005)
13. Giannelli, C., Jüttler, B., Speleers, H.: THB-splines: the truncated basis for hierarchical splines. Comput. Aided Geom. Des. **29**, 485–498 (2012)
14. Giannelli, C., Jüttler, B., Speleers, H.: Strongly stable bases for adaptively refined multilevel spline spaces. Adv. Comput. Math. **40**, 459–490 (2014)
15. Kiss, G., Giannelli, C., Zore, U., Jüttler, B., Großmann, D., Barner, J.: Adaptive CAD model (re-)construction with THB-splines. Graph. Models **76**, 273–288 (2014)
16. Kraft, R.: Adaptive and linearly independent multilevel B-splines. In: Le Méhauté, A., Rabut, C., Schumaker, L.L. (eds.) Surface Fitting and Multiresolution Methods, pp. 209–218. Vanderbilt University Press, Nashville (1997)
17. Lai, M.J., Schumaker, L.L.: A domain decomposition method for computing bivariate spline fits of scattered data. SIAM J. Numer. Anal. **47**, 911–928 (2009)
18. Lee, B.G., Lyche, T., Mørken, K.: Some examples of quasi–interpolants constructed from local spline projectors. In: T. Lyche, L.L. Schumaker (eds.), Mathematical Methods for Curves and Surfaces, Oslo 2000. pp. 243–252. Vanderbilt University Press (2001)
19. Lyche, T., Schumaker, L.L.: Local spline approximation methods. J. Approx. Theory **15**, 294–325 (1975)
20. Mourrain, B.: On the dimension of spline spaces on planar T-meshes. Math. Comput. **83**, 847–871 (2014)
21. Patrizi, F., Dokken, T.: Linear dependence of bivariate Minimal Support and Locally Refined B-splines over LR-meshes. Comput. Aided Geom. Des. **77**, 101803 (2020)
22. Prautzsch, H., Boehm, W., Paluszny, M.: Bézier and B-Spline Techniques. Springer, Berlin (2002)

23. Sederberg, T.W., Zheng, J., Bakenov, A., Nasri, A.: T-splines and T-NURCCs. ACM Trans. Graph. **22**, 477–484 (2003)
24. Speleers, H.: Hierarchical spline spaces: quasi-interpolants and local approximation estimates. Adv. Comput. Math. **43**, 235–255 (2017)
25. Speleers, H., Manni, C.: Effortless quasi-interpolation in hierarchical spaces. Numer. Math. **132**, 155–184 (2016)

Part II
Contributed Papers

Chapter 4
A Progressive Construction of Univariate Spline Quasi-Interpolants on Uniform Partitions

Abdelaziz Abbadi and María José Ibáñez

Abstract We present a method for constructing differential and integral spline quasi-interpolants defined on uniform partitions of the real line. It is based on a expression to the quasi-interpolation error for a enough regular function and involves the errors for the non-reproduced monomials. From it, a minimization problem is proposed whose solution is calculated progressively. It is characterized in terms of some splines which do not depend on the linear functional defining the quasi-interpolation operators. The resulting quasi-interpolants are compared with other well-known schemes to show the efficiently of this construction.

Keywords Differential quasi-interpolant · Integral quasi-interpolant · Quasi-interpolation error · B-spline

4.1 Introduction

Quasi-interpolation is a widely used method to efficiently approximate functions or data. Several techniques based on Appell sequences, Neumann series, or Fourier transform have been developed to define quasi-interpolants from the translates of a compactly support function (see [5, pp. 121–128], [11, pp. 68–78], [6–10, 21] and [15, pp. 359–363]). The first method proposed in [11, pp. 72] illustrates how to construct the linear functional λ whose associated (differential) quasi-interpolation operator is exact on the space $\mathcal{S} := \text{span}\left(\phi\,(\cdot - i)\right)_{i\in\mathbb{Z}^s}$, ϕ being a box xpline. For that, only the values at 0 of the polynomials that form the Appell sequence associated with the function ϕ are needed.

A. Abbadi (✉)
University Mohammed I, Team ANAA, LANO Laboratory, Oujda, Morocco
e-mail: abzabbadi@yahoo.fr

M. J. Ibáñez
University of Granada, Department of Applied Mathematics, Granada, Spain
e-mail: mibanez@ugr.es

D. Barrera et al. (eds.), *Mathematical and Computational Methods for Modelling, Approximation and Simulation*, SEMA SIMAI Springer Series 29,
https://doi.org/10.1007/978-3-030-94339-4_4

Some results in the bivariate case are presented in [17–20]. The construction of bivariate discrete quasi-interpolants with nearly optimal approximation orders and small norms is considered in [1–4, 13].

Once the quasi-interpolation operator Q is constructed, a standard argument (see e.g. [12, pp. 144]) shows that $\|f - Qf\|_\infty \leq (1 + \|Q\|_\infty)\,\text{dist}(f, \mathcal{R})$, where $\mathcal{R}$ is the polynomial space reproduced by Q, i.e. $Qg = g$ for all $g \in \mathcal{R}$. Therefore, by enforcing certain conditions, an estimation of the quasi-interpolation error is obtained, but the constant appearing in this error bound is not optimal. In fact, it is independent of the class of the functions to be approximated.

A progressive construction of differential quasi-interpolants (abbr. DQIs) based on cardinal B-splines is proposed to improve the quasi-interpolation error. The linear functional λ defining the DQI will evaluate the function at integer points near 0. For any enough regular function, it is possible to provide a formula to the quasi-interpolation error involving a term measuring how well the quasi-interpolation operator approximates some non-reproduced monomials. Firstly, the term depends on the sequence γ of coefficients defining the quasi-interpolation operator Q and it is quite natural to minimize it, and for then we drive the progressive problem of minimisation which allows us to determine some equations from the next non-reproduced monomials.

More precisely, (a) we impose some constraints on γ, yielding the exactness of Q on some polynomial space $\mathbb{P}(\phi) \subset \mathcal{S}$ by using the values at 0 of the polynomials in the Appell sequence; (b) we express the leading term of the quasi-interpolation error by means the uniform norms of some splines which do not depend on λ on the unit square; and (c) we define an appropriate minimization problem which will will be solved progressively.

The paper is organized as follows. In Sect. 4.2, we set some notations and we recall some results needed for deriving conditions that guarantee the exactness of the DQI. In Sect. 4.3, we derive a formula for the quasi-interpolation error. Using this formula, we propose a new class of DQIs which is obtained by solving a minimization problem. The progressive solution of this problem is detailed in Sect. 4.4. Finally, in Sect. 4.5, we give some examples of DQIs to show the efficiently of this new construction of quasi-interpolants.

4.2 Notations and Preliminaries

Let ϕ be the cardinal B-spline M_n of order $n \geq 2$ centered at zero, and define the quasi-interpolation operator (QIO for short) Q by

$$Q(f) = \sum_{i \in \mathbb{Z}} \lambda f(\cdot + i) M_n(\cdot - i), \quad f \in C(\mathbb{R}), \tag{4.1}$$

where λ is a functional having one of the two expressions

$$\lambda f := \sum_{k=0}^{l} \sum_{j \in J_k} \gamma_{j,k} f^{(k)}(-j), \tag{4.2}$$

$$\lambda f := \sum_{j \in J} \gamma_j \langle f, M_n(\cdot + j) \rangle, \tag{4.3}$$

with $0 \le l \le n-1$, and finite sets $(J_k)_{0 \le k \le l}$ and J.

The parameters $\gamma_{j,k}$, $0 \le k \le l$, and γ_j are real numbers, and

$$\langle f, M_n(\cdot + j) \rangle = \int_{\mathbb{R}} f(x) M_n(x+j) dx.$$

For $h > 0$, let Q_h be the scaled QIO associated with Q. It is defined by $Q_h := \sigma_h Q \sigma_{\frac{1}{h}}$, where $\sigma_h f := f(\frac{\cdot}{h})$.

Note that the functional λ provides a discrete QIO when $l = 0$.

To characterize the exactness of Q on the space $\mathbb{P}_{n-1}$ of all polynomials of degree $\le n-1$, let us consider the sequence $(g_\alpha)_{\alpha \ge 0}$ of Appell's polynomials. If $m_\alpha(x) = \frac{1}{\alpha!} x^\alpha$, $\alpha \ge 0$ is the normalized monomial of degree α, then the Appell's polynomials as defined recursively as follows:

$$g_0(x) = 1,$$

$$g_\alpha(x) = m_\alpha(x) - \sum_{j \in \mathbb{Z}} M_n(j) \sum_{\beta=0}^{\alpha-1} m_{\alpha-\beta}(-j) g_\beta(x), \ \alpha \ge 1. \tag{4.4}$$

Then, we have the following results [11].

Lemma 4.1 *It holds*

$$m_\alpha = \sum_{i \in \mathbb{Z}} g_\alpha(i) M_n(\cdot - i), \ 0 \le \alpha \le n-1, \tag{4.5}$$

$$g_\alpha = \sum_{\beta=0}^{\alpha} g_{\alpha-\beta}(0) m_\beta, \ \alpha \ge 0. \tag{4.6}$$

Now, the exactness conditions are deduced from the two following results [6].

Proposition 4.1 *The QIO Q given by (4.1) and one of the two expressions (4.2) or (4.3) is exact on the space $\mathbb{P}_{n-1}$ if the functional λ coincides on $\mathbb{P}_{n-1}$ with the differential form δ defined by*

$$\delta(f) := \delta f = \sum_{\beta \ge 0} g_\beta(0) f^{(\beta)}(0).$$

Equivalently, such a exactness is achieved if and only if for all $\alpha = 0, \ldots, n-1$ *it holds*

$$\lambda m_\alpha = g_\alpha(0)\,. \tag{4.7}$$

4.3 A Minimization Problem

The problem that we consider here is the construction of a new class of quasi-interpolants minimizing the error in the uniform norm.

Let $Q(f)$ be the quasi-interpolant given by (4.1) and (4.2). For $n \geq 3$, and taking into account of the parity of n, $n = 2\nu$ or $n = 2\nu - 1$, $\nu \geq 2$, it is straightforward to verify that for $x = (d+\xi)h \in [dh, (d+1)h]$, $d \in \mathbb{Z}$, $\xi \in [0,1]$, the quasi-interpolant $Q_h f(x)$ can be written only in terms of the values

$$f^{(k)}(h(d - r_1 - \nu + 1)), \ldots, f^{(k)}(h(d - r_0 + \nu)),\ k = 0, \ldots, l,$$

where $r_0 := \min J$ and $r_1 = \max J$, with $J = \bigcup_{k=0}^{l} J_k$.

Proposition 4.2 *For* $f \in C^l(\mathbb{R})$, *it holds*

$$|f(x) - Q_h f(x)| \leq \frac{h^n}{(n-1)!} \|f^{(n)}\|_{\infty, I_{h,J,d}} \int_{-r_1-\nu+1}^{-r_0+\nu} |K(\xi,\tau)|d\tau,$$

where $I_{h,J,d} = [h(d - r_1 - \nu + 1), h(d - r_0 + \nu)]$ *and*

$$K(\xi,\tau) = (\xi - \tau)_+^{n-1} - Q[(\cdot - \tau)_+^{n-1}](\xi).$$

Proof Since $f(x) - Q_h f(x)$ vanishes on the space of polynomials of degree $\leq n-1$, from Peano's theorem (see [16]) it follows that for $f \in C^n(\mathbb{R})$

$$f(x) - Q_h f(x) = \frac{1}{(n-1)!} \int_{h(d-r_1-\nu+1)}^{h(d-r_0+\nu)} K_h(x,t) f^{(n)}(t)dt,$$

where

$$K_h(x,t) = (x-t)_+^{n-1} - Q_h[(\cdot - t)_+^{n-1}](x).$$

If $x := (d+\xi)h$ with $\xi \in [0,1]$, then after a variable change we get

$$f(x) - Q_h f(x) = \frac{h^n}{(n-1)!} \int_{-r_1-\nu+1}^{-r_0+\nu} K(\xi,t) f^{(n)}(h(d+\tau))d\tau,$$

where

$$K(\xi,\tau) = (\xi-\tau)_+^{n-1} - Q[(\cdot-\tau)_+^{n-1}](\xi).$$

Then,

$$|f(x) - Q_h f(x)| \le \frac{h^n}{(n-1)!}\|f^{(n)}\|_{\infty, I_{h,J,d}} \int_{-r_1-\nu+1}^{-r_0+\nu} |K(\xi,\tau)|d\tau.$$

and the proof is complete. □

In order to improve the error above, we seek to construct Q so that $K(\xi,\tau)$ be as small as possible. To do this, for $\tau \in \big[-r_1-\nu+1, r_0+\nu\big]$ consider the best uniform approximation

$$p^*(\xi) := \sum_{s=0}^{n+r} a_s(\tau) m_s(\xi)$$

of $(\cdot-\tau)_+^{n-1}$ by polynomials of degree $\le n+r$, $\quad r \ge 0$, i.e.,

$$\big\|p^* - (\cdot-\tau)_+^{n-1}\big\|_{\infty,[0,1]} = \inf_{p\in\mathbb{P}_{n+r}} \big\|p - (\cdot-\tau)_+^{n-1}\big\|_{\infty,[0,1]}.$$

Since Q is exact on $\mathbb{P}_{n-1}$, we get

$$Qp^* = \sum_{s=0}^{n-1} a_s(\tau) m_s + \sum_{s=n}^{n+r} a_s(\tau) Q m_s,$$

so that,

$$p^* - Qp^* = \sum_{s=n}^{n+r} a_s(\tau)(m_s - Qm_s).$$

Then, $\sum_{s=n}^{n+r} |a_s(\tau)| \big\|m_s - Qm_s\big\|_{\infty,[0,1]}$ is an approximation of $|K(\cdot,\tau)|$. Hence, following the idea in [14], it is possible to construct Q by minimizing the errors

$$|m_s - Qm_s\|_{\infty,[0,1]}, \; n \le s \le n+r.$$

There are many degrees of freedom in determining the univariate DQIs, therefore we propose to minimize the errors associated with the different non reproduced monomials, not only the first one. Let

$$V = \left\{\gamma = (\gamma_{j,k})_{j\in J_k, 0\le k\le l} : \lambda m_\alpha = g_\alpha(0), 0 \le \alpha \le n\right\},$$

and consider the following minimization problem.

Problem 4.1 Given $r \geq 0$, minimize on V the errors $\|m_{n+s} - Qm_{n+s}\|_{\infty,[0,1]}$, $0 \le s \le r$ associated with the QIO Q.

This problem will be solved progressively.

4.4 Solving the Minimization Problem

4.4.1 Case r = 0

The problem is to minimize $\|m_n - Qm_n\|_{\infty,[0,1]}$ subject to the exactness of Q on $\mathbb{P}_{n-1}$, i.e.,

$$\lambda m_\alpha = g_\alpha(0), \quad 0 \le \alpha \le n-1.$$

For this, we use the Schoenberg operator $\mathcal{S}$ defined by

$$\mathcal{S}f = \sum_{i\in\mathbb{Z}} f(i)\, M_n(\cdot - i). \tag{4.8}$$

The operator $\mathcal{S}$ is exact on $\mathbb{P}_1$.

Lemma 4.2 *It holds*

$$m_n - Qm_n = G_{n,0} - \left(\lambda m_n - g_n(0)\right),$$

where

$$G_{n,0} := m_n - \mathcal{S}g_n. \tag{4.9}$$

Proof Since

$$m_n(x+i) = \sum_{\ell=0}^{n} m_\ell(i)\, m_{n-\ell}(x),$$

then

$$Qm_n = \sum_{i\in\mathbb{Z}} \Big(\sum_{\ell=0}^{n} m_\ell(i)\, \lambda m_{n-\ell} \Big) M_n(\cdot - i)$$

$$= \lambda m_n + \sum_{i\in\mathbb{Z}} \Big(\sum_{\ell=1}^{n} m_\ell(i)\, \lambda m_{n-\ell} \Big) M_n(\cdot - i)\,.$$

The exactness of Q on $\mathbb{P}_{n-1}$ gives

$$Qm_n = \lambda m_n + \sum_{i\in\mathbb{Z}} \Big(\sum_{\ell=1}^{n} m_\ell(i)\, \lambda m_{n-\ell} \Big) M_n(\cdot - i)$$

$$= \lambda m_n + \sum_{i\in\mathbb{Z}} \Big(\sum_{\ell=1}^{n} g_{n-\ell}(0)\, m_\ell(i) \Big) M_n(\cdot - i)$$

$$= \lambda m_n - g_n(0) + \sum_{i\in\mathbb{Z}} \Big(\sum_{\ell=0}^{n} g_{n-\ell}(0)\, m_\ell(i) \Big) M_n(\cdot - i)\,.$$

By (4.6), we get

$$Qm_n = \lambda m_n - g_n(0) + \sum_{i\in\mathbb{Z}} g_n(i)\, M_n(\cdot - i)$$

$$= \lambda m_n - g_n(0) + \mathcal{S} g_n$$

and the proof is complete. □

The previous lemma allows us to characterize the solution of Problem 4.1.

Proposition 4.3 *γ is a solution of Problem 4.1 if and only if $\lambda m_n - g_n(0)$ is the best uniform approximation of $G_{n,0}$ by constants over the interval $[0, 1]$.*

As a consequence, we have the following result.

Corollary 4.1 *For $r = 0$, γ is a solution of Problem (4.1) if and only if*

$$\lambda m_n - g_n(0) = \frac{1}{2}\Big(\max_{[0,1]} G_{n,0} + \min_{[0,1]} G_{n,0} \Big).$$

Proof It follows from Lemma (4.2) and a classical result of the best uniform approximation by polynomials [22]. □

Therefore, Problem 4.1 has at least one solution if $\sum_{k=0}^{l} \operatorname{card} J_k \geq n+1$. The solution is unique if and only if $\sum_{k=0}^{l} \operatorname{card} J_k = n+1$.

Note that the spline $G_{n,0} := m_n - \mathcal{S}g_n$ does not depend on the functional λ, so that the parameters γ which define the operator $Q$are not involved.

Assume that we have solved the problem in the case $r = 0$. The next step is to solve Problem 4.1 progressively for any $r \geq 0$.

4.4.2 Case r > 0

Firstly, we compute the quasi-interpolation error for the normalized monomial of degree $n + 1$.

Lemma 4.3 *It holds*

$$m_{n+1} - Qm_{n+1} = G_{n,1} - \left(\lambda m_{n+1} - g_{n+1}(0)\right),$$

with

$$G_{n,1} := m_{n+1} - \mathcal{S}g_{n+1} - (\lambda m_n - g_n)\mathcal{S}m_1. \tag{4.10}$$

Proof Since

$$m_{n+1}(x+i) = \sum_{\ell=0}^{n+1} m_\ell(i)\, m_{n+1-\ell}(x),$$

we get,

$$\begin{aligned}
Qm_{n+1} &= \sum_{i\in\mathbb{Z}} \Big(\sum_{\ell=0}^{n+1} m_\ell(i)\, \lambda m_{n+1-\ell} \Big) M_n(\cdot - i) \\
&= \lambda m_{n+1} + \sum_{i\in\mathbb{Z}} \Big(\sum_{\ell=1}^{n+1} m_\ell(i)\, \lambda m_{n+1-\ell} \Big) M_n(\cdot - i) \\
&= \lambda m_{n+1} + \sum_{i\in\mathbb{Z}} \Big(\sum_{\ell=2}^{n+1} m_\ell(i)\, \lambda m_{n+1-\ell} \Big) M_n(\cdot - i) \\
&\quad + \sum_{i\in\mathbb{Z}} \big(m_1(i)\, \lambda m_n \big) M_n(\cdot - i) \\
&= \lambda m_{n+1} + \sum_{i\in\mathbb{Z}} \Big(\sum_{\ell=2}^{n+1} m_\ell(i)\, \lambda m_{n+1-\ell} \Big) M_n(\cdot - i) + \lambda m_n \mathcal{S}m_1.
\end{aligned}$$

Being Q exact on $\mathbb{P}_{n-1}$, then

$$
\begin{aligned}
Qm_{n+1} &= \lambda m_{n+1} + \sum_{i\in\mathbb{Z}} \Big(\sum_{\ell=2}^{n+1} g_{n+1-\ell}\,(0)\, m_\ell(i)\Big) M_n\,(\cdot - i) + \lambda m_n \mathcal{S} m_1 \\
&= \lambda m_{n+1} + \sum_{i\in\mathbb{Z}} \Big(\sum_{\ell=0}^{n+1} g_{n+1-\ell}\,(0)\, m_\ell(i)\Big) M_n\,(\cdot - i) - g_{n+1}(0) \\
&\quad - g_n\,(0) \sum_{i\in\mathbb{Z}} m_1(i) M_n\,(\cdot - i) + \lambda m_n \mathcal{S} m_1 .
\end{aligned}
$$

By using (4.6), we get

$$
Qm_{n+1} = \big(\lambda m_{n+1} - g_{n+1}\,(0)\big) + \mathcal{S} g_{n+1} + \big(\lambda m_n - g_n\,(0)\big)\, \mathcal{S} m_1 ,
$$

and the result follows. □

Now, the solution of Problem 4.1 is characterized as in the case $r = 0$.

Corollary 4.2 *For $r = 1$, γ is a solution of Problem 4.1 if and only if*

$$
\lambda m_{n+1} - g_{n+1}\,(0) = \frac{1}{2}\Big(\max_{[0,1]} G_{n,1} + \min_{[0,1]} G_{n,1}\Big).
$$

In the general case, for $r \geq 1$, we define the spline function

$$
G_{n,r} = \big(\lambda m_{n+r} - \mathcal{S} g_{n+r}\big) - \sum_{s=1}^{r} \Big(\lambda m_{n+r-s} - g_{n+r-s}(0)\Big) \mathcal{S} m_s ,
$$

where the values $(\lambda m_{n+r-s})_{1\leq s\leq r}$ are determined by solving Problem 4.1 for $0 \leq s \leq r$. The following result holds.

Corollary 4.3 *For $r \geq 1$, γ is a solution of Problem 4.1 if and only if*

$$
\lambda m_{n+r} - g_{n+r}\,(0) = \frac{1}{2}\Big(\max_{[0,1]} G_{n,r} + \min_{[0,1]} G_{n,r}\Big).
$$

Proof We have

$$
m_{n+r}\,(x + i) = \sum_{\ell=0}^{n+r} m_\ell\,(i)\, m_{n+r-\ell}\,(x) ,
$$

so that,

$$
\begin{aligned}
Qm_{n+r} &= \sum_{i\in\mathbb{Z}}\sum_{\ell=0}^{n+r} m_\ell(i)\,\lambda m_{n+r-\ell} M_n(\cdot - i)\\
&= \lambda m_{n+r} + \sum_{i\in\mathbb{Z}}\sum_{\ell=1}^{n+r} m_\ell(i)\,\lambda m_{n+r-\ell} M_n(\cdot - i)\\
&= \lambda m_{n+r} + \sum_{i\in\mathbb{Z}}\sum_{\ell=r+1}^{n+r} m_\ell(i)\,\lambda m_{n+r-\ell} M_n(\cdot - i)\\
&+ \sum_{i\in\mathbb{Z}}\sum_{\ell=1}^{r} m_\ell(i)\,\lambda m_{n+r-\ell} M_n(\cdot - i)\\
&= \lambda m_{n+r} + \sum_{i\in\mathbb{Z}}\sum_{\ell=r+1}^{n+r} g_{n+r-\ell}(0)\, m_\ell(i)\, M_n(\cdot - i)\\
&+ \sum_{i\in\mathbb{Z}}\sum_{\ell=1}^{r} m_\ell(i)\,\lambda m_{n+r-\ell} M_n(\cdot - i)\\
&= \lambda m_{n+r} + \sum_{i\in\mathbb{Z}}\sum_{\ell=0}^{n+r} g_{n+r-\ell}(0)\, m_\ell(i)\, M_n(\cdot - i)\\
&- \sum_{i\in\mathbb{Z}}\sum_{\ell=0}^{r} g_{n+r-\ell}(0)\, m_\ell(i)\, M_n(\cdot - i)\\
&+ \sum_{i\in\mathbb{Z}}\sum_{\ell=1}^{r} m_\ell(i)\,\lambda m_{n+r-\ell} M_n(\cdot - i)\\
&= \lambda m_{n+r} + \sum_{i\in\mathbb{Z}} g_{n+r}(i)\, M_n(\cdot - i) - g_{n+r}(0)\\
&+ \sum_{\ell=1}^{r} \lambda m_{n+r-\ell} - g_{n+r-\ell}(0) \sum_{i\in\mathbb{Z}} m_\ell(i)\, M_n(\cdot - i)\\
&= \lambda m_{n+r} - g_{n+r}(0) + \mathcal{S} g_{n+r} + \sum_{\ell=1}^{r} \lambda m_{n+r-\ell} - g_{n+r-\ell}(0)\, \mathcal{S} m_\ell .
\end{aligned}
$$

From the definition of $G_{n,r}$, we get

$$m_{n+r} - Qm_{n+r} = G_{n,r} - \left(\lambda m_{n+r} - g_{n+r}(0)\right).$$

The rest of the proof is similar to that of Corollary 4.1. □

4.5 Some Examples of Differential Quasi-Interpolants

In order to show the performance of construction proposed above, we derive the expressions of some quadratic and cubic differential quasi-interpolants. We first give the explicit expressions of necessary polynomials g_α, $\alpha = 0, \ldots, n$. In the quadratic case, they are

$$g_0(x) = 1,\ g_1(x) = x,\ g_2(x) = \tfrac{1}{2}x^2 - \tfrac{1}{8},\ g_3(x) = \tfrac{1}{6}x^3 - \tfrac{1}{8}x,$$
$$g_4(x) = \tfrac{1}{24}x^4 - \tfrac{1}{16}x^2 + \tfrac{1}{192},$$

and in the cubic one

$$g_0(x) = 1,\quad g_1(x) = x,\quad g_2(x) = \tfrac{1}{2}x^2 - \tfrac{1}{6},$$
$$g_3(x) = \tfrac{1}{6}x^3 - \tfrac{1}{6}x,\ g_4(x) = \tfrac{1}{24}x^4 - \tfrac{1}{12}x^2 + \tfrac{1}{72},\ g_5(x) = \tfrac{1}{120}x^5 - \tfrac{1}{36}x^3 + \tfrac{1}{72}x.$$

4.5.1 *Quadratic Differential Quasi-Interpolants*

Let $J_0 = J_1 = \{-1, 0, 1\}$. For the sake of simplicity, we use the notations a_j and b_j instead of $\gamma_{0,j}$ and $\gamma_{1,j}$ respectively. Let Q be the differential quasi-interpolant (DQI)

$$Qf = \sum_{i\in\mathbb{Z}} \lambda f(\cdot + i) M_3(\cdot - i),$$

where

$$\lambda f = a_{-1}f(1) + a_0 f(0) + a_1 f(-1) + b_{-1}f'(1) + b_0 f'(0) + b_1 f'(-1).$$

The exactness of Q on $\mathbb{P}_2$ yields the system

$$\sum_{j\in J_0} a_j m_\alpha(-j) + \sum_{j\in J_1} b_j m'_\alpha(-j) = g_\alpha(0), \quad \alpha \le 2. \tag{4.11}$$

The natural choices $J_0 = \{0\}$ and $J_1 = \{-1, 0\}$ give the DQI $\widetilde{Q}$

$$\widetilde{Q}f := \sum_{i\in\mathbb{Z}} \Big(f(i) + \frac{1}{8}\,(f'(i) - f'(i+1)) \Big) M_3(\cdot - i).$$

The choice $J_0 = \{-1, 0\}$ and $J_1 = \{-1, 0, 1\}$ involve five parameters, while the exactness on $\mathbb{P}_2$ provides only three equations. So, we construct a DQI Q^* by adding two other equations from the progressive minimization of $\|m_3 - Q^* m_3\|_{\infty,[0,1]}$ and $\|m_4 - Q^* m_4\|_{\infty,[0,1]}$. We have the following result.

Proposition 4.4 *A DQI Q exact on $\mathbb{P}_2$ minimizes $\|m_3 - Qm_3\|_{\infty,[0,1]}$ if and only if*

$$\sum_{j\in J_0} a_j m_3(-j) + \sum_{j\in J_1} b_j m_2(-j) = 0. \tag{4.12}$$

Proof According to Corollary 4.1, we have

$$\lambda m_3 - g_3\,(0) = \frac{1}{2}\Big(\max_{[0,1]} G_{3,0} + \min_{[0,1]} G_{3,0} \Big).$$

Thus (see Fig. 4.1),

$$\begin{aligned} G_{3,0}(x) &= m_3(x) - \mathcal{S}g_3(x) \\ &= m_3(x) - \sum_{i=-1}^{2} g_3(i) M_3(x - i) \end{aligned}$$

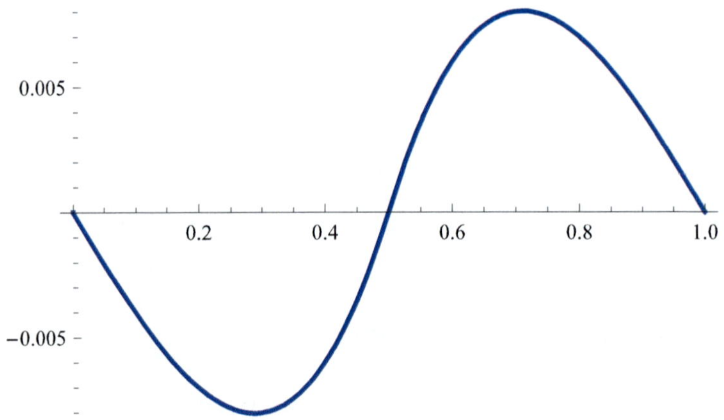

Fig. 4.1 Graph of $G_{3,0}$

$$= \begin{cases} \frac{1}{6}x^3 - \frac{1}{24}x, & 0 \le x \le \frac{1}{2}, \\ \frac{1}{6}x^3 - \frac{1}{2}x^2 + \frac{11}{24}x - \frac{1}{8}, & 0 \le x \le \frac{1}{2}. \end{cases}$$

Since

$$\max_{[0,1]} G_{3,0} = G_{3,0}\Big(\frac{6-\sqrt{3}}{6}\Big) = \frac{\sqrt{3}}{216},$$

$$\min_{[0,1]} G_{3,0} = G_{3,0}\Big(\frac{\sqrt{3}}{6}\Big) = -\frac{\sqrt{3}}{216},$$

we deduce that

$$\lambda m_3 = g_3(0) + \frac{1}{2}\Big(\max_{[0,1]} G_{3,0} + \min_{[0,1]} G_{3,0}\Big) = 0,$$

which completes the proof. □

Now we ask for the minimization also of the error for the quartic normalized monomial.

Proposition 4.5 *Let Q be a DQI which is exact on $\mathbb{P}_2$ and minimizes $\|m_3 - Qm_3\|_{\infty,[0,1]}$. Then, it minimizes $\|m_4 - Qm_4\|_{\infty,[0,1]}$ if and only if*

$$\sum_{j\in J_0} a_j m_4(-j) + \sum_{j\in J_1} b_j m_3(-j) = -\frac{1}{24}. \tag{4.13}$$

Proof Using Corollary 4.2, the proof runs as in Proposition 4.4. □

The solution of system (4.11)–(4.13) provides the DQI Q^*.

$$Q^* f := \sum_{i\in\mathbb{Z}} \Big(\frac{3}{2}f(i) - \frac{1}{2}f(i+1) + \frac{1}{3}f'(i) + \frac{7}{48}f'(i+1) + \frac{1}{48}f'(i-1)\Big) M_3(\cdot - i).$$

Next, we give error bounds of the approximation by the quadratic DQIs $\widetilde{Q}$ and Q^* defined above.

Proposition 4.6 *Let Q be one of the two DQIs $\widetilde{Q}$ and Q^*. Then, for all $f \in C^3(\mathbb{R})$, we have*

$$\|D^\alpha(f - Qf)\|_\infty \le c_\alpha(Q)\|f^{(3)}\|_\infty, \alpha = 0, 1,$$

where $c_0(\widetilde{Q}) = 0.0794922$, $c_0(Q^*) = 0.0609918$, $c_1(\widetilde{Q}) = 0.5034708$ *and* $c_1(Q^*) = 0.5520833$.

Proof Let us consider the DQI $\widetilde{Q}$ resulting from the choices $J_0 = \{0\}$ and $J_1 = \{-1, 0\}$. With $x = (d + \xi)h, \xi \in [0, 1], forh > 0$ and some $d \in \mathbb{Z}$, we have

$$|f(x) - \widetilde{Q}_h f(x)| \leq \frac{h^3}{2} \|f^{(3)}\|_{\infty, I_{h,J,d}} \int_{-1}^{3} |\widetilde{K}(\xi, t)| dt,$$

where $\widetilde{K}(\xi, t) = (\xi - t)_+^2 - \widetilde{Q}[(\cdot - t)_+^2](\xi)$ and $I_{h,J,d} = [h(d-1), h(d+3)]$. Let $\widetilde{K}(\xi) = \frac{1}{2} \int_{-1}^{3} |\widetilde{K}(\xi, t)| dt = \sum_{i=1}^{4} I_i(\xi)$, where

$$I_1(\xi) = \tfrac{1}{2} \int_{-1}^{0} |\widetilde{K}(\xi, t)| dt, \ I_2(\xi) = \tfrac{1}{2} \int_{0}^{1} |\widetilde{K}(\xi, t)| dt,$$
$$I_3(\xi) = \tfrac{1}{2} \int_{1}^{2} |\widetilde{K}(\xi, t)| dt, \ I_4(\xi) = \tfrac{1}{2} \int_{2}^{3} |\widetilde{K}(\xi, t)| dt.$$

Using the Bernstein-Bèzier representation of $\widetilde{K}(\xi, t)$ in every interval $[i, i+1]$, $i = -1, \ldots 2$, it holds that $\widetilde{K}(\xi)$ is bounded by

$$\widetilde{P}(\xi) =: \frac{407}{5120} - \frac{37}{768}\xi - \frac{13}{256}\xi^2 + \frac{7}{32}\xi^3 + \frac{23}{96}\xi^4 - \frac{401}{960}\xi^5.$$

Then, the claim follows with$c_0(\widetilde{Q}) = \|\widetilde{P}\|_{\infty,[0,1]} = \widetilde{P}(0) = \frac{407}{5120} \approx 0.0794922$.

The constants $c_0(Q^*)$, $c_1(\widetilde{Q})$ and $c_1(Q^*)$ are obtained by using the same technique. □

Figure 4.2 shows the test function $f(x) = \sin(1 - \cos(x^2))$ and $\widetilde{Q}f$ (resp. $Q_h^* f$(red)) for $h = \frac{1}{2^n}, 2 \leq n \leq 4$.

Figure 4.3 shows the graphs of the quasi-interpolation errors for the scaled quasi-interpolants associated with $\widetilde{Q}_h$ and Q_h^* with $h = \frac{1}{64}$.

Figure 4.3 shows that Q^* produces a better result than $\widetilde{Q}$.

4.5.2 Cubic Differential QIs

Let Q be the DQI

$$Qf = \sum_{i \in \mathbb{Z}} \lambda f(\cdot + i) M_4(\cdot - i),$$

with

$$\lambda f = a_{-1} f(1) + a_0 f(0) + a_1 f(-1) + b_{-1} f'(1) + b_0 f'(0) + b_1 f'(-1).$$

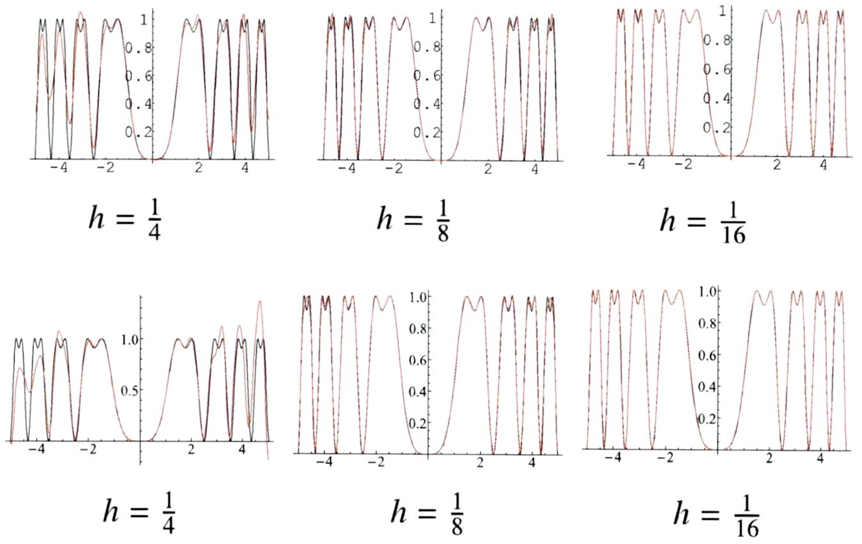

Fig. 4.2 Graphs of f (black) and $\widetilde{Q}f$ (resp. $Q_h^* f$)(red) for $h = \frac{1}{2^n}, 2 \le n \le 4$

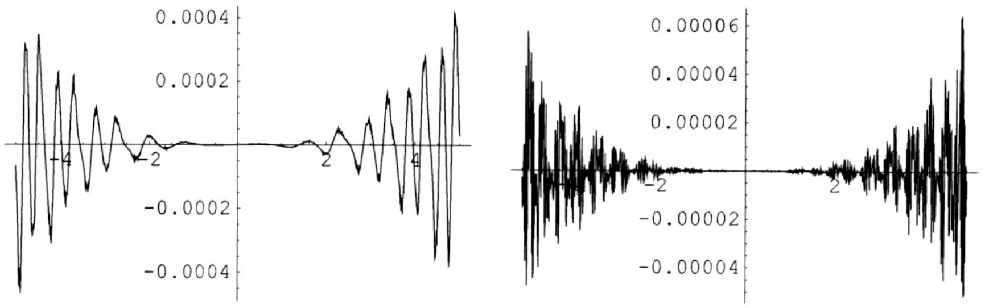

Fig. 4.3 Errors for $\widetilde{Q}f$ (left) and Q_h^* with $h = \frac{1}{64}$

It is exact on $\mathbb{P}_3$ if and only if the parameters $a_j,\ j \in J_0 = \{-1, 0, 1\}$ and $b_j,\ j \in J_1 = \{-1, 0, 1\}$ satisfy the constraints

$$\sum_{j \in J_0} a_j m_\alpha(-j) + \sum_{j \in J_1} b_j m'_\alpha(-j) = g_\alpha(0),\ \alpha \le 3. \tag{4.14}$$

The two natural choices $J_0 = J_1 = \{-1, 0\}$ and $J_0 = J_1 = \{0, 1\}$ give, respectively, the following DQIs Q_1 and Q_2:

$$Q_1 f := \sum_{i \in \mathbb{Z}} \Big(2f(i) - f(i+1) + \frac{2}{3} f'(i) + \frac{1}{3} f'(i+1) \Big) M_4(\cdot - i),$$

$$Q_2 f := \sum_{i \in \mathbb{Z}} \Big(- f(i-1) + 2f(i) - \frac{1}{3} f'(i-1) - \frac{2}{3} f'(i) \Big) M_4(\cdot - i).$$

The choice $J_0 = J_1 = \{-1, 0, 1\}$ provides 6 parameters a_j and b_j, so two additional equations are needed in order to obtain a unique DQI. Then, we have to minimize $\|m_4 - Qm_4\|_{\infty,[0,1]}$ and $\|m_5 - Qm_5\|_{\infty,[0,1]}$.

Proposition 4.7 *The DQI Q which is exact on $\mathbb{P}_3$ minimizes $\|m_4 - Qm_4\|_{\infty,[0,1]}$ if and only if*

$$\sum_{j\in J_0} a_j m_4(-j) + \sum_{j\in J_1} b_j m_3(-j) = \frac{35}{2304}.$$

Proof The proof is similar to that of Proposition 4.4. □

Proposition 4.8 *Let Q be a DQI which is exact on $\mathbb{P}_3$ and minimizes $\|m_4 - Qm_4\|_{\infty,[0,1]}$. Then, Q minimizes $\|m_5 - Qm_5\|_{\infty,[0,1]}$ if and only if*

$$\sum_{j\in J_0} a_j m_5(-j) + \sum_{j\in J_1} b_j m_4(-j) = -\frac{\sqrt{541500 - 90895\sqrt{30}}}{864000}.$$

Proof Using Corollary 4.2, we have

$$\lambda m_5 - g_5(0) = \frac{1}{2}\Big(\max_{[0,1]} G_{4,1} + \min_{[0,1]} G_{4,1}\Big).$$

On the other hand, on [0, 1] the spline $G_{4,1}$ can be written (see Fig. 4.4) as

$$\begin{aligned} G_{4,1}(x) &= m_5(x) - Sg_5(x) - \big(\lambda m_4 - g_4(0)\big)Sm_1(x) \\ &= \tfrac{1}{5!}x^5 - \sum_{i=-1}^{2}\Big(\frac{1}{5!}i^5 - \frac{1}{6}\frac{1}{3!}i^3 + \frac{1}{72}i\Big)M_4(x-i) - \frac{1}{867}\sum_{i=-1}^{2} iM_4(x-i) \\ &= \tfrac{1}{120}x^5 - \tfrac{1}{72}x^3 + \tfrac{49}{11520}x. \end{aligned}$$

Since

$$\max_{[0,1]} G_{4,1} = \frac{(9\sqrt{30} + 20\sqrt{71})\sqrt{60 - \sqrt{2130}}}{864000},$$

$$\min_{[0,1]} G_{4,1} = \frac{(9\sqrt{30} - 20\sqrt{71})\sqrt{60 + \sqrt{2130}}}{864000},$$

we have

$$\frac{1}{2}\Big(\max_{[0,1]} G_{4,1} + \min_{[0,1]} G_{4,1}\Big) = -\frac{\sqrt{541500 - 90895\sqrt{30}}}{864000}.$$

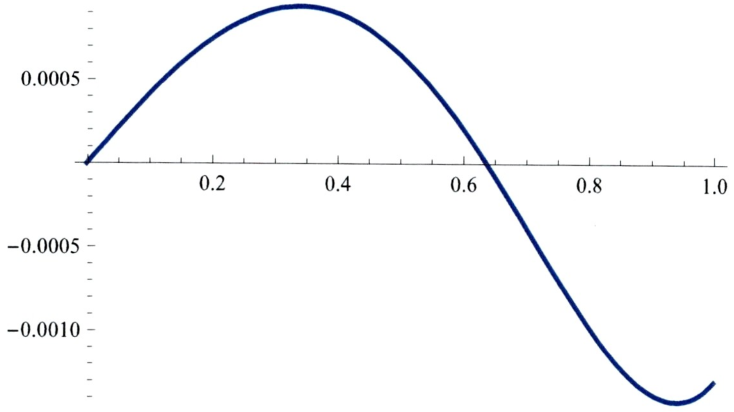

Fig. 4.4 Graph of $G_{4,1}$

Then,

$$\lambda m_5 = g_5(0) - \frac{\sqrt{541500 - 90895\sqrt{30}}}{864000} = -\frac{\sqrt{541500 - 90895\sqrt{30}}}{864000},$$

and this completes the proof. □

For the choice $J_0 = J_1 = \{-1, 0, 1\}$ the exactness conditions (4.7) and the new conditions stated in the two last propositions give a linear system of equations with the unique solution

$$\begin{aligned}
a_{-1} &= \frac{-4950 + \sqrt{5(108300 - 18179\sqrt{30})}}{9600} \approx -0.493862,\\
a_0 &= \tfrac{65}{32} = 2.03125,\\
a_1 &= \frac{-4950 - \sqrt{5(108300 - 18179\sqrt{30})}}{9600} \approx -0.537388,\\
b_{-1} &= \frac{5025 - \sqrt{5(108300 - 18179\sqrt{30})}}{28800} \approx 0.167225,\\
b_0 &= -\frac{\sqrt{5(108300 - 18179\sqrt{30})}}{7200} \approx -0.0290167,\\
b_1 &= \frac{-5025 - \sqrt{5(108300 - 18179\sqrt{30})}}{28800} \approx -0.181733.
\end{aligned}$$

Denote by Q^* the corresponding DQI. Next, error estimates are derived for the DQIs Q_1, Q_2 and Q^*. Its proof runs as in Proposition 4.6.

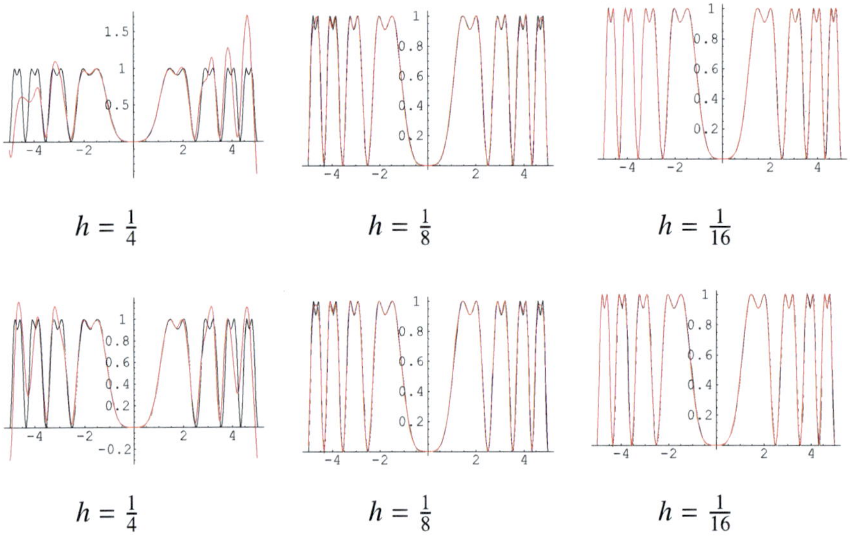

Fig. 4.5 Graphs of f (black) and $Q_{1,h}f$ (resp. Q^*f) (red) for $h = \frac{1}{2^n}, 2 \le n \le 4$

Proposition 4.9 *Let Q be one of the three DQI Q_1, Q_2 and Q^*. Then, for all $f \in C^4(\mathbb{R})$ we have*

$$\|D^\alpha(f - Qf)\|_\infty \le c_\alpha(Q)\|f^{(4)}\|_\infty \, \alpha = 0, 1,$$

where

$$c_0(Q_1) = c_0(Q_2) = 0.01609, \quad c_0(Q^*) = 0.014229,$$
$$c_1(Q_1) = c_1(Q_2) = 0.0446358, \; c_1(Q^*) = 0.04210.$$

Figure 4.5 shows the test function f used in the quadratic case and its quasi-interpolants $Q_{1,h}f$ (resp. $Q_h^* f$) for $h = \frac{1}{2^n}, 2 \le n \le 4$.

Figure 4.6 shows the quasi-interpolation errors for the scaled quasi-interpolantion operators $Q_{1,h}$ and Q_h^* with $h = \frac{1}{64}$.

4.6 Conclusion

In this paper, we have presented a new method for constructing univariate differential and integral quasi-interpolants on a uniform partition of $\mathbb{R}$ The basic idea consists of the minimization of the errors for the non-reproduced monomials. The solution of the problem at every step is characterized in terms of some splines independent on the linear functional defining the quasi-interpolation operator. The

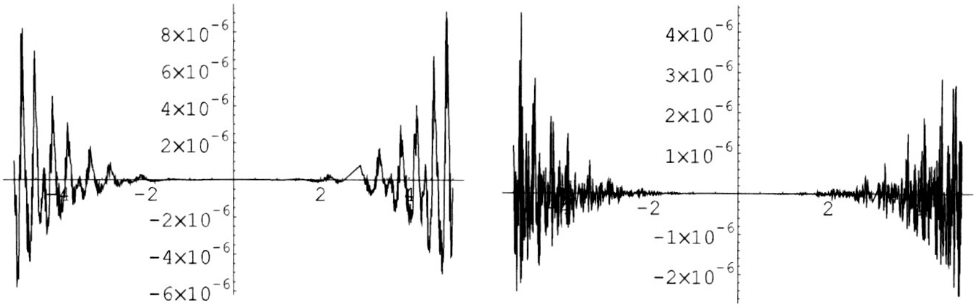

Fig. 4.6 Errors for $Q_{1,h}f$ (left) and Q_h^* with $h = \frac{1}{64}$

differencial case is analyzed in detail and some new differential quasi-interpolants are given. They are compared with the classical operators and it is shown the good performance of the new quasi-interpolants.

References

1. Barrera, D., Ibáñez, M.J.: Minimizing the quasi-interpolation error for bivariate discrete quasi-interpolants. J. Comput. Appl. Math. **224**, 250–268 (2009)
2. Barrera, D., Ibáñez, M.J., Sablonnière, P.: Near-best discrete quasi-interpolants on uniform and nonuniform partitions. In: Cohen, A., Merrien, J.-L., & Schumaker, L.L. (eds.), Curve and Surface Fitting: Saint-Malo 2002. Nashboro Press, Brentwood, pp. 31–40 (2003)
3. Barrera, D., Ibáñez, M.J., Sablonnière, P., Sbibih, D.: Near minimally normed spline quasi-interpolants on uniform partitions. J. Comput. Appl. Math. **181**, 211–233 (2005)
4. Barrera, D., Ibáñez, M.J., Sablonnière, P., Sbibih, D.: Near-best quasi-interpolants associated with H-splines on a three-direction mesh. J. Comput. Appl. Math. **183**, 133–152 (2005)
5. Chui, C.K.: Multivariate Splines. SIAM, Philadelphia (1988)
6. Chui, C.K., Diamond, H.: A natural formulation of quasi-interpolation by multivariate splines. Proc. Amer. Soc. **99**, 643–646 (1987)
7. Chui, C.K., Diamand, H.: A characterization of multivariate quasi-interpolation formulas and its applications. Number. Math. **57**, 105–121 (1990)
8. Chui, K., Lai, M.J.: A multivariate analog of Marsden's identity and quasi-interpolation scheme. Constr. Approx. **3**, 111–122 (1987)
9. Dahmen, W., Micchelli, C.K.: Translates of multivariate splines. Linear Algebra Appl. **52–53**, 217–234 (1983)
10. de Boor, C.: The polynomials in the linear span of integer translates of a compactly supported function. Constr. Approx. **3**, 199–208 (1987)
11. de Boor, C., Höllig, K., Riemenschneider, S.: Box Splines. Springer, New York (1993)
12. DeVore, R.A., Lorentz, G.G.: Constructive Approximation. Springer, Berlin (1993)
13. Ibáñez Pérez, M.J.: Quasi-interpolantes spline discretos de norma casi mínima. Teoría y aplicaciones. Doctoral Dissertation, University of Granada (2003)
14. Ibáñez Pérez, M.J.: On Chebyshev-type discrete quasi-interpolants. Math. Comput. Simul. **77**, 218–227 (2008)
15. Lai, M.-J., Schumaker, L.L.: Spline Functions on Triangulations. Cambridge University Prss, Cambridge (2007)
16. Phillips, G.M.: Interpolation and Approximation by Polynomials. Springer, New-York (2003)

17. Sablonnière, P.: Quasi-interpolants associated to H-splines on a three-direction mesh. J. Comput. Appl. Math. **66**, 433–442 (1996)
18. Sablonnière, P.: On some families of B-splines on the uniform four-direction mesh of the plan. In: Conference on Multivariate Approximation and Interpolation with Applications in CAGD, Signal and Image Processing, Eilat, Israel, Sept. 7–11 (unpublished) (1998)
19. Sablonnière, P.: Quasi-interpolantes splines sobre particiones uniformes. In: First Meeting in Approximation Theory, Úbeda (Spain), July 2000. Prépublication IRMAR 00-38 (2000)
20. Sablonnière, P.: H-splines and quasi-interpolants on a three directional mesh. In: Buhamann, M.D., Mache, D.H. (eds.), Advanced Problems in Constructive Approximation. Internationl Series of Numerical Mathematics vol. 142, pp. 187–201. Birkhäuser, Basel (2002)
21. Ward, J.D.: Polynomial reproducing formulas and the commutator of a locally supported spline. In: Chui, C.K., Schumaker, L. L., Utreras, F.I. (eds.), Topics in Multivariate Approximation. Academic Press, San Diego, pp. 255–263 (1987)
22. Watson, G.A.: Approximation Theory and Numerical Methods. Wiley, Chichester (1980)

Chapter 5
Richardson Extrapolation of Nyström Method Associated with a Sextic Spline Quasi-Interpolant

Chafik Allouch, Ikram Hamzaoui, and Driss Sbibih

Abstract In this paper, we analyse the *Nyström* method based on a sextic spline quasi-interpolant for approximating the solution of a linear *Fredholm* integral equation of the second kind. For a sufficiently smooth kernel the method is shown to have convergence of order 8 and the *Richardson* extrapolation is used to further improve this order to 9. Numerical examples are given to confirm the theoretical estimates.

Keywords Spline quasi-interpolant · *Fredholm* integral equation · *Nyström* method · *Richardson* extrapolation

5.1 Introduction

Consider the *Fredholm* integral equation defined on $\mathbb{E} = \mathscr{C}[0, 1]$ by

$$u(s) - \int_0^1 \kappa(s,t)u(t)dt = f(s), \quad 0 \leq s \leq 1, \tag{5.1}$$

where κ is a smooth kernel, $f \in \mathbb{E}$ is a real-valued continuous function and u denotes the unknown function. The *Nyström* method (see [5]) for solving (5.1) consists in replacing the integral in (5.1) by a numerical formula and it has been widely studied in the literature. A general framework for the method in the case of interpolatory projection is presented in [1, 2]. In [6] the method using a quartic spline quasi-interpolant is proposed. A superconvergent version of the *Nyström*

C. Allouch (✉) · I. Hamzaoui
The Multidisciplinary Faculty of Nador, Team of Modeling and Scientific Computing, Nador, Morocco
e-mail: c.allouch@ump.ma

D. Sbibih
University Mohammed I, FSO, LANO Laboratory, Oujda, Morocco

D. Barrera et al. (eds.), *Mathematical and Computational Methods for Modelling, Approximation and Simulation*, SEMA SIMAI Springer Series 29,
https://doi.org/10.1007/978-3-030-94339-4_5

method based on spline quasi-interpolants of degree $d \geq 2$ is analysed in [7]. In this paper we construct a quadrature formula based on integrating a sextic spline quasi-interpolant and this formula is used for the numerical solution of the *Fredholm* integral equation (5.1). We show that the convergence order of the approximate solution to the exact solution is the same as that of the quadrature rule. We show that the approximate solution of (5.1) has an asymptotic error expansion and one step of the *Richardson* extrapolation further improves the order of convergence.

The paper is divided into five sections. In Sect. 5.2, we set the notation and the sextic spline quasi-interpolant $\mathcal{Q}_n$ is constructed. In Sect. 5.3, we introduce the quadrature rule based on $\mathcal{Q}_n$ and we establish an expression of the error estimate. In Sect. 5.4, the *Nyström* method for the approximate solution of (5.1) is analysed and asymptotic series expansion for the proposed solution is obtained. Numerical examples are given in Sect. 5.5.

5.2 Sextic Spline Quasi-Interpolant

5.2.1 B-splines

Definition 5.1 Let $d \in \mathbb{N}$ and let

$$x_{-d} \leq \ldots \leq x_{-1} \leq 0 = x_0 < \ldots < x_n = 1 \leq x_{n+1} \leq \ldots \leq x_{n+d}$$

be an extended partition of the interval $I = [0, 1]$. The normalized B-spline of degree d associated with the knots $x_i \ldots, x_{i+d+1}$ is defined by

$$B_{i,d}(x) = (x_{i+d+1} - x_i)[x_i, \ldots, x_{i+d+1}](. - x)_+^d,$$

where $[x_i, \ldots, x_{i+d+1}](. - x)_+^d$ is the divided difference of $t \longrightarrow (t - x)_+^d$ with respect to the $d + 2$ points $x_i, \ldots, x_{i+d+1}$.

By using the definition of the divided differences, we obtain

$$B_{i,d}(x) = [x_{i+1}, \ldots, x_{i+d+1}](. - x)_+^d - [x_i, \ldots, x_{i+d}](. - x)_+^d. \tag{5.2}$$

Thus, from the above formula, we get

$$B_{i,0}(x) = (x_{i+1} - x)_+^0 - (x_i - x)_+^0$$

which is the characteristic function on the interval $[x_i, x_{i+1}[$, i.e.

$$B_{i,0}(x) = \begin{cases} 1, & \text{if} \quad x_i \leq x < x_{i+1} \\ 0, & \text{otherwise.} \end{cases} \tag{5.3}$$

The B-splines of higher degree ($d \geq 1$) can be evaluated by using the following recursion formula (see [3, Chap.4]):

$$B_{i,d}(x) = w_{i,d} B_{i,d-1} + (1 - w_{i+1,d}) B_{i+1,d-1}, \tag{5.4}$$

with

$$w_{i,d}(x) = \begin{cases} \frac{x - x_i}{x_{i+d} - x_i} & \text{if} \quad x_i < x_{i+d}, \\ 0 & \text{otherwise.} \end{cases}$$

5.2.2 *Construction of the Discrete Spline Quasi-Interpolant*

Let $\mathbb{X}_n = \{x_k = \frac{k}{n},\ 0 \leq k \leq n\}$ denote the uniform partition of the interval I onto n equal subintervals $I_k = [x_{k-1}, x_k],\ 1 \leq k \leq n$ with meshlength $h = \frac{1}{n}$. Let $S_6(I, \mathbb{X}_n)$ be the space of $\mathscr{C}^5$ sextic splines on this partition. Its canonical basis is formed by the $n + 6$ normalized B-splines $\{B_k \equiv B_{k-7,6},\ k \in J_n\}$ where $J_n = \{1, \ldots, n + 6\}$. The support of B_k is $[x_{k-7}, x_k]$ if we add multiple knots at the endpoints

$$x_{-6} = x_{-5} = \ldots = x_0 = 0 \quad \text{and} \quad x_n = x_{n+1} = \ldots = x_{n+6} = 1.$$

For $7 \leq k \leq n$, we have $B_k(x) = \bar{B}(\frac{x}{h} - k)$, where $\bar{B}$ is the cardinal B-spline associated with the knots $\{0, 1, 2, 3, 4, 5, 6, 7\}$ and defined by

$$\bar{B}(x) = \begin{cases} \frac{1}{720}x^6, & 0 \leq x \leq 1, \\ -\frac{7}{720} + \frac{7}{120}x - \frac{7}{48}x^2 + \frac{7}{36}x^3 - \frac{7}{48}x^4 + \frac{7}{120}x^5 - \frac{1}{120}x^6, & 1 \leq x \leq 2, \\ \frac{1337}{720} - \frac{133}{24}x + \frac{329}{48}x^2 - \frac{161}{36}x^3 + \frac{77}{48}x^4 - \frac{7}{24}x^5 + \frac{1}{48}x^6, & 2 \leq x \leq 3, \\ -\frac{12089}{360} + \frac{196}{3}x - \frac{1253}{24}x^2 + \frac{196}{9}x^3 - \frac{119}{24}x^4 + \frac{7}{12}x^5 - \frac{1}{36}x^6, & 3 \leq x \leq 4, \\ \frac{59591}{360} - \frac{700}{3}x + \frac{3227}{24}x^2 - \frac{364}{9}x^3 + \frac{161}{24}x^4 - \frac{7}{12}x^5 + \frac{1}{48}x^6, & 4 \leq x \leq 5, \\ -\frac{208943}{720} + \frac{7525}{24}x - \frac{6671}{48}x^2 + \frac{1169}{36}x^3 - \frac{203}{48}x^4 + \frac{7}{24}x^5 - \frac{1}{120}x^6, & 5 \leq x \leq 6, \\ \frac{1}{720}(7 - x)^6, & 6 \leq x \leq 7, \\ 0, & \text{elsewhere.} \end{cases}$$

We recall (see [10, Theorem 4.21 & Remark 4.1]) the representation of monomials using symmetric functions of the interior knots $N_k = \{x_{k-6}, \ldots, x_{k-1}\}$ in the support of B_k, which are defined by $\sigma_0(N_k) = 1$ and for $1 \leq r \leq 6$:

$$\sigma_r(N_k) = \sum_{1 \leq \ell_1 < \cdots < \ell_r \leq 6} x_{k-\ell_1} \ldots x_{k-\ell_r}.$$

For $0 \le r \le 6$, let $m_r(x) = x^r$. Then, we have

$$m_r(x) = \sum_{k \in J_n} (-1)^{6-r} \frac{r!}{6!} D^{6-r} \psi_k(0) B_k(x) = \sum_{k \in J_n} \theta_k^{(r)} B_k(x),$$

where

$$\psi_k(t) = \prod_{\ell=1}^{6} (x_{k-\ell} - t).$$

Hence

$$\theta_k^{(r)} = \binom{6}{r}^{-1} \sigma_r(N_k), \quad 0 \le r \le 6.$$

For $r = 0$, we have $\theta_k^{(0)} = 1$, for all $k \in J_n$, since $\sum_{k \in J_n} B_k(x) = 1$.

For $r = 1$, we have $\binom{6}{r}^{-1} = \frac{1}{6}$ and $\sigma_1(N_k) = \sum_{1 \le \ell \le 6} x_{k-\ell} = x_{k-1} + \ldots + x_{k-6}$. Thus, we obtain the *Greville* abscissae:

$$\theta_k = \theta_k^{(1)} = \frac{1}{6} \sum_{\ell=1}^{6} x_{k-\ell},$$

which are the coefficients of $m_1(x) = \sum_{k \in J_n} \theta_k B_k(x)$.

The sextic discrete spline quasi-interpolant (abbr. dQI) used here (see [8]) is the following spline operator

$$\mathcal{Q}_n f = \sum_{k \in J_n} \mu_k(f) B_k,$$

whose coefficients are linear combinations of discrete values of f on a set of data points $\mathbb{T}_n = \{t_j, \; j \in \Gamma_n\}$ where $\Gamma_n = \{j = 1, 2, \ldots, n+2\}$. The elements of $\mathbb{T}_n$ are defined by

$$t_1 = 0, \quad t_{n+2} = 1, \quad t_j = \frac{x_{j-2} + x_{j-1}}{2}, \quad 2 \le j \le n+1.$$

The dQI is constructed to be exact on Π_6, where Π_6 is the space of polynomials of degree at most 6, that means $\mathcal{Q}_n m_r = m_r$ for $0 \le r \le 6$ and therefore

$$\mu_k(m_r) = \theta_k^{(r)}, \quad k \in J_n, \quad 0 \le r \le 6.$$

For $7 \le k \le n$, the functionals μ_k use values of f in a neighbourhood of the support of B_k, thus it is natural to express μ_k in the following way

$$\mu_k(f) = \sum_{i=1}^{7} \alpha_i f_{k-i+2},$$

where $f_k = f(t_k)$. This leads us to solve the system of linear equations

$$\sum_{i=1}^{7} \alpha_i t_{k-i+2}^r = \theta_k^{(r)}, \quad 0 \le r \le 6.$$

For $1 \le k \le 6$ and $n+1 \le k \le n+6$ we write respectively

$$\mu_k(f) = \sum_{i=1}^{7} \beta_{i,k} f_i \quad \text{and} \quad \mu_k(f) = \sum_{i=1}^{7} \gamma_{i,k} f_{n-i+3}, \tag{5.5}$$

which is equivalent to the systems of linear equations

$$\sum_{i=1}^{7} \beta_{i,k} t_i^r = \theta_k^{(r)} \quad \text{and} \quad \sum_{i=1}^{7} \gamma_{i,k} t_{n-i+3}^r = \theta_k^{(r)}, \quad 0 \le r \le 6.$$

All these systems have *Vandermonde* determinants and since the $(t_j)_{j \in \Gamma_n}$ are distinct, they have unique solutions, whence the existence and unicity of the dQI. The functional coefficients are respectively defined by the following formulas:

$$\mu_1(f) = f_1,$$

$$\mu_2(f) = \frac{3887}{10395} f_1 + \frac{231}{256} f_2 - \frac{385}{768} f_3 + \frac{231}{640} f_4 - \frac{165}{896} f_5 + \frac{385}{6912} f_6 - \frac{21}{2816} f_7,$$

$$\mu_3(f) = -\frac{5689}{22275} f_1 + \frac{27631}{19200} f_2 - \frac{9151}{34560} f_3 + \frac{1091}{9600} f_4 - \frac{79}{1920} f_5 + \frac{997}{103680} f_6 - \frac{221}{211200} f_7,$$

$$\mu_4(f) = -\frac{20959}{155925} f_1 + \frac{3089}{9600} f_2 + \frac{5015}{3456} f_3 - \frac{4811}{4800} f_4 + \frac{3277}{6720} f_5 - \frac{7381}{51840} f_6 + \frac{1961}{105600} f_7,$$

$$\mu_5(f) = \frac{5821}{31185} f_1 - \frac{1193}{1920} f_2 + \frac{26737}{17280} f_3 + \frac{1}{320} f_4 - \frac{1217}{6720} f_5 + \frac{4001}{51840} f_6 - \frac{83}{7040} f_7,$$

$$\mu_6(f) = -\frac{2159}{31185} f_1 + \frac{2957}{11520} f_2 - \frac{30451}{34560} f_3 + \frac{13673}{5760} f_4 - \frac{33727}{40320} f_5 + \frac{17977}{103680} f_6 - \frac{2159}{126720} f_7,$$

$$\mu_k(f) = -\frac{2159}{138240}(f_{k+1} + f_{k-5}) + \frac{751}{4608}(f_k + f_{k-4}) - \frac{37003}{46080}(f_{k-1} + f_{k-3}) + \frac{79879}{34560} f_{k-2},$$
$$(7 \le k \le n)$$

and for $n+1 \le k \le n+6$, $\mu_k(f)$ is given by the second formula in (5.5) with $\gamma_{i,k} = \beta_{i,k}$, $1 \le i \le 7$. Since $\mathcal{Q}_n$ reproduces Π_6 it is easy to show that for $f \in$

$\mathscr{C}^7[0,1]$, we have

$$\|f - \mathscr{Q}_n f\|_\infty \leq c_1 h^7 \|f^{(7)}\|_\infty, \quad (5.6)$$

where c_1 is a constant independent of n.

It is more convenient to write the quasi-interpolant $\mathscr{Q}_n$ under the quasi-*Lagrange* form

$$\mathscr{Q}_n f = \sum_{j\in\Gamma_n} f_j L_j, \quad (5.7)$$

where the quasi-*Lagrange* functions L_j are linear combinations of seven B-splines. For example, using the value f_1 are $\{\mu_1, \mu_2, \mu_3, \mu_4, \mu_5, \mu_6\}$, therefore we have

$$L_1 = B_1 + \frac{3887}{10395} B_2 - \frac{5689}{22275} B_3 - \frac{20959}{155925} B_4 + \frac{5821}{31185} B_5 - \frac{2159}{31185} B_6.$$

For $8 \leq k \leq n-5$, we have

$$L_k = -\frac{2159}{13824}(B_{k-1} + B_{k+5}) + \frac{751}{4608}(B_k + B_{k+4}) - \frac{37003}{46080}(B_{k+1} + B_{k+3}) + \frac{79879}{34560} B_{k+2}.$$

This representation is used in Sects. 5.3 and 5.4 below.

5.3 Quadrature Formula Associated with $\mathscr{Q}_n$

By integrating $\mathscr{Q}_n f$ in the quasi-*Lagrange* form (5.7) we obtain as in [9], the following quadrature formula

$$\int_0^1 f(x)dx = I(f) \simeq I_n(f) = \int_0^1 \mathscr{Q}_n f(x)dx = h \sum_{j\in\Gamma_n} \omega_j f_j,$$

with weights $\omega_j = \frac{1}{h}\int_0^1 L_j(x)dx$. Using the fact that $\int_0^1 B_j(x) = \frac{x_j - x_{j-7}}{7}$, we get

$$\begin{aligned} I_n(f) = h\sum_{j=8}^{n-5} f_j + h\Bigg[&\frac{101}{735}(f_1 + f_{n+2}) + \frac{113221}{138240}(f_2 + f_{n+1}) + \frac{1035241}{967680}(f_3 + f_n) \\ &+ \frac{464651}{483840}(f_4 + f_{n-1}) + \frac{3446899}{3386880}(f_5 + f_{n-2}) + \frac{962903}{967680}(f_6 + f_{n-3}) \\ &+ \frac{193657}{193536}(f_7 + f_{n-4})\Bigg]. \end{aligned} \quad (5.8)$$

Since $\mathscr{Q}_n$ is exact on Π_6, and the weights and knots $(t_j)_{j\in\Gamma_n}$ are symmetric with respect to the midpoint of I, we deduce that the quadrature rule (5.8) is exact on Π_7.

Therefore, the error $E_n(f) = I(f) - I_n(f)$ is a $\mathcal{O}(h^8)$, when $f \in \mathcal{C}^8[0,1]$. Now, according to the *Peano* kernel theorem (see [4, Chap.3]), we have

$$E_n(f) = \frac{1}{7!}\int_0^1 K(t) f^{(8)}(t)dt,$$

where $K(t)$ is the *Peano* kernel defined by

$$K(t) = \int_0^1 (s-t)_+^7 ds - h\sum_{j\in\Gamma_n}\omega_j(t_j - t)_+^7.$$

Theorem 5.1 *The* Peano *kernel $K(t)$ is negative in the intervals $J_1 = [0, t_0]$ and $J_3 = [1-t_0, 1]$ and positive in $J_2 = [t_0, 1-t_0]$ with $t_0 = \tau_0 h$ and $\tau_0 = 1.38135$ (Fig. 5.1).*

Proof Using

$$K(t) = \frac{1}{8}(1-t)^8 - h\sum_{j\in\Gamma_n}\omega_j(t_j - t)_+^7,$$

we see immediately that $K(0) = K(1) = 0$. In fact, since for all $j \in \Gamma_n$, $(t_j - 1)_+^7 = 0$, we obtain $K(1) = 0$. On the other hand, $K(0) = 0$ is due to the fact that the polynomial $p(x) = x^7$, belongs to Π_7, hence it is exactly integrated by the quadrature rule I_n and therefore

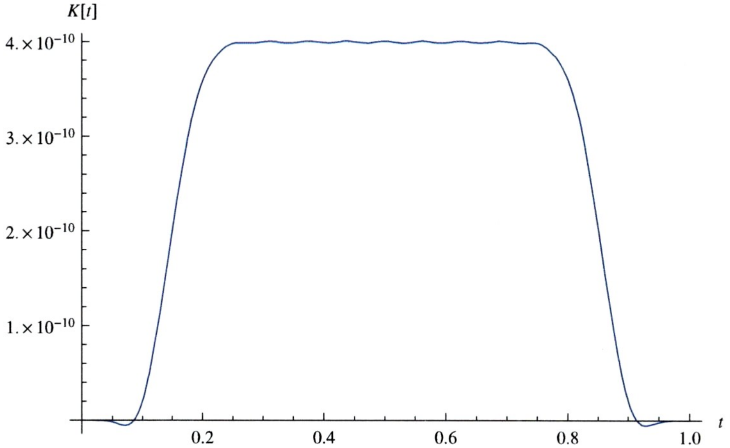

Fig. 5.1 Graph of the *Peano* kernel with $n = 16$

$$I(p) = \frac{1}{8} = h \sum_{j \in \Gamma_n} \omega_j t_j^7.$$

We need to study the sign of $K(t)$ as it is done for the quartic case, (see [6]). Now, setting $t = \tau h, \ \tau \in [0, n]$ and $x = \xi h$, we obtain

$$\frac{1}{h^8} K(t) = \int_0^n (\xi - \tau)_+^7 d\xi - \sum_{j \in \Gamma_n} \omega_j (\tau_j - \tau)_+^7,$$

where

$$\tau_1 = 0, \quad \tau_{n+2} = n \quad \text{and} \quad \tau_j = j - \frac{3}{2} \quad \text{for} \quad j = 2, \ldots, n+1.$$

We have also

$$\int_0^n (\xi - \tau)_+^7 d\xi = \int_\tau^n (\xi - \tau)^7 d\xi = \left[\frac{(\xi - \tau)^8}{8} \right]_\tau^n = \frac{(n - \tau)^8}{8},$$

which gives

$$\frac{1}{h^8} K(t) = \frac{(n - \tau)^8}{8} - \sum_{j \in \Gamma_n} \omega_j (\tau_j - \tau)_+^7 = p(\tau).$$

Consequently, $K(t)$ and $p(\tau)$ have the same sign. By using the symmetry of nodes and weights, it is easy to verify that $p(\tau) = p(n - \tau)$. Then

$$p(\tau) = \frac{\tau^8}{8} - \sum_{j \in \Gamma_n} \omega_j (\tau_j - \tau)_+^7.$$

Now let study the sign of $p(\tau)$ on $[0, n]$. Let $p_j \equiv p|_{[\tau_j, \tau_{j+1}]}, \ j = 1, \ldots, n+1$.

- In the interval $[\tau_1, \tau_2] = [0, \frac{1}{2}]$:

$$p_1(\tau) = \frac{\tau^7}{8} \left(\tau - \frac{808}{735} \right) \leq 0,$$

  which admits $\tau = 0$ as root on this interval.
- In the interval $[\tau_2, \tau_3] = [\frac{1}{2}, \frac{3}{2}]$:

$$p_2(\tau) = p_1(\tau) - \frac{267}{326} \left(\tau - \frac{1}{2} \right)^7,$$

which admits $\tau_0 = \frac{1985}{1437} = 1.38135$ as root in the interval $[\tau_2, \tau_3]$. We can check numerically that $p(\tau) \leq 0$ in $[\tau_2, \tau_0]$ and $p(\tau) \geq 0$ in $[\tau_0, \tau_3]$.

- In the interval $[\tau_3, \tau_4] = [\frac{3}{2}, \frac{5}{2}]$:

$$p_3(\tau) = p_2(\tau) - \frac{996}{931}\left(\tau - \frac{3}{2}\right)^7 \geq 0,$$

which does not admits any roots in the interval $[\tau_3, \tau_4]$.

- In the interval $[\tau_4, \tau_5] = [\frac{5}{2}, \frac{7}{2}]$:

$$p_4(\tau) = p_3(\tau) - \frac{339}{353}\left(\tau - \frac{5}{2}\right)^7 \geq 0,$$

which does not admits any roots in the interval $[\tau_4, \tau_5]$.

- In the interval $[\tau_5, \tau_6] = [\frac{7}{2}, \frac{9}{2}]$:

$$p_5(\tau) = p_4(\tau) - \frac{402}{395}\left(\tau - \frac{7}{2}\right)^7 \geq 0,$$

which does not admits any roots in the interval $[\tau_5, \tau_6]$.

- In the interval $[\tau_6, \tau_7] = [\frac{9}{2}, \frac{11}{2}]$:

$$p_6(\tau) = p_5(\tau) - \frac{1411}{1418}\left(\tau - \frac{9}{2}\right)^7 \geq 0,$$

which does not admits any roots in the interval $[\tau_6, \tau_7]$.

- In the interval $[\tau_7, \tau_8] = [\frac{11}{2}, \frac{13}{2}]$:

$$p_7(\tau) = p_6(\tau) - \frac{1600}{1599}\left(\tau - \frac{11}{2}\right)^7 \geq 0,$$

which does not admits any roots in the interval $[\tau_7, \tau_8]$.

- In the interval $[\tau_8, \tau_9] = [\frac{13}{2}, \frac{15}{2}]$:

$$p_8(\tau) = p_7(\tau) - \left(\tau - \frac{13}{2}\right)^7 \geq 0.$$

- In the interval $[\tau_i, \tau_{i+1}]$, $8 \leq i \leq n-5$ it can be shown by induction that $p_i(\tau) = p_{i-1}(\tau - 1)$.
- In the last seven intervals, since $p(\tau) = p(n - \tau)$, we get

$$p_{n+3-j}(\tau) = p_j(n - \tau), \quad \tau \in [\tau_{n+3-j}, \tau_{n+4-j}], \quad 1 \leq j \leq 7,$$

which means that the behaviour of p is symmetrical of that one in the first seven intervals. This completes the proof.

□

Using the above theorem, the following asymptotic error formula can be proved.

Proposition 5.1 *For any function* $f \in \mathscr{C}^8[0,1]$*, there exist a point* $\tau \in [0,1]$ *such that*

$$E_n(f) = I(f) - I_n(f) = c_0 h^8 f^{(8)}(\tau) + \mathscr{O}(h^9), \tag{5.9}$$

where $c_0 = \frac{1107467}{3251404800} \simeq 3.41 \times 10^{-4}$.

Proof The proof is similar to the proof of Theorem 2 in [6]. □

5.4 The Nyström Method

By using the quadrature scheme (5.8) to approximate the integral in (5.1), we obtain a new equation

$$u_n(s) - h \sum_{j \in \Gamma_n} \omega_j \kappa(s, t_j) u_n(t_j) = f(s), \quad s \in [0,1], \tag{5.10}$$

where the unknowns are $\{u_n(t_j),\ j \in \Gamma_n\}$ and they can be evaluated by solving the following linear system of size $n+2$

$$u_n(t_i) - h \sum_{j \in \Gamma_n} \omega_j \kappa(t_i, t_j) u_n(t_j) = f(t_i), \quad i \in \Gamma_n. \tag{5.11}$$

From (5.10), the approximate solution $u_n(s)$ is completely determined by their values at the nodes $(t_i)_{i \in \Gamma_n}$. In fact,

$$u_n(s) = f(s) + h \sum_{j \in \Gamma_n} \omega_j k(s, t_j) u_n(t_j), \quad s \in [0,1].$$

Now let $\mathscr{K}$ be the integral operator defined by

$$(\mathscr{K} u)(s) = \int_0^1 \kappa(s,t) u(t) dt,$$

and let $\mathscr{K}_n$ be the following *Nyström* approximation

$$(\mathscr{K}_n u)(s) = h \sum_{j \in \Gamma_n} \omega_j \kappa(s, t_j) u(t_j).$$

The following theorem gives a complete information for analyzing the convergence of the *Nyström* method.

Theorem 5.2 *Let $\kappa(s,t)$ be a continuous kernel for $s,t \in [0,1]$. Assume that the quadrature scheme (5.8) is convergent for all continuous functions on $[0,1]$. Further, assume that the integral equation (5.1) is uniquely solvable for a given $f \in \mathscr{C}[0,1]$. Then, for all sufficiently large n, say $n \geq N$, the operators $(I-\mathscr{K}_n)^{-1}$ exist and are uniformly bounded,*

$$\|(I-\mathscr{K}_n)^{-1}\|_\infty \leq \frac{1+\|(I-\mathscr{K})^{-1}\|_\infty\|\mathscr{K}_n\|_\infty}{1-\|(I-\mathscr{K})^{-1}\|_\infty\|(\mathscr{K}-\mathscr{K}_n)\mathscr{K}_n\|_\infty} \leq c, \quad n \geq N$$

with a suitable constant $c < \infty$. For the equations $(I-\mathscr{K})u = f$ and $(I-\mathscr{K}_n)u_n = f$,

$$\|u-u_n\|_\infty \leq \|(I-\mathscr{K})^{-1}\|_\infty\|(\mathscr{K}-\mathscr{K}_n)u\|_\infty, \tag{5.12}$$

$$\leq c\|(\mathscr{K}-\mathscr{K}_n)u\|_\infty, \quad n \geq N. \tag{5.13}$$

Proof See *Atkinson* [1, Theorem 4.1.2]. □

Theorem 5.3 *Let u be the exact solution of (5.1). Assume that $\kappa(s,.)u(.) \in \mathscr{C}^8[0,1]$ for all $s \in [0,1]$. Then, for a sufficiently large n,*

$$\|u-u_n\|_\infty = \mathscr{O}(h^8). \tag{5.14}$$

Proof The estimation (5.13) shows that $\|u-u_n\|_\infty$ and $\|(\mathscr{K}-\mathscr{K}_n)u\|_\infty$ converges to zero with the same speed. By (5.9), we have for $s \in [0,1]$ the asymptotic integration error

$$(\mathscr{K}u)(s) - (\mathscr{K}_n u)(s) = c_0 h^8 \left[\frac{\partial^8}{\partial t^8}\kappa(s,t)u(t)\right]_{t=\tau} + \mathscr{O}(h^9). \tag{5.15}$$

Hence, from (5.13) and (5.15), the *Nyström* method converges with an order of $\mathscr{O}(h^8)$, provided $\kappa(s,t)u(t)$ is eight times continuously differentiable with respect to t, uniformly in s. □

An asymptotic series expansion for the *Nyström* solution u_n is obtained below.

Theorem 5.4 *Under the assumption of Theorem 5.3, we have*

$$u_n - u = c_0[(I-\mathscr{K})^{-1}\mathscr{W}u]h^8 + \mathscr{O}(h^9), \tag{5.16}$$

with

$$(\mathscr{W}u)(s) = \left[\frac{\partial^8}{\partial t^8}\kappa(s,t)u(t)\right]_{t=\tau}, \quad s \in [0,1].$$

Proof Since

$$(I - \mathcal{K}_n)(u - u_n) = (\mathcal{K} - \mathcal{K}_n)u,$$

we can write as in [1, Chap.4]

$$u - u_n = e_n + R_n,$$

where

$$\begin{aligned} e_n &= (I - \mathcal{K})^{-1}(\mathcal{K} - \mathcal{K}_n)u, \\ R_n &= [(I - \mathcal{K}_n)^{-1} - (I - \mathcal{K})^{-1}](\mathcal{K} - \mathcal{K}_n)u, \\ &= (I - \mathcal{K}_n)^{-1}(\mathcal{K}_n - \mathcal{K})(I - \mathcal{K})^{-1}(\mathcal{K} - \mathcal{K}_n)u. \end{aligned}$$

Using the asymptotic expansion (5.15), we get

$$e_n = [(I - \mathcal{K})^{-1}\mathcal{W}u]c_0h^8 + \mathcal{O}(h^9)$$

and

$$\begin{aligned} (I - \mathcal{K}_n)R_n &= (\mathcal{K}_n - \mathcal{K})e_n, \\ &= -c_0h^8\left[\frac{\partial^8}{\partial t^8}\kappa(.,t)e_n(t)\right]_{t=\tau_n'} + \mathcal{O}(h^9), \\ &= -c_0h^{16}\left[\frac{\partial^8}{\partial t^8}\kappa(.,t)c(t)\right]_{t=\tau_n'} + \mathcal{O}(h^9), \end{aligned}$$

where $c(t) = c_0[(I - \mathcal{K})^{-1}\mathcal{W}u](t)$ and $\tau_n \in [0, 1]$. Letting S_n be the solution of equation

$$(I - \mathcal{K}_n)S_n = -c_0h^7\left[\frac{\partial^8}{\partial t^8}\kappa(.,t)c(t)\right]_{t=\tau_n'},$$

we deduce that

$$R_n(t) = (S_n(t) - S(t))h^9 + S(t)h^9,$$

where S satisfies

$$(I - \mathcal{K})S = -c_0h^7\left[\frac{\partial^8}{\partial t^8}\kappa(.,t)c(t)\right]_{t=\tau_n'}.$$

Taking into account that $S - S_n \simeq e_n$, we finally obtain

$$R_n(t) \simeq S(t)h^9.$$

This completes the proof. □

One step of *Richardson* extrapolation can be used to further improve the order of convergence of u_n. Let u_{2n} be the solution associated with a uniform partition of $[0, 1]$ with $2n$ intervals and norm $\frac{h}{2}$. Define

$$u_{2n}^R = \frac{2^8 u_{2n} - u_n}{2^8 - 1}.$$

Theorem 5.5 *If* $\kappa(s, .)u(.) \in \mathscr{C}^8[0, 1]$ *for all* $s \in [0, 1]$, *then, we have*

$$\|u - u_{2n}^R\|_\infty = \mathscr{O}(h^9). \tag{5.17}$$

Proof From Theorem 5.4 we obtain

$$u_{2n} - u = c_0[(I - \mathscr{K})^{-1}\mathscr{W}u]\left(\frac{h}{2}\right)^8 + \mathscr{O}(h^9). \tag{5.18}$$

The estimate (5.17) follows from (5.16) and (5.18). □

5.5 Numerical Results

Example 1 Consider the following linear *Fredholm* integral equation of the second kind

$$u(s) - \int_0^1 s^{\frac{1}{2}} t^8 u(t)dt = f(s), \quad s \in [0, 1],$$

where the exact solution is $u(s) = s$ and f is chosen accordingly. The errors

$$\|u - u_n\|_\infty = \mathscr{O}(h^\alpha) \quad \text{and} \quad \|u - u_{2n}^R\|_\infty = \mathscr{O}(h^\beta)$$

were approximated respectively by

$$\max\left\{|u(\frac{i}{100}) - u_n(\frac{i}{100})|, \ i = 0, 1, \ldots, 100\right\}$$

and

$$\max\left\{|u(\frac{i}{100}) - u_{2n}^R(\frac{i}{100})|,\ i = 0, 1, \ldots, 100\right\}.$$

Using two successive values of n, the values of α and β are computed and are listed in Table 5.1.

From the above table it can be seen that the computed orders of convergence match well with the expected values.

Example 2 Consider the following *Fredholm* integral equation quoted from [6]

$$u(s) - \int_0^1 (s+1)e^{-st}u(t)dt = f(s), \quad s \in [0, 1],$$

where f is chosen so that $u(s) = \cos(s)$. The results are given in Table 5.2.

We denote by u_n^Q and u_n^{QB} the approximated solutions given by *Nyström* method based on the integration of a quartic spline quasi-interpolant and *Nyström* method associated with the extrapolated quadrature formula I_{QB} in [6] respectively. The errors

$$\|u - u_n^Q\|_\infty = \mathcal{O}(h^\gamma) \quad \text{and} \quad \|u - u_n^{QB}\|_\infty = \mathcal{O}(h^\delta)$$

which are listed in Table 5.3, are quoted from [6]. Note that the predicted values of γ and δ are respectively, 6 and 7. The numerical algorithm was run on a PC with Intel Core i5 1.60 GHz CPU, 8GB RAM, and the programs were compiled by using Wolfram Mathematica.

Table 5.1 Nyström and extrapolated Nyström methods for example 1

n	$\|u - u_n\|_\infty$	α	$\|u - u_{2n}^R\|_\infty$	β
16	1.11×10^{-8}			
32	5.31×10^{-11}	7.71	9.69×10^{-12}	
64	2.27×10^{-13}	7.87	1.89×10^{-14}	9.00
128	9.22×10^{-16}	7.94	3.69×10^{-17}	9.00
256	3.67×10^{-18}	7.97	7.22×10^{-20}	9.00
512	1.45×10^{-20}	8.00	1.42×10^{-22}	8.99

Table 5.2 Nyström and extrapolated Nyström methods for example 2

n	$\|u - u_n\|_\infty$	α	$\|u - u_{2n}^R\|_\infty$	β
16	1.05×10^{-12}			
32	4.97×10^{-15}	7.72	1.21×10^{-18}	
64	2.11×10^{-17}	7.88	2.34×10^{-21}	9.01
128	8.57×10^{-20}	7.94	4.55×10^{-24}	9.01
256	3.41×10^{-22}	7.97	8.86×10^{-27}	9.00

Table 5.3 Nyström method based on I_Q and I_{QB} for example

n	$\|u - u_n^Q\|_\infty$	α	$\|u - u_n^{QB}\|_\infty$	β
8	2.33×10^{-8}		9.96×10^{-9}	
16	4.75×10^{-10}	5.61	7.61×10^{-11}	7.03
32	8.30×10^{-12}	5.84	5.96×10^{-13}	6.99
64	1.36×10^{-13}	5.92	4.44×10^{-15}	7.07
128	2.22×10^{-15}	5.94	3.41×10^{-17}	7.02

It can be seen from Tables 5.2 and 5.3 that the approximation u_{2n}^R with $n = 32$ is better than the approximation u_n^{QB} with $n = 128$.

5.6 Conclusion

The results, which are displayed in Table 5.1, show that a very high accuracy is obtained even for a kernel which is only continuous with respect to the variable s. On the other hand, we obtained significant performances in comparison with those of the quadrature rules in [6] and this is due to the fact that the order of convergence of the proposed method is higher. Note that the size of the corresponding linear system is $n + 2$. It can be shown that to solve the present problem by a piecewise polynomial interpolation scheme, a linear system of size at least $4n$ will need to be solved to obtain accuracy of comparable order.

References

1. Atkinson, K.E.: The Numerical Solution of Integral Equations of the Second Kind. Cambridge University Press, Cambridge (1997)
2. Atkinson, K.E., Han, W.: Theoretical Numerical Analysis, 2nd edn. Springer, Berlin (2005)
3. de Boor, C.: A Practical Guide to Splines. Springer, Berlin (1978)
4. Engels, H.: Numerical Quadrature and Cubature. Academic, New York (1980)
5. Nyström, E.: Uber die praktische Auflosung von Integralgleichungen mit Anwendungen auf Randwertaufgaben. Acta Math. **54**, 185–204 (1930)
6. Sablonnière, P., Sbibih, D., Tahrichi, M.: Error estimate and extrapolation of a quadrature formula derived from a quartic spline quasi-interpolant. BIT Numer. Math. **50**, 843–862 (2010)
7. Sablonnière, P., Sbibih, D., Tahrichi, M.: High-order quadrature rules based on spline quasi-interpolants and application to integral equations. Appl. Numer. Math. **62**, 507–520 (2012)
8. Sablonnière, P.: Spline quasi-interpolants and applications to numerical Analysis. Rend. Sem. Univ. Pol. Torino **63**(2), 211–222 (2005)
9. Sablonnière, P.: A quadrature formula associated with a univariate quadratic spline quasi-interpolant. BIT Numer. Math. **47**, 825–837 (2007)
10. Schumaker, L.L.: Spline Functions: Basic Theory. Wiley, New York (1973)

Chapter 6
Superconvergent Methods Based on Cubic Splines for Solving Linear Integral Equations

Azzeddine Bellour, Driss Sbibih, and Ahmed Zidna

Abstract In this paper we propose two collocation methods, based on superconvergent cubic spline interpolant and quasi-interpolant, for approximating the solution of the second kind Fredholm integral equations. Convergence analysis is established. Some numerical examples are given to show the validity of the presented methods.

Keywords PGS Cubic spline · Collocation method · Superconvergence Cubic Quasi-interpolant · Fredholm integral equation

6.1 Introduction

We consider the following Fredholm integral of the second kind equation:

$$x(t) = f(t) + \int_a^b k(t, s)x(s)ds, t \in I = [a, b], \tag{6.1}$$

where the functions f and k are sufficiently smooth. The existence and the uniqueness of the smooth solution of (6.1) are given, for example, in [14, 15, 25].

These types of equations appear in many fields of science and engineering, for instance in elasticity, heat and mass transfer, fluid dynamics, electromagnetism, economics, etc. Given that in the most cases, it is not possible to find the closed form of the solution, different numerical methods have been introduced in the literature

A. Bellour
Department of Mathematics, Ecole Normale Supérieure de Constantine, Constantine, Algeria

D. Sbibih (✉)
University Mohammed I, URAC05, FSO, ANO Laboratory, Oujda, Morocco

A. Zidna
LGIPM Laboratory, University of Lorraine, Metz, France
e-mail: ahmed.zidna@univ-lorraine.fr

D. Barrera et al. (eds.), *Mathematical and Computational Methods for Modelling, Approximation and Simulation*, SEMA SIMAI Springer Series 29,
https://doi.org/10.1007/978-3-030-94339-4_6

for approximating the solution of Eq. (6.1), for more details see, for example, [1–3, 6, 7, 9, 11, 19].

The Galerkin/collocation, Nyström and degenerate kernel methods are the commonly used techniques for this purpose. They have been extensively studied in the literature (see the monograph [6] and the references therein). In [16], Kulkarni introduced an efficient method, based on projections, that provides a high order of convergence in comparison with the Galerkin/collocation method.

Over the past decade, another class of methods based on quasi-interpolating operators, which are not projectors but reproduce polynomial spaces, was introduced in the literature (see e.g. [2–5, 21]). In [3], for example, the authors studied two methods based on left and right partial approximations of the kernel by spline quasi-interpolants with respect to one variable, the other remaining free. The associated numerical tests, listed in [3], show that these techniques are better than those provided by projection methods [6]. In numerical analysis, the superconvergence is a phenomenon where the order of convergence at certain special points is higher than the order of convergence over the whole definition domain. These special points are called superconvergence points. Superconvergence is an important property in theory which can be advantageously exploited in practice. It can be used in collocation finite element method for differential equation (see [13, 24], for instance), in boundary integral methods or integral equations (see [17, 18, 22, 26], for example) and to construct good spline interpolants and quasi-interpolants with an optimal approximation order. According to this last fact, Sablonnière introduced in [20] a superconvergent interpolation method by C^2 cubic splines, called pretty good slopes (abbr. PGS), which gives an overall fourth order approximation of the first derivative at uniform spaced knots. In [10], Boujraf et al. presented a method to construct a new cubic spline quasi-interpolant with the same evaluation points as the classical one but superconvergent at the knots of the partition, i.e., more accurate locally which leads to a significant improvement of the global approximation error.

Following the seminal works cited above, we have recently developed in [8] two methods which allow to approximate the solution of (6.1) by spline functions of class C^2 and degree 3. In this paper, we use superconvergent cubic splines for improving the numerical convergence rates of the methods described in [8].

The paper is organized as follows. In Sect. 6.2, we present the first collocation method based on the PGS cubic spline interpolation, and we prove that it has an approximation order four on the whole interval $[a, b]$. In Sect. 6.3, we describe two numerical schemes, based on a superconvergent cubic spline quasi-interpolant, for approximating the above equation. We show that they have both an approximation order four. Section 6.4 deals with some numerical examples that confirm the theoretical results. Finally, in Sect. 6.5, we give a conclusion and a comparison with the collocation methods developed in [8].

6.2 PGS Cubic Spline Approximation

Let $\tau_n = \{t_i = a + ih, i = 0, 1, .., n\}$ be a uniform partition of the interval I, where the stepsize $h = \frac{b-a}{n}$, and let $t_{-3} = t_{-2} = t_{-1} = t_0$, $t_{n+3} = t_{n+2} = t_{n+1} = t_n$ some additional multiple knots. We denote by $S_3^2(I, \tau_n)$ the space of C^2 cubic splines with knot set τ_n and multiple set at the endpoints. It has dimension $n + 3$ and the cubic B-splines $\{B_i, i = 1, \ldots, n+3\}$ with support $[t_{i-4}, t_i]$ form a basis of this space.

Let $s \in S_3^2(I, \tau_n)$ be a cubic spline, generated by the PGS conditions given by Sablonnière in [20], interpolating the $(n + 1)$ values $x_i = x(t_i)$, $i = 0, \ldots, n$.

Let $d_i = s'(t_i)$, $i = 0, \ldots, n$. Then, the restrictions of s to the intervals $\sigma_i = [t_{i-1}, t_i]$, $i = 1, \ldots, n$, can be written in the form:

$$s_i(t) = x_{i-1}\left(\frac{t_i - t}{h}\right)^3 + \left(3x_{i-1} + hd_{i-1}\right)\left(\frac{t - t_{i-1}}{h}\right)\left(\frac{t_i - t}{h}\right)^2 + (3x_i - hd_i) \times \left(\frac{t - t_{i-1}}{h}\right)^2\left(\frac{t_i - t}{h}\right) + x_i\left(\frac{t - t_{i-1}}{h}\right)^3. \tag{6.2}$$

The values of d_i are the solutions of the following linear system:

$$d_{i-1} + 4d_i + d_{i+1} = \frac{3}{h}\left(x_{i+1} - x_{i-1}\right), i = 1, \ldots, n-1, \tag{6.3}$$

to which we must add two conditions at the endpoints in order to get $n+1$ equations for the $n + 1$ unknowns d_i, $i = 0, \ldots, n$. Sablonnière [20] suggests to take the following additional equations:

$$\begin{aligned} d_0 + 5d_1 + 2d_2 &= \frac{1}{6h}(-19x_0 - 9x_1 + 27x_2 + x_3), \\ 2d_{n-2} + 5d_{n-1} + d_n &= \frac{1}{6h}(-19x_{n-3} - 9x_{n-2} + 27x_{n-1} + x_n). \end{aligned} \tag{6.4}$$

These two equations derive from some imposed conditions that allow to get forth order approximation of the end first derivatives, for more details, see [20].

Combining (6.4) with the first and the last equation from (6.3) respectively, we get the following two equations:

$$\begin{aligned} -\frac{1}{3}d_0 + d_1 &= \frac{1}{18h}(-17x_0 + 9x_1 + 9x_2 - x_3), \\ d_{n-1} - \frac{1}{3}d_n &= \frac{1}{18h}(17x_n - 9x_{n-1} - 9x_{n-2} + x_{n-3}). \end{aligned} \tag{6.5}$$

In the matrix notation, (6.3) and (6.5) can be written as:

$$Md = \frac{1}{h} Nx, \tag{6.6}$$

where $d = (d_0, \ldots, d_n)^T$, $x = (x_0, \ldots, x_n)^T$ and M (resp. N) are the matrices given by

$$M = \begin{bmatrix} -\frac{1}{3} & 1 & 0 & 0 & \ldots\ldots & 0 & 0 & 0 & 0 \\ 1 & 4 & 1 & 0 & \ldots\ldots & 0 & 0 & 0 & 0 \\ 0 & 1 & 4 & 1 & \ldots\ldots & 0 & 0 & 0 & 0 \\ \vdots & \vdots & \vdots & \vdots & \vdots & \vdots & \vdots & \vdots & \vdots \\ \vdots & \vdots & \vdots & \vdots & \vdots & \vdots & \vdots & \vdots & \vdots \\ 0 & 0 & 0 & 0 & \ldots\ldots & 1 & 4 & 1 & 0 \\ 0 & 0 & 0 & 0 & \ldots\ldots & 0 & 1 & 4 & 1 \\ 0 & 0 & 0 & 0 & \ldots\ldots & 0 & 0 & 1 & -\frac{1}{3} \end{bmatrix} \text{ and}$$

$$N = \begin{bmatrix} -\frac{17}{18} & \frac{1}{2} & \frac{1}{2} & -\frac{1}{18} & 0 & \ldots & 0 & 0 & 0 \\ -3 & 0 & 3 & 0 & 0 & \ldots & 0 & 0 & 0 \\ 0 & -3 & 0 & 3 & 0 & \ldots & 0 & 0 & 0 \\ \vdots & \vdots & \vdots & \vdots & \vdots & \vdots & \vdots & \vdots & \\ \vdots & \vdots & \vdots & \vdots & \vdots & \vdots & \vdots & \vdots & \\ 0 & 0 & 0 & 0 & \ldots & -3 & 0 & 3 & 0 \\ 0 & 0 & 0 & 0 & \ldots & 0 & -3 & 0 & 3 \\ 0 & 0 & 0 & 0 & \ldots & \frac{1}{18} & -\frac{1}{2} & -\frac{1}{2} & \frac{17}{18} \end{bmatrix}$$

It is well known that for a smooth enough x, the approximation order by s is equal to 4, i.e. $x - s = O(h^4)$. Thus, from (6.1), we obtain for all $i = 0, \ldots, n$

$$\begin{aligned} x_i &= f_i + \sum_{j=1}^{n} \int_{t_{j-1}}^{t_j} k(t_i, r) s_j(r) dr + O(h^4) \\ &= f_i + \sum_{j=1}^{n} x_{j-1} \underbrace{\int_{t_{j-1}}^{t_j} k(t_i, r) \left(\frac{t_j - r}{h}\right)^3 dr}_{a_{i,j-1}} \\ &\quad + \sum_{j=1}^{n} (3x_{j-1} + hd_{j-1}) \underbrace{\int_{t_{j-1}}^{t_j} k(t_i, r) \left(\frac{r - t_{j-1}}{h}\right) \left(\frac{t_j - r}{h}\right)^2 dr}_{b_{i,j-1}} \end{aligned}$$

$$+\sum_{j=1}^{n}(3x_j - hd_j)\underbrace{\int_{t_{j-1}}^{t_j} k(t_i,r)\left(\frac{r-t_{j-1}}{h}\right)^2\left(\frac{t_j-r}{h}\right)dr}_{c_{i,j}}$$

$$+\sum_{j=1}^{n}x_j\underbrace{\int_{t_{j-1}}^{t_j} k(t_i,r)\left(\frac{r-t_{j-1}}{h}\right)^3 dr}_{e_{i,j}} + O(h^4)$$

$$= f_i + \sum_{j=0}^{n}x_j a_{i,j} + 3\sum_{j=0}^{n}x_j b_{i,j} + h\sum_{j=0}^{n}d_j b_{i,j} + 3\sum_{j=0}^{n}x_j c_{i,j} - h\sum_{j=0}^{n}d_j c_{i,j}$$

$$+\sum_{j=0}^{n}x_j e_{i,j} + O(h^4) \tag{6.7}$$

such that, $a_{i,n} = b_{i,n} = c_{i,0} = e_{i,0} = 0$ for $i = 0, \ldots, n$.

Now, we approximate x_i by $\widehat{x_i}$ and d_i by $\widehat{d_i}$ such that $\widehat{x_i}$ and $\widehat{d_i}$ $(i = 0, \ldots, n)$ satisfy the system (6.6) and for all $i = 0, \ldots, n$, we have

$$\widehat{x_i} = f_i + \sum_{j=0}^{n}\widehat{x_j} a_{i,j} + 3\sum_{j=0}^{n}\widehat{x_j} b_{i,j} + h\sum_{j=0}^{n}\widehat{d_j} b_{i,j} + 3\sum_{j=0}^{n}\widehat{x_j} c_{i,j} - h\sum_{j=0}^{n}\widehat{d_j} c_{i,j}$$

$$+\sum_{j=0}^{n}\widehat{x_j} e_{i,j}.$$

In the matrix notation, we have

$$\widehat{x} = F + A\widehat{x} + 3B\widehat{x} + hB\widehat{d} + 3C\widehat{x} - hC\widehat{d} + E\widehat{x},$$

where $\widehat{x} = (\widehat{x_0}, \ldots, \widehat{x_n})^T, \widehat{d} = (\widehat{d_0}, \ldots, \widehat{d_n})^T, F = (f_0, \cdots, f_n)^T, A = (a_{i,j}), B = (b_{i,j}), C = (c_{i,j})$ and $E = (e_{i,j})$.

If we denote by N the right hand side matrix of the linear system (6.6), it follows that

$$(I - A - 3B - 3C - E)\widehat{x} = F + h(B - C)\widehat{d} = F + (B - C)M^{-1}N\widehat{x}.$$

Hence,

$$\left(I - A - 3B - 3C - E - (B - C)M^{-1}N\right)\widehat{x} = F. \tag{6.8}$$

Finally, by replacing the unknown values $x_i, i = 0, \ldots, n$ in the formula of s_i (Eq. (6.2)) by the approximate values $\widehat{x_i}, i = 0, \ldots, n$, which are computed by solving the linear system (6.8), we obtain an approximate solution $\widehat{s} \in S_3^2(I, \Pi_n)$ of the exact solution x such that $\widehat{s} = \widehat{s}_i$ on σ_i, $i = 1, \ldots, n$, where

$$\widehat{s}_i(t) = \widehat{x}_{i-1}\left(\frac{t_i - t}{h}\right)^3 + \left(3\widehat{x}_{i-1} + h\widehat{d}_{i-1}\right)\left(\frac{t - t_{i-1}}{h}\right)\left(\frac{t_i - t}{h}\right)^2$$
$$+ \left(3\widehat{x}_i - h\widehat{d}_i\right)\left(\frac{t - t_{i-1}}{h}\right)^2\left(\frac{t_i - t}{h}\right) + \widehat{x}_i\left(\frac{t - t_{i-1}}{h}\right)^3. \quad (6.9)$$

6.2.1 *Analysis of Convergence*

The following results are needed for computing the convergence order of the method.

Lemma 6.1 *Let B be $n \times n$ matrix such that $\|B\|_\infty < 1$. Then, the matrix $(I - B)$ is invertible. Moreover, $\|(I - B)^{-1}\|_\infty \leq \frac{1}{1 - \|B\|_\infty}$.*

By using the inversion of general tridiagonal matrices [23], it is easy to prove the following result.

Lemma 6.2 *The matrix M is invertible, and the inverse $M^{-1} = \left(u_{i,j}\right)_{1 \leq i,j \leq n+1}$ is symmetric and for $i \geq j$, we have*

$$u_{j,i} = u_{i,j} = (-1)^{i+j}\frac{\alpha_{j-1}\alpha_{n+1-i}}{\alpha_{n+1}},$$

such that the sequence $(\alpha_i)_{0 \leq i \leq n+1}$ is defined by

$$\begin{cases} \alpha_i &= \frac{7+3\sqrt{3}}{6\sqrt{3}}(2 - \sqrt{3})^i - \frac{7-3\sqrt{3}}{6\sqrt{3}}(2 + \sqrt{3})^i, i = 0, 1, \ldots, n, \\ \alpha_{n+1} &= -\frac{1}{3}\alpha_n - \alpha_{n-1}. \end{cases}$$

Proof For any symmetric invertible tridiagonal matrix (see [23]) of the form

$$A_{n+1} = \begin{bmatrix} b_1 & c_1 & 0 & 0 & \ldots\ldots & 0 & 0 & 0 & 0 \\ a_2 & b_2 & c_2 & 0 & \ldots\ldots & 0 & 0 & 0 & 0 \\ 0 & a_3 & b_3 & c_3 & \ldots\ldots & 0 & 0 & 0 & 0 \\ \vdots & \vdots & \vdots & \vdots & \vdots \quad \vdots & \vdots & \vdots & \vdots & \\ \vdots & \vdots & \vdots & \vdots & \vdots \quad \vdots & \vdots & \vdots & \vdots & \\ 0 & 0 & 0 & 0 & \ldots\ldots & a_{n-1} & b_{n-1} & c_{n-1} & 0 \\ 0 & 0 & 0 & 0 & \ldots\ldots & 0 & a_n & b_n & c_n \\ 0 & 0 & 0 & 0 & \ldots\ldots & 0 & 0 & a_{n+1} & b_{n+1} \end{bmatrix}$$

the inverse $A_{n+1}^{-1} = \left(u_{i,j}\right)_{1 \leq i,j \leq n+1}$ is symmetric and for $i \geq j$, we have

$$u_{j,i} = u_{i,j} = \begin{cases} (-1)^{i+j} a_{j+1} a_{j+2} \dots a_i \frac{\alpha_{j-1}\phi_{i+1}}{\alpha_{n+1}} & \text{if } i > j, \\ \frac{\alpha_{i-1}\phi_{i+1}}{\alpha_{n+1}} & \text{if } i = j, \end{cases} \tag{6.10}$$

such that the sequences $(\alpha_i)_{0 \leq i \leq n+1}$ and $(\phi_i)_{1 \leq i \leq n+2}$ are defined by

$$\begin{cases} \alpha_i = b_i \alpha_{i-1} - a_i c_{i-1} \alpha_{i-2}, i = 2, \dots, n+1, \\ \alpha_0 = 1, \alpha_1 = b_1, \end{cases}$$

and

$$\begin{cases} \phi_i \quad = b_i \phi_{i+1} - a_{i+1} c_i \phi_{i+2}, i = n, \dots, 1, \\ \phi_{n+2} = 1, \phi_{n+1} = b_{n+1}. \end{cases}$$

Hence, for the matrix M, the sequences $(\alpha_i)_{0 \leq i \leq n+1}$ and $(\phi_i)_{1 \leq i \leq n+2}$ are given by

$$\begin{cases} \alpha_i \quad = 4\alpha_{i-1} - \alpha_{i-2}, i = 2, \dots, n, \\ \alpha_{n+1} = \frac{-1}{3}\alpha_n - \alpha_{n-1}, \\ \alpha_0 \quad = 1, \alpha_1 = \frac{-1}{3}, \end{cases} \tag{6.11}$$

and

$$\begin{cases} \phi_i \quad = 4\phi_{i+1} - \phi_{i+2}, i = n, \dots, 2, \\ \phi_1 \quad = \frac{-1}{3}\phi_2 - \phi_3, \\ \phi_{n+2} = 1, \phi_{n+1} = \frac{-1}{3}. \end{cases} \tag{6.12}$$

It is easy to see that the solution of the difference equation (6.11) is given by

$$\begin{cases} \alpha_i \quad = \frac{7+3\sqrt{3}}{6\sqrt{3}}(2-\sqrt{3})^i - \frac{7-3\sqrt{3}}{6\sqrt{3}}(2+\sqrt{3})^i, i = 0, 1, \dots, n, \\ \alpha_{n+1} = -\frac{1}{3}\alpha_n - \alpha_{n-1}. \end{cases}$$

On the other hand, by using (6.11), we can see that the sequence $(\alpha_{n+2-i})_{1 \leq i \leq n+2}$ satisfies (6.12). Since $(\phi_i)_{1 \leq i \leq n+2}$ is the unique solution of (6.12), we deduce that

$$\phi_i = \alpha_{n+2-i}, i = 1, \dots, n+2.$$

Which implies, by using (6.10), that for $i \geq j$

$$u_{j,i} = u_{i,j} = (-1)^{i+j} \frac{\alpha_{j-1}\alpha_{n+1-i}}{\alpha_{n+1}},$$

which completes the proof of Lemma 6.2. □

Remark 6.1 It is easy to see that the matrix M^{-1} is symmetric. Moreover, $\forall\, i,\, j = 1, 2, \ldots, n+1$,

$$u_{n-i+2,n-j+2} = u_{i,j} = (-1)^{i+j} \frac{\alpha_{\min(i,j)-1}\alpha_{n+1-\max(i,j)}}{\alpha_{n+1}}.$$

Lemma 6.3 *For all $i = 2, \ldots, n$, we have*

- $\alpha_i = 4\alpha_{i-1} - \alpha_{i-2}$.
- $\alpha_i < 0$ and $\alpha_{n+1} > 0$.

Proof Let $i = 2, \ldots, n$, we have by Lemma 6.2

$$\begin{aligned} 4\alpha_{i-1} - \alpha_{i-2} =& \frac{7+3\sqrt{3}}{6\sqrt{3}} \left(4(2-\sqrt{3})^{i-1} - (2-\sqrt{3})^{i-2}\right) - \frac{7-3\sqrt{3}}{6\sqrt{3}} \times \\ & \left(4(2+\sqrt{3})^{i-1} - (2+\sqrt{3})^{i-2}\right) \\ =& \frac{7+3\sqrt{3}}{6\sqrt{3}}(2-\sqrt{3})^{i-2}(7-4\sqrt{3}) - \frac{7-3\sqrt{3}}{6\sqrt{3}}(2+\sqrt{3})^{i-2} \times \\ & (7+4\sqrt{3}) = \alpha_i. \end{aligned}$$

and

$$2i > \frac{\ln(\frac{7+3\sqrt{3}}{7-3\sqrt{3}})}{\ln(2+\sqrt{3})} \cong 1.45 \Longleftrightarrow (2+\sqrt{3})^{2i} > \frac{7+3\sqrt{3}}{7-3\sqrt{3}} \Longleftrightarrow \alpha_i < 0.$$

Thus, $\alpha_{n+1} = -\frac{1}{3}\alpha_n - \alpha_{n-1} > 0$. □

Lemma 6.4 *$\|M^{-1}\|_\infty$ is bounded. Moreover, for all $n \geq 10$, we have*

$$2.2718528 \cong \frac{6\sqrt{3}-3}{76-42\sqrt{3}} \leq \|M^{-1}\|_\infty \leq \frac{1118577}{492362} \cong 2.2718589.$$

Proof Let $L_i = \sum_{j=1}^{n+1} |u_{i,j}|$, $i = 1, \ldots, n+1$. The proof is split into three steps.

Step 1 Compute L_i for $i = 2, \ldots, n$. We have, by Lemma 6.2

$$\begin{aligned}\alpha_{n+1} L_i =& \alpha_{n+1}\left(\sum_{j=1}^{i} |u_{i,j}| + \sum_{j=i+1}^{n+1} |u_{i,j}|\right)\\ =& \sum_{j=1}^{i} |\alpha_{j-1}\alpha_{n+1-i}| + \sum_{j=i+1}^{n+1} |\alpha_{i-1}\alpha_{n+1-j}|.\\ =& |\alpha_{n+1-i}| \sum_{j=0}^{i-1} |\alpha_j| + |\alpha_{i-1}| \sum_{j=0}^{n-i} |\alpha_j|,\end{aligned}$$

which implies, by taking into account Lemma 6.3, that

$$\alpha_{n+1} L_i = -2\alpha_{n+1-i} + \alpha_{n+1-i} \sum_{j=0}^{i-1} \alpha_j - 2\alpha_{i-1} + \alpha_{i-1} \sum_{j=0}^{n-i} \alpha_j. \tag{6.13}$$

On the other hand,

$$\begin{aligned}-2\alpha_{n+1-i} - 2\alpha_{i-1} =& -\frac{2+\sqrt{3}}{3\sqrt{3}}\left[(7+3\sqrt{3}) - (7-3\sqrt{3})(2+\sqrt{3})^n\right] \times\\ &\left((2-\sqrt{3})^i + (2-\sqrt{3})^{n+2}(2+\sqrt{3})^i\right),\end{aligned} \tag{6.14}$$

and

$$\begin{aligned}\alpha_{n+1-i} \sum_{j=0}^{i-1} \alpha_j + \alpha_{i-1} \sum_{j=0}^{n-i} \alpha_j =& -\left(\frac{7+3\sqrt{3}}{6\sqrt{3}}\right)^2 \sqrt{3}(2-\sqrt{3})^n + \left(\frac{7-3\sqrt{3}}{6\sqrt{3}}\right)^2 \times\\ &\sqrt{3}(2+\sqrt{3})^n + \frac{5}{18\sqrt{3}}\left((7+3\sqrt{3})(2-\sqrt{3})^{n+1} - (7-3\sqrt{3})(2-\sqrt{3})\right) \times\\ &(2+\sqrt{3})^i - \frac{5}{18\sqrt{3}}\left((7-3\sqrt{3})(2+\sqrt{3})^{n+1} - (7+3\sqrt{3})(2+\sqrt{3})\right) \times\\ &(2-\sqrt{3})^i.\end{aligned} \tag{6.15}$$

By substituting (6.14) and (6.15) into (6.13), we obtain for $i = 2, \ldots, n$

$$\alpha_{n+1} L_i = \frac{1}{36\sqrt{3}} \left((7 - 3\sqrt{3})^2 (2 + \sqrt{3})^n - (7 + 3\sqrt{3})^2 (2 - \sqrt{3})^n \right) + \frac{2+\sqrt{3}}{18\sqrt{3}} \times$$
$$\left((7 - 3\sqrt{3})(2 + \sqrt{3})^n - (7 + 3\sqrt{3}) \right) \left((2 - \sqrt{3})^{n+2} (2 + \sqrt{3})^i + (2 - \sqrt{3})^i \right). \tag{6.16}$$

Step 2 Show that $L_i \leq L_2 = L_n$ for $i = 2, \ldots, n$. We have by Remark 6.1,

$$L_2 = \sum_{j=1}^{n+1} |u_{2,j}| = \sum_{j=1}^{n+1} |u_{n,n-j+2}| = \sum_{j=1}^{n+1} |u_{n,j}| = L_n.$$

Moreover, by (6.16), we have for $i = 3, \ldots, n-1$

$$\begin{aligned}
\alpha_{n+1}(L_i - L_2) &= \frac{2+\sqrt{3}}{18\sqrt{3}} \left((7 - 3\sqrt{3})(2 + \sqrt{3})^n - (7 + 3\sqrt{3}) \right) \times \\
&\left((2 - \sqrt{3})^n \left((2 + \sqrt{3})^{i-2} - 1 \right) + (2 - \sqrt{3})^2 \left((2 - \sqrt{3})^{i-2} - 1 \right) \right) \\
&\leq \frac{2+\sqrt{3}}{18\sqrt{3}} \left((7 - 3\sqrt{3})(2 + \sqrt{3})^n - (7 + 3\sqrt{3}) \right) \times \\
&\left((2 - \sqrt{3})^n \left((2 + \sqrt{3})^{n-3} - 1 \right) + (2 - \sqrt{3})^2 \left((2 - \sqrt{3}) - 1 \right) \right) \\
&\leq \frac{2+\sqrt{3}}{18\sqrt{3}} \left((7 - 3\sqrt{3})(2 + \sqrt{3})^n - (7 + 3\sqrt{3}) \right) \times \\
&\left((2 - \sqrt{3})^3 + (2 - \sqrt{3})^2 \left((2 - \sqrt{3}) - 1 \right) \right) \\
&= \frac{2+\sqrt{3}}{18\sqrt{3}} \left((7 - 3\sqrt{3})(2 + \sqrt{3})^n - (7 + 3\sqrt{3}) \right) (2 - \sqrt{3})^2 (3 - 2\sqrt{3}) < 0.
\end{aligned}$$

This implies that $L_i \leq L_2 = L_n$ for $i = 2, \ldots, n$.

Step 3 Show that $\|M^{-1}\|_\infty$ is bounded. We have by Remark 6.1,

$$L_1 = \sum_{j=1}^{n+1} |u_{1,j}| = \sum_{j=1}^{n+1} |u_{n+1,n-j+2}| = \sum_{j=1}^{n+1} |u_{n+1,j}| = L_{n+1}.$$

Moreover,

$$\alpha_{n+1}L_1 = \sum_{j=1}^{n+1} |\alpha_{n+1-j}| = \sum_{j=0}^{n} |\alpha_j| = 2 - \sum_{j=0}^{n} \alpha_j$$
$$= \frac{1}{3} + \frac{15\sqrt{3}-24}{18\sqrt{3}}\left(2+\sqrt{3}\right)^{n+1} + \frac{15\sqrt{3}+24}{18\sqrt{3}}\left(2-\sqrt{3}\right)^{n+1} \tag{6.17}$$

and by (6.16), we have

$$\alpha_{n+1}L_2 = \frac{1}{9} + \frac{23-13\sqrt{3}}{18\sqrt{3}}\left(2+\sqrt{3}\right)^{n} - \frac{23+13\sqrt{3}}{18\sqrt{3}}\left(2-\sqrt{3}\right)^{n}.$$

Hence,

$$\alpha_{n+1}(L_1 - L_2) = \frac{2}{9} + \frac{19\sqrt{3}-26}{18\sqrt{3}}\left(2+\sqrt{3}\right)^{n} + \frac{28\sqrt{3}+47}{18\sqrt{3}}\left(2-\sqrt{3}\right)^{n} > 0.$$

Thus, by Step 2, we deduce that $\|M^{-1}\|_\infty = \max_{1\le i\le n+1}(L_i) = L_1$.
Now, by Lemma 6.3, we have

$$\alpha_{n+1} = -\frac{1}{3}\alpha_n - \alpha_{n-1} = \frac{(76-42\sqrt{3})}{18\sqrt{3}}\left(2+\sqrt{3}\right)^{n} - \frac{(76+42\sqrt{3})}{18\sqrt{3}}\left(2-\sqrt{3}\right)^{n}.$$

Thus, by (6.17), we obtain

$$\|M^{-1}\|_\infty = L_1 = \frac{6\sqrt{3} + \left(6\sqrt{3}-3\right)\left(2+\sqrt{3}\right)^{n} + \left(6\sqrt{3}+3\right)\left(2-\sqrt{3}\right)^{n}}{\left(76-42\sqrt{3}\right)\left(2+\sqrt{3}\right)^{n} - \left(76+42\sqrt{3}\right)\left(2-\sqrt{3}\right)^{n}}.$$

It is easy to see that the sequence

$$u_n = \frac{6\sqrt{3} + \left(6\sqrt{3}-3\right)\left(2+\sqrt{3}\right)^{n} + \left(6\sqrt{3}+3\right)\left(2-\sqrt{3}\right)^{n}}{\left(76-42\sqrt{3}\right)\left(2+\sqrt{3}\right)^{n} - \left(76+42\sqrt{3}\right)\left(2-\sqrt{3}\right)^{n}}, n \in \mathbb{N}$$

is decreasing and converges to $\frac{6\sqrt{3}-3}{76-42\sqrt{3}}$.
This implies that $\|M^{-1}\|_\infty$ is bounded. Moreover, for all $n \ge 10$, we have

$$2.2718528 \cong \frac{6\sqrt{3}-3}{76-42\sqrt{3}} \le \|M^{-1}\|_\infty \le u_{10} = \frac{1118577}{492362} \cong 2.2718589.$$

□

The following theorem describes the convergence of the PGS cubic spline approximation method.

Theorem 6.1 *Let* $f \in C^4(I)$, $k \in C^4(I \times I)$ *such that* $\frac{1610939}{492362}\|k\|(b-a) < 1$. *Then Eqs. (6.9) define a unique approximation* $\widehat{s}$ *and the resulting error* $\varepsilon := x - \widehat{s}$ *satisfies*

$$\|\varepsilon\| \leq \beta h^4,$$

where β *is a finite constant independent of* h.

Proof We have $\|A\| \leq \frac{\|k\|(b-a)}{4}$, $\|B\| \leq \frac{\|k\|(b-a)}{12}$, $\|C\| \leq \frac{\|k\|(b-a)}{12}$, $\|E\| \leq \frac{\|k\|(b-a)}{4}$ and $\|N\| = 6$. Hence, by Lemma 6.4

$$\begin{aligned}\|A + 3B + 3C + E + (B - C)M^{-1}N\| &\leq \|k\|(b-a) + \|k\|(b-a)\|M^{-1}\| \\ &\leq \frac{1610939}{492362}\|k\|(b-a)\| < 1.\end{aligned}$$

Then, by Lemma 6.1, Eq. (6.8) has a unique solution $\widehat{x}$.

It follows that Eqs. (6.9) define a unique approximation $\widehat{s}$.

Now, let $\widehat{e} = x - \widehat{x} = (x_0 - \widehat{x}_0, \ldots, x_n - \widehat{x}_n)^T$. Then, from Eq. (6.7),

$$\Big(I - A - 3B - 3C - E - (B - C)M^{-1}N\Big)\widehat{e} = O(h^4).$$

Hence, $\widehat{e} = \Big(I - A - 3B - 3C - E - (B - C)M^{-1}N\Big)^{-1} O(h^4)$.

Which implies, by Lemma 6.1, that there exists $\beta_1 > 0$ such that

$$\|\widehat{e}\| \leq \underbrace{\frac{\beta_1}{1 - \frac{1610939}{492362}\|k\|(b-a)}}_{\beta_2} h^4.$$

On the other hand, from (6.6), we have $(d - \widehat{d}) = \frac{1}{h}M^{-1}N\widehat{e}$.

Hence, $\|d - \widehat{d}\| \leq \frac{4832817}{246181}\beta_2 h^3$.

Then, for all $i = 1, \ldots, n$ and $t \in \sigma_i$, we have

$$\begin{aligned}|s_i(t) - \widehat{s}_i(t)| &\leq 4|x_{i-1} - \widehat{x}_{i-1}| + h|d_{i-1} - \widehat{d}_{i-1}| + 4|x_i - \widehat{x}_i| + h|d_i - \widehat{d}_i| \\ &\leq \frac{11635082}{246181}\beta_2 h^4.\end{aligned}$$

It follows that $\|x - \widehat{s}\| \leq \|x - s\| + \|s - \widehat{s}\| \leq Ch^4 + \frac{11635082}{246181}\beta_2 h^4$.

Thus, the proof is completed by taking $\beta = C + \frac{11635082}{246181}\beta_2$. □

6.3 Superconvergent Cubic Spline Quasi-Interpolant Method

In this section we apply the superconvergent discrete cubic spline quasi-interpolant, introduced in [10], for approximating the solution of Eq. (6.1). This linear operator, noted Q_3, is defined by $Q_3 x = \sum_{j=1}^{n+3} \lambda_j(x) B_j$, where the coefficients $\lambda_j(x)$ are linear combinations of some values of x on the partition τ_n and computed such that the quasi-interpolant Q_3 satisfies the two following properties:

- Q_3 is exact on polynomials of degree 3, i.e. $Q_3 p = p$ for all polynomials p of degree ≤ 3.
- Q_3 is more accurate on the knots (t_i), i.e. $Q_3 x(t_i) - x_i = O(h^5)$, $\forall\ i = 0, 1, \ldots, n$.

Then, these coefficients are given by:

$$
\begin{aligned}
&\lambda_1(x) = x_0,\\
&\lambda_2(x) = \tfrac{11}{36}x_0 + \tfrac{4}{3}x_1 - x_2 + \tfrac{4}{9}x_3 - \tfrac{1}{12}x_4,\\
&\lambda_3(x) = -\tfrac{5}{36}x_0 + \tfrac{11}{9}x_1 - \tfrac{1}{9}x_3 + \tfrac{1}{36}x_4,\\
&\lambda_j(x) = \tfrac{1}{36}x_{j-4} - \tfrac{5}{18}x_{j-3} + \tfrac{3}{2}x_{j-2} - \tfrac{5}{18}x_{j-1} + \tfrac{1}{36}x_j, \text{ for } j = 4, \cdots, n,\\
&\lambda_{n+1}(x) = \tfrac{1}{36}x_{n-4} - \tfrac{1}{9}x_{n-3} + \tfrac{11}{9}x_{n-1} - \tfrac{5}{36}x_n,\\
&\lambda_{n+2}(x) = -\tfrac{1}{12}x_{n-4} + \tfrac{4}{9}x_{n-3} - x_{n-2} + \tfrac{4}{3}x_{n-1} + \tfrac{11}{36}x_n,\\
&\lambda_{n+3}(x) = x_n.
\end{aligned}
$$

The exactness of Q_3 on polynomials of degree ≤ 3 allows to prove, (see e.g. [12]), that there exist constants $C_k, k = 0, 1, 2$, independent of n, such that for any function $f \in C^4(I)$, we have:

$$
\|Q_3^{(k)} f - f^{(k)}\|_\infty \leq C_k h^{4-k} \|f^{(4-k)}\|_\infty.
$$

In addition, if $f \in C^5(I)$, we have $|Q_3 x(t_i) - x_i| = O(h^5)$ for all $i \in 0, 1, \ldots, n$ (see [10]).

There are at least two manners for approximating the solution of the equation (6.1) by using the above superconvergent quasi-interpolant Q_3:

1. Quasi-interpolation method 1 which consists in replacing the solution x in the right hand side of (1.1) by Q_3.
2. Quasi-interpolation method 2 which is based on the approximation of the univariate left section of the kernel k: $t \to k(t, s)$ by the quasi-interpolant Q_3.

6.3.1 Quasi-Interpolation Method 1

Let us describe the first technique. As $x - Qx = O(h^4)$ for x smooth enough, we obtain for all $i = 0, \dots, n$

$$x_i = f_i + \sum_{j=1}^{n+3} \lambda_j(x) \underbrace{\int_a^b B_j(s) k(t_i, s) ds}_{B_{i+1,j}} + O(h^4). \tag{6.18}$$

Now, we approximate x_i by $\widetilde{x}_i$, for $i = 0, \dots, n$, such that

$$\widetilde{x}_i = f_i + \sum_{j=1}^{n+3} \widetilde{\lambda}_j(x) B_{i+1,j}, \tag{6.19}$$

where $\widetilde{\lambda}_j(x)$ are obtained by replacing x_i by $\widetilde{x}_i$ in the expressions of $\lambda_j(x)$ previously defined.

In the matrix notation, (6.18) can be written in the form:

$$\widetilde{X} = F + B\widetilde{\lambda}(x) \tag{6.20}$$

such that $\widetilde{X} = (\widetilde{x}_0, \widetilde{x}_1, \dots, \widetilde{x}_n)^T$, $F = (f_0, \cdots, f_n)^T$, $\widetilde{\lambda}(x) = (\widetilde{\lambda}_0(x), \widetilde{\lambda}_1(x), \dots, \widetilde{\lambda}_{n+3}(x))^T$, $B = (B_{i,j})$ for $i = 1, \cdots, n+1$ and $j = 1, \cdots, n+3$.

On the other hand, we have

$$\widetilde{\lambda}(x) = \begin{pmatrix} 1 & 0 & 0 & 0 & 0 & 0 & \dots & 0 & 0 & \\ \frac{11}{36} & \frac{4}{3} & -1 & \frac{4}{9} & -\frac{1}{12} & 0 & 0 & \dots & 0 & 0 \\ -\frac{5}{36} & \frac{11}{9} & 0 & -\frac{1}{9} & \frac{1}{36} & 0 & 0 & \dots & 0 & 0 \\ \frac{1}{36} & -\frac{5}{18} & \frac{3}{2} & -\frac{5}{18} & \frac{1}{36} & 0 & 0 & \dots & 0 & 0 \\ 0 & \frac{1}{36} & -\frac{5}{18} & \frac{3}{2} & -\frac{5}{18} & \frac{1}{36} & 0 & \dots & 0 & 0 \\ \vdots & \vdots & \vdots & \vdots & \vdots & \vdots & \vdots & \vdots & \vdots & \vdots \\ 0 & 0 & \dots & 0 & 0 & \frac{1}{36} & -\frac{5}{18} & \frac{3}{2} & -\frac{5}{18} & \frac{1}{36} \\ 0 & 0 & \dots & 0 & 0 & 0 & \frac{1}{36} & -\frac{1}{9} & \frac{11}{9} & \frac{5}{36} \\ 0 & 0 & \dots & 0 & 0 & -\frac{1}{12} & \frac{4}{9} & -1 & -\frac{4}{3} & \frac{11}{36} \\ 0 & 0 & \dots & 0 & 0 & 0 & 0 & 0 & 0 & 1 \end{pmatrix} \widetilde{X}.$$

We denote by A the right hand side matrix. Then, the system (6.20) becomes:

$$(I - BA)\widetilde{X} = F \tag{6.21}$$

The following theorem describes the convergence of the first superconvergent cubic spline quasi-interpolant technique.

Theorem 6.2 *Let $f \in C^4(I)$ and $k \in C^4(I \times I)$ such that $\frac{19}{6}\| \int_a^b |k(.,s)|ds\| < 1$. Then Eq. (6.21) defines a unique approximation $\widetilde{Q}_3 x = \sum_{j=1}^{n+3} \widetilde{\lambda}_j(x) B_j$ and the resulting error $\widehat{e} := x - \widetilde{Q}_3 x$ satisfies*

$$\|\widehat{e}\| \leq \gamma h^4,$$

where γ is a finite constant independent of h.

Proof We have, $\|A\| = \frac{19}{6}$ and $\|BA\| \leq \frac{19}{6}\| \int_a^b |k(.,s)|ds\|$.

Then, by Lemma 6.1, Eq. (6.21) has a unique solution $\widetilde{X}$.

On the other hand, there exists a constant $c > 0$ such that $\|x - Q_3 x\| \leq ch^4$.

Then, from (6.18) and (6.19), we have for $X = (x_0, x_1, \ldots, x_n)^T$

$$\|X - \widetilde{X}\| \leq c\| \int_a^b |k(.,s)|ds\| \|(I - BA)^{-1}\| h^4,$$

which implies that $\|X - \widetilde{X}\| \leq \frac{c\| \int_a^b |k(.,s)|ds\|}{1 - \frac{19}{6}\| \int_a^b |k(.,s)|ds\|} h^4$.

We deduce that

$$\begin{aligned}
\|\widehat{e}\| &\leq \|x - Qx\| + \|Qx - \widetilde{Q}x\| \\
&\leq ch^4 + \|A\| \|X - \widetilde{X}\| \\
&\leq ch^4 + \frac{\frac{19}{6} c\| \int_a^b |k(.,s)|ds\|}{1 - \frac{19}{6}\| \int_a^b |k(.,s)|ds\|} h^4.
\end{aligned}$$

Thus, the proof is completed by taking $\gamma = c + \frac{\frac{19}{6} c\| \int_a^b |k(.,s)|ds\|}{1 - \frac{19}{6}\| \int_a^b |k(.,s)|ds\|}$. □

Remark 6.2 The condition on the kernel k, given in the above theorem, can be always satisfied by considering the integral equation with a kernel $\tilde{k} = \lambda.k$, where λ is a constant which can be chosen such that $\lambda\| \int_a^b |k(.,s)|ds\| < 1$.

6.3.2 Quasi-Interpolation Method 2

Now, in order to present the second technique, we first need to write the quasi-interpolant Q_3 under the quasi-Lagrange form:

$$Q_3 x = \sum_{i=0}^{n} x_i L_i$$

where the quasi-Lagrange functions L_i are linear combinations of some B-splines B_j. More precisely, they are given by:

$$
\begin{aligned}
L_0 &= B_1 + \tfrac{11}{36}B_2 - \tfrac{5}{36}B_3 + \tfrac{1}{36}B_4,\\
L_1 &= \tfrac{4}{3}B_2 + \tfrac{11}{9}B_3 - \tfrac{5}{18}B_4 + \tfrac{1}{36}B_5,\\
L_2 &= -B_2 + \tfrac{3}{2}B_4 - \tfrac{5}{18}B_5 + \tfrac{1}{36}B_6,\\
L_3 &= \tfrac{4}{9}B_2 - \tfrac{1}{9}B_3 - \tfrac{5}{18}B_4 + \tfrac{3}{2}B_5 - \tfrac{5}{18}B_6 + \tfrac{1}{36}B_7,\\
L_4 &= -\tfrac{1}{12}B_2 + \tfrac{1}{36}B_3 + \tfrac{1}{36}B_4 - \tfrac{5}{18}B_5 + \tfrac{3}{2}B_6 - \tfrac{5}{18}B_7 + \tfrac{1}{36}B_8,\\
L_i &= \tfrac{1}{36}B_i - \tfrac{5}{18}B_{i+1} + \tfrac{3}{2}B_{i+2} - \tfrac{5}{18}B_{i+3} + \tfrac{1}{36}B_{i+4}, \text{ for } i = 5, \cdots, n-5,\\
L_{n-4} &= -\tfrac{1}{12}B_{n+2} + \tfrac{1}{36}B_{n+1} + \tfrac{1}{36}B_n - \tfrac{5}{18}B_{n-1} + \tfrac{3}{2}B_{n-2} - \tfrac{5}{18}B_{n-3} + \tfrac{1}{36}B_{n-4},\\
L_{n-3} &= \tfrac{4}{9}B_{n+2} - \tfrac{1}{9}B_{n+1} - \tfrac{5}{18}B_n + \tfrac{3}{2}B_{n-1} - \tfrac{5}{18}B_{n-2} + \tfrac{1}{36}B_{n-3},\\
L_{n-2} &= -B_{n+2} + \tfrac{3}{2}B_n - \tfrac{5}{18}B_{n-1} + \tfrac{1}{36}B_{n-2},\\
L_{n-1} &= \tfrac{4}{3}B_{n+2} + \tfrac{11}{9}B_{n+1} - \tfrac{5}{18}B_n + \tfrac{1}{36}B_{n-1},\\
L_n &= B_{n+3} + \tfrac{11}{36}B_{n+2} - \tfrac{5}{36}B_{n+1} + \tfrac{1}{36}B_n.
\end{aligned}
$$

This form of Q_3 in terms of quasi-Lagrange functions allows to easily compute its infinity norm. Indeed, it is well known that $\|Q_3\|_\infty$ is equal to the Chebyshev norm of the Lebesgue function:

$$\Lambda_3 = \sum_{i=0}^{n} \|L_i\|.$$

Hence, it was shown in [10] that

$$\|Q_3\|_\infty = \Lambda_3 = 2.24236.$$

Then, the approximation of the univariate function $t \to k(t, s)$ by the above quasi-interpolant has the form:

$$k(t, s) \approx \tilde{k}(t, s) = \sum_{i=0}^{n} k(t_i, s)L_i(t).$$

Therefore, the approximate solution $\tilde{x}_n(t)$ satisfies the following equation:

$$
\begin{aligned}
\tilde{x}_n(t) &= f(t) + \textstyle\sum_{i=0}^{n} L_i(t) \displaystyle\int_a^b k(t_i, s)\tilde{x}_n(s)ds\\
&= f(t) + \textstyle\sum_{i=0}^{n} \alpha_i L_i(t).
\end{aligned}
$$

By substituting the above expression of $\tilde{x}_n(t)$ in the approximate integral equation of (1.1), we obtain:

$$\sum_{i=0}^{n} \alpha_i L_i(t) = \sum_{i=0}^{n} \left[\int_a^b k(t_i, s)f(s)ds + \sum_{j=0}^{n} \alpha_j \int_a^b k(t_i, s)L_j(s)ds \right] L_i(t).$$

As the quasi-Lagrange splines (L_i) are linearly independent, we deduce that:

$$\alpha_i - \sum_{j=0}^{n} \alpha_j \int_a^b k(t_i, s) L_j(s) ds = \int_a^b k(t_i, s) f(s) ds, \ \forall\, 0 \le i \le n.$$

In the matrix form, the above system becomes:

$$(I - A)\alpha = d, \tag{6.22}$$

where $A = (a_{i,j})_{0 \le i,j \le n}$, $\alpha = (\alpha_0, \cdots, \alpha_n)^T$, $d = (d_0, \cdots, d_n)^T$, with $a_{i,j} = \int_a^b k(t_i, s) L_j(s) ds$ and $d_i = \int_a^b k(t_i, s) f(s) ds$.

Now we have the following result.

Theorem 6.3 *Assume that $k \in C^4(I \times I)$ and $\Lambda_3 \| \int_a^b |k(., s)| ds \|_\infty < 1$. Then Eq. (6.22) defines a unique approximation $\tilde{x}_n(t) = f(t) + \sum_{i=0}^{n} \alpha_i L_i(t)$ of $x(t)$, and the resulting error $\tilde{e} := x - \tilde{x}_n$ satisfies*

$$\|\tilde{e}\| \le C h^4,$$

where C is a finite constant independent of h.

Proof It is simple to verify that $\|A\|_\infty = \Lambda_3 \| \int_a^b |k(., s)| ds \|_\infty$. Then, according to the above assumption, Eq. (6.22) has a unique solution.

On the other hand, as $\tilde{k}(., s) = Q_3 k(., s)$, we have:

$$|k(t, s) - \tilde{k}(., s)| \le C_0 h^4 \|D^{4,0} k\|_{\infty,\Omega}, \ \ \Omega = I \times I, \tag{6.23}$$

where C_0 is a constant independent of h, and $D^{4,0}$ denotes the forth derivative of k with respect to the first variable.

Using the fact that $\sum_{i=0}^{n} L_i(t) = 1$, we can easily get:

$$x(t) - \tilde{x}_n(t) = \sum_{j=0}^{n} L_j(t) \left[\int_a^b k(t, s)(x(s) - \tilde{x}_n(s)) ds \right.$$
$$\left. + \int_a^b (k(t, s) - k(t_j, s)) \tilde{x}_n(s) ds \right].$$

Then we obtain

$$\|x - \tilde{x}_n\|_\infty \leq \Lambda_3 \| \int_a^b |k(., s)| ds \|_\infty \|x - \tilde{x}_n\|_\infty + C_1 h^4,$$

where $C_1 = \Lambda_3 C_0 (b - a) \|\tilde{x}_n\|_\infty \|D^{4,0} k\|_{\infty,\Omega}$.

From the expression of $\tilde{x}_n$, we can easily prove that

$$\|\tilde{x}_n\|_\infty \leq \frac{\|f\|_\infty}{1 - \Lambda_3 \| \int_a^b |k(., s)| ds \|_\infty}.$$

Hence, the proof is completed by taking $C = \dfrac{\Lambda_3 C_0 (b - a) \|f\|_\infty \|D^{4,0} k\|_{\infty,\Omega}}{(1 - \Lambda_3 \| \int_a^b |k(., s)| ds \|_\infty)^2}$. □

6.4 Numerical Examples

In this section, we present an illustration of the obtained theoretical results. More precisely, we give three numerical examples, where the exact solutions x are already known. In each example, we calculate the error between x and the approximate solution for $t = 0, 0.1, 0.2, 0.3, 0.4, 0.5, 0.6, 0.7, 0.8, 0.9, 1$ and $n = 10, 20$, by using both methods developed in the previous sections. We use the same examples as in [8] for comparing the new numerical results with the ones described in [8].

In order to present the absolute errors, we use the abbreviation **PGS** for the Pretty Good Slopes spline cubic interpolation method and **SSQI1** (resp. **SSQI2**) for the first (resp. second) method based on the Superconvergent Spline Quasi-Interpolant Q_3.

Example 6.1 Consider the following integral equation

$$x(t) = f(t) + \int_0^1 k(t, s) x(s) ds, \ t \in [0, 1].$$

with $k(t, s) = \frac{1}{12} \frac{st-1}{1+t^2}$ and $f(t)$ is chosen so that the exact solution of this equation is $x(t) = \sin(t) + 1$.

Example 6.2 Consider the following integral equation

$$x(t) = f(t) + \int_0^1 k(t, s) x(s) ds, \ t \in [0, 1].$$

with $k(t,s) = \frac{1}{11}\sin(st+3s)$ and $f(t)$ is chosen so that the exact solution of this equation is $x(t) = 2\cos(t) + 2$.

Example 6.3 Consider the following integral equation

$$x(t) = f(t) + \int_0^1 k(t,s)x(s)ds, \; t \in [0,1].$$

with $k(t,s) = \frac{(s-t)\sin(s-t)}{12(1+t)}$ and $f(t)$ is chosen so that the exact solution of this equation is $x(t) = \frac{t^5}{20} - \frac{t^4}{12}$.

The numerical results related to the above examples are illustrated in Tables 6.1, 6.2, and 6.3 below where, for different values of n, we give the absolute errors $|x(t_i) - Px(t_i)|$, $t_i = ih, 0 \le i \le 10$ and $h = \frac{1}{10}$. Here, P denotes one of the considered three operators **PGS**, **SSQI1** or **SSQI2**.

Table 6.1 Absolute errors for Example 6.1

t	PGS	SSQI1	SSQI2	PGS	SSQI1	SSQI2
0	0.42×10^{-7}	0.54×10^{-8}	0.28×10^{-7}	0.23×10^{-8}	0.32×10^{-9}	0.67×10^{-9}
0.1	0.40×10^{-7}	0.51×10^{-8}	0.27×10^{-7}	0.31×10^{-8}	0.17×10^{-8}	0.21×10^{-7}
0.2	0.37×10^{-7}	0.49×10^{-7}	0.58×10^{-6}	0.27×10^{-8}	0.25×10^{-9}	0.12×10^{-8}
0.3	0.32×10^{-7}	0.29×10^{-8}	0.21×10^{-6}	0.28×10^{-8}	0.22×10^{-9}	0.37×10^{-8}
0.4	0.28×10^{-7}	0.19×10^{-8}	0.23×10^{-6}	0.14×10^{-8}	0.18×10^{-9}	0.37×10^{-8}
0.5	0.24×10^{-7}	0.10×10^{-8}	0.15×10^{-6}	0.14×10^{-8}	0.14×10^{-9}	0.23×10^{-8}
0.6	0.21×10^{-7}	0.22×10^{-9}	0.57×10^{-7}	0.79×10^{-9}	0.10×10^{-9}	0.72×10^{-9}
0.7	0.16×10^{-7}	0.57×10^{-9}	0.12×10^{-7}	0.71×10^{-9}	0.76×10^{-10}	0.35×10^{-9}
0.8	0.14×10^{-7}	0.31×10^{-7}	0.19×10^{-6}	0.29×10^{-9}	0.47×10^{-10}	0.83×10^{-9}
0.9	0.10×10^{-7}	0.16×10^{-8}	0.11×10^{-7}	0.17×10^{-8}	0.84×10^{-9}	0.45×10^{-8}
1	0.92×10^{-8}	0.13×10^{-8}	0.94×10^{-8}	0.68×10^{-9}	0.60×10^{-10}	0.15×10^{-9}

Table 6.2 Absolute errors for Example 6.2

t	PGS	SSQI1	SSQI2	PGS	SSQI1	SSQI2
0	0.15×10^{-7}	0.15×10^{-7}	0.10×10^{-9}	0.55×10^{-9}	0.94×10^{-9}	0.60×10^{-11}
0.1	0.15×10^{-7}	0.15×10^{-7}	0.10×10^{-9}	0.12×10^{-8}	0.12×10^{-8}	0.66×10^{-10}
0.2	0.14×10^{-7}	0.37×10^{-7}	0.20×10^{-8}	0.13×10^{-8}	0.10×10^{-8}	0.74×10^{-11}
0.3	0.13×10^{-7}	0.22×10^{-7}	0.14×10^{-9}	0.43×10^{-9}	0.10×10^{-8}	0.72×10^{-11}
0.4	0.13×10^{-7}	0.21×10^{-7}	0.12×10^{-9}	0.15×10^{-8}	0.98×10^{-9}	0.71×10^{-11}
0.5	0.12×10^{-5}	0.20×10^{-7}	0.10×10^{-9}	0.52×10^{-9}	0.94×10^{-9}	0.69×10^{-11}
0.6	0.12×10^{-7}	0.19×10^{-7}	0.89×10^{-10}	0.80×10^{-9}	0.90×10^{-9}	0.67×10^{-11}
0.7	0.11×10^{-7}	0.18×10^{-7}	0.71×10^{-10}	0.63×10^{-9}	0.86×10^{-9}	0.64×10^{-11}
0.8	0.10×10^{-7}	0.52×10^{-7}	0.19×10^{-8}	0.12×10^{-9}	0.82×10^{-8}	0.62×10^{-11}
0.9	0.15×10^{-7}	0.10×10^{-7}	0.82×10^{-10}	0.23×10^{-8}	0.15×10^{-8}	0.56×10^{-10}
1	0.10×10^{-7}	0.96×10^{-8}	0.80×10^{-10}	0.53×10^{-8}	0.65×10^{-9}	0.68×10^{-11}

Table 6.3 Absolute errors for Example 6.3

t	PGS	SSQI1	SSQI2	PGS	SSQI1	SSQI2
0	0.87×10^{-8}	0.31×10^{-8}	0.76×10^{-11}	0.98×10^{-9}	0.45×10^{-9}	0.43×10^{-12}
0.1	0.23×10^{-8}	0.21×10^{-8}	0.44×10^{-11}	0.25×10^{-9}	0.83×10^{-8}	0.10×10^{-9}
0.2	0.10×10^{-7}	0.27×10^{-6}	0.18×10^{-8}	0.63×10^{-9}	0.22×10^{-9}	0.17×10^{-10}
0.3	0.32×10^{-7}	0.93×10^{-9}	0.69×10^{-9}	0.14×10^{-9}	0.14×10^{-9}	0.10×10^{-10}
0.4	0.58×10^{-7}	0.69×10^{-9}	0.41×10^{-9}	0.85×10^{-9}	0.83×10^{-10}	0.60×10^{-11}
0.5	0.89×10^{-7}	0.63×10^{-9}	0.25×10^{-9}	0.34×10^{-8}	0.39×10^{-10}	0.36×10^{-11}
0.6	0.12×10^{-6}	0.70×10^{-9}	0.16×10^{-9}	0.43×10^{-8}	0.99×10^{-11}	0.22×10^{-11}
0.7	0.15×10^{-6}	0.90×10^{-9}	0.11×10^{-9}	0.52×10^{-8}	0.79×10^{-11}	0.13×10^{-11}
0.8	0.18×10^{-6}	0.27×10^{-6}	0.22×10^{-9}	0.62×10^{-8}	0.15×10^{-10}	0.73×10^{-12}
0.9	0.22×10^{-6}	0.15×10^{-8}	0.10×10^{-10}	0.83×10^{-8}	0.86×10^{-8}	0.34×10^{-11}
1	0.24×10^{-6}	0.19×10^{-8}	0.10×10^{-10}	0.83×10^{-8}	0.58×10^{-11}	0.53×10^{-12}

Remark 6.3 The Fredholm integral equations corresponding to the numerical examples, described above, are chosen so that their associated exact solutions are known. Then, for the comparison of the studied methods, one can use in this case either absolute or relative error. Although the relative error is more accurate, we prefer to use the absolute error in order to compare the obtained results with those published in [8].

In Figs. 6.1 and 6.2 below, we present a comparison between the errors, corresponding to Example 4.3, obtained by using similar operators described here and in [10]. Through these figures, we remark that the superconvergent approximants are better than the classical ones.

6.5 Conclusion

In this paper, we have proposed two main methods for approximating the solution of Fredholm linear integral equations based on superconvergent cubic splines which have been recently introduced in the papers [10] and [20]. We notice that these methods are of great interest for solving many numerical analysis problems. Their efficiency lies in their simplicity and robustness. These properties derive from the simplicity, stability and high order of convergence of cubic splines associated with the superconvergence phenomenon. According to the above theoretical and numerical results, the two methods based on the quasi-interpolation are simple and easy to implement. Moreover, the numerical tests show that they are better in comparison with the first method which has a global character. On the other hand, although the superconvergence is satisfied only at the knots, the constructed operators improve the rate of convergence on the whole interval of study. This fact is confirmed by comparing our results published recently in [8] with the above ones.

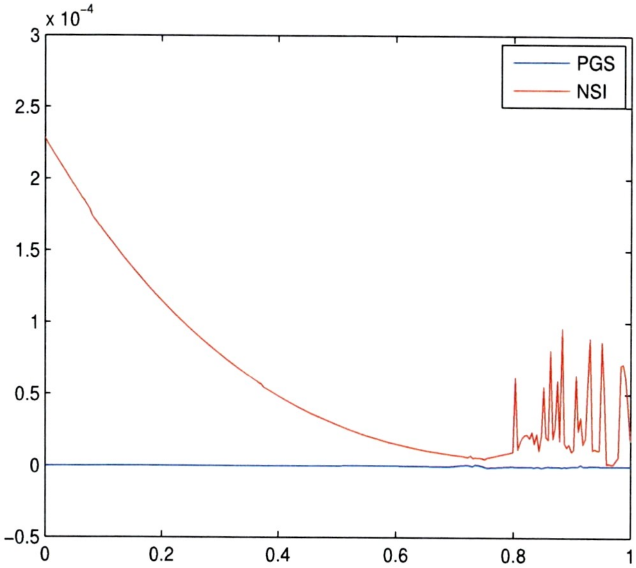

Fig. 6.1 Errors corresponding to the natural (NSI) (resp. pretty good slopes (PGS)) cubic spline interpolation methods

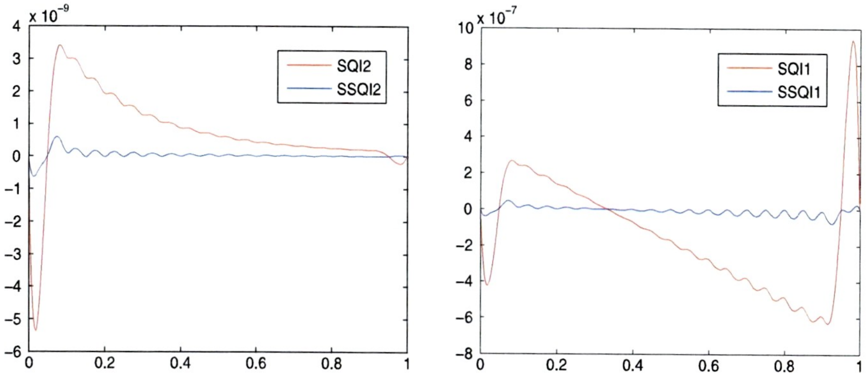

Fig. 6.2 Errors corresponding to the classical (SQI1 and SQI2) and the superconvergent (SSQI1 and SSQI2) quasi-interpolation methods

Acknowledgment This work was partially supported by the PHC Toubkal Project 2020 No 43594NB.

References

1. Abbasbandy, S., Shivanian, E.: A new analytical technique to solve Fredholm integral equations. Numer. Algorithms **56**, 27–43 (2011)
2. Allouch, C., Sablonnière, P.: Iteration methods for Fredholm integral equations of the second kind based on spline quasi-interpolants. Math. Comput. Simul. **99**, 19–27 (2014)
3. Allouch, C., Sablonnière, P., Sbibih, D.: Solving Fredholm integral equations by approximating kernels by spline quasi-interpolants. Numer. Algorithms **56**, 437–453 (2011)
4. Allouch, C., Sbibih, D., Sablonnière, P.: A modified Kulkarni's method based on a discrete spline quasi-interpolant. Math. Comput. Simul. **81**, 1991–2000 (2011)
5. Allouch, C., Sbibih, D., Sablonnière, P.: A collocation method for the numerical solution of a two dimensional integral equation using a quadratic spline quasiinterpolant. Numer. Algorithms **62**, 445–468 (2013)
6. Atkinson, K.E.: The Numerical Solution of Integral Equations of the Second Kind. Cambridge University Press, Cambridge (1997)
7. Atkinson, K., Graham, I., Sloan, I.: Piecewise continuous collocation for integral equations. SIAM J. Numer. Anal. **20**, 172–186 (1983)
8. Bellour, A., Sbibih, D., Zidna, A.: Two cubic spline methods for solving Fredholm integral equations. Appl. Math. Comput. **276**, 1–11 (2016)
9. Bica, A.M.: The numerical method of successive interpolations for Fredholm functional integral equations. Numer. Algorithms **58**, 351–377 (2011)
10. Boujraf, A., Sbibih, D., Tahrichi, M., Tijini, A.: A superconvergent cubic spline quasi-interpolant and application. Afr. Mat. **26**, 1531–1547 (2015)
11. Brutman, L.: An application of the generalized alternating polynomials to the numerical solution of Fredholm integral equations. Numer. Algorithms **5**, 437–442 (1993)
12. DeVore, R.A., Lorentz, G.G.: Constructive Approximation. Springer, Berlin (1993)
13. Fangy, Q., Yamamoto, T.: Superconvergence of finite difference approximations for convection diffusion problems. Numer. Linear Algebra Appl. **8**, 99–110 (2001)
14. Hochstadt, H.: Integral Equations. Wiley, New York (1993)
15. Kress, R.: Linear Integral Equations. Springer, New York (1991)
16. Kulkarni, R.P.: A superconvergence result for solutions of compact operator equations. Bull. Austral. Math. Soc. **68**, 517–528 (2003)
17. Liu, J., Zhang, X., Yu, D.: Superconvergence and ultraconvergence of Newton-Cotes rules for supersingular integrals. J. Comput. Appl. Math. **233**, 2841–2854 (2010)
18. Liu, D., Wu, J., Yu, D.: The superconvergence of the Newton-Cotes rule for Cauchy principal value integrals. J. Comput. Appl. Math. **235**, 696–707 (2010)
19. Maleknejad, K., Mahmoudi, Y.: Numerical solution of linear Fredholm integral equation by using hybrid Taylor and block-pulse functions. Appl. Math. Comput. **149**, 799–806 (2004)
20. Sablonnière, P.: Gradient approximation on uniform meshes by finite differences and cubic spline interpolation. In: Hancock, E.R., Martin, R.R., Sabin, M.A. (eds.) Mathematics of Surfaces XIII. NCS, vol. 5654, pp. 322–334. Springer, Berlin (2009)
21. Sablonnière, P., Sbibih, D., Tahrichi, M.: High-order quadrature rules based on spline quasi-interpolants and application to integral equations. Appl. Numer. Math. **62**, 507–520 (2012)
22. Sloan, J.H., Wendland, W.L.: Superconvergence of the Galerkin iterates for integral equations of the second kind. J. Integral Equ. **9**, 1–23 (1985)
23. Usmani, R.: Inversion of a tridiagonal Jacobi matrix. Linear Algebra Appl. **212/213**, 413–414 (1994)
24. Wahlbin, L.B.: Local behaviour in finite element methods. In: Ciarlet, P.G., Lione, J.L. (eds.) Handbook of Numerical Analysis, Vol II, Finite Element Methods (Part 1), pp. 355–522. Elsevier, North Holland (1991)
25. Zeidler, E.: Nonlinear Functional Analysis and Its Applications. Springer, Berlin (1991)
26. Zhang, X., Wu, J., Yu, D.: Superconvergence of the composite Simpson's rule for a certain finite-part integral and its applications. J. Comput. Appl. Math. **223**, 598–613 (2009)

Chapter 7
The Completely Discretized Problem of the Dual Mixed Formulation for the Heat Diffusion Equation in a Polygonal Domain by the Crank-Nicolson Scheme in Time

Reda Korikache and Luc Paquet

Abstract The purpose of this paper is to prove *a priori* error estimates for the completely discretized problem of the dual mixed formulation for the heat diffusion equation in a polygonal domain. We complete the discretization of the problem (Farhloul et al., Functional Analysis and Evolution Equations. The Günter Lumer Volume, p. 240, Birkhäuser, Basel, 2007) in time by using the Crank-Nicolson scheme and we show the existence, the stability and a priori error estimates for the solution of the completely discretized problem.

Keywords Dual mixed finite element method · Heat diffusion equation · Singularities · Grids refinements · A priori error estimates

2010 MSC: 35K05, 58J35

7.1 Introduction

The aim of this paper is to study the completely discretized problem of the dual mixed formulation for the heat evolution equation in a polygonal domain Ω of $\mathbb{R}^2$. Let us note that we have dealt in [2] with the mixed dual semi-discretized formulation in space. In this work, we will complete the discretization of the problem in time by using the Crank-Nicolson scheme. For this completely

R. Korikache (✉)
LANO, ESTO, University Mohammed 1, Oujda, Morocco
e-mail: r.korikache@ump.ac.ma

L. Paquet
LAMAV, Polytechnic University Hauts-De-France, Valenciennes, France
e-mail: Luc.Paquet@uphf.fr

D. Barrera et al. (eds.), *Mathematical and Computational Methods for Modelling, Approximation and Simulation*, SEMA SIMAI Springer Series 29,
https://doi.org/10.1007/978-3-030-94339-4_7

discretized problem, we are going to demonstrate the existence and the uniqueness of the solution. Then, we will show the stability of this scheme. Finally, we will prove, under some conditions of meshing refinement in the reentrant corner of the polygonal domain, a priori error estimates of order 1 in space, and 2 in time for the solution of the completely discretized problem when using the Crank-Nicolson implicit scheme.

7.2 The Model Problem

Let Ω be a bounded polygonal domain of $\mathbb{R}^2$. In particular the boundary of $\Omega : \partial\Omega \equiv \Gamma = \cup_{j=1}^{N} \overline{\Gamma}_j$ for some $N \in \mathbb{N}$, where Γ_j is an open segment of a straight line of the plane $\mathbb{R}^2$, $\forall\ j = 1, 2, \ldots, N$. As it is well known, the geometric singularities of the domain (the angles) induce in general singularities on the solution of the Cauchy problem with Dirichlet boundary condition for the heat diffusion equation (see for example the books of P. Grisvard [3, 4]). As shown in [4] and [3], we may suppose without harming to the generality that Ω has only one nonconvex angle, in other words one reentrant corner, and that its vertex is located at the origin. In the following, we denote by ω ($\omega > \pi$) the measure of that angle. For a fixed $T > 0$, let us set $Q := \Omega\times]0, T[$ and let us denote by $\Sigma := \Gamma\times]0, T[$ the lateral boundary of the cylinder Q. We introduce the following weighted Sobolev space (see [3], definition 8.4.1.1 and lemma 8.4.1.2 p. 388):

$$H^{2,\alpha}(\Omega) = \{v \in H^1(\Omega);\ r^\alpha D^\beta v \in L^2(\Omega),\ \forall\beta \in \mathbb{N}^2 : |\beta| = 2\},$$

which is a Hilbert space for the norm

$$\|v\|_{2,\alpha,\Omega} = (\|v\|_{1,\Omega}^2 + |v|_{2,\alpha,\Omega}^2)^{1/2},$$

where the semi-norm $|.|_{2,\alpha,\Omega}$ is defined by

$$|v|_{2,\alpha,\Omega} = \left(\sum_{|\beta|=2} \left\| r^\alpha D^\beta v \right\|_{0,\Omega}^2\right)^{1/2},$$

r denotes the distance to the origin of $\mathbb{R}^2$. Let us recall that:

$$\mathring{H}^1(\Omega) = \{v \in H^1(\Omega);\ v_{/\partial\Omega} = 0\}.$$

Let us consider the Cauchy problem for the heat diffusion equation in Ω up to time T: given the right-hand side $f \in L^2\left(0, T; L^2(\Omega)\right)$ and the initial condition $g \in$

$\overset{\circ}{H}{}^1(\Omega)$, find $u \in H^1\left(0, T; L^2(\Omega)\right) \cap L^2(0, T; \overset{\circ}{H}{}^1(\Omega))$ weak solution of:

$$\begin{cases} u_t(x,t) - \Delta u(x,t) = f(x,t) \text{ in } Q, \\ u(x,t) = 0 \text{ on } \Sigma, \\ u(x,0) = g(x)\,, \quad \text{for } x \in \Omega. \end{cases} \tag{7.1}$$

We have seen in Proposition 3 of [2], that Problem (7.1) admits a unique solution:

$$u \in H^1(0, T; L^2(\Omega)) \cap L^2(0, T; H^{2,\alpha}(\Omega)) \cap L^2(0, T; \overset{\circ}{H}{}^1(\Omega)),$$

with $\alpha \in \left]1 - \frac{\pi}{\omega}, 1\right[$.

In the following, we will consider the additional unknown heat flux $\mathbf{p} := \nabla u$ (rigorously speaking, the heat flux density vector is the opposite [10]). We have proved in [2, Theorem 4] that the couple $(\mathbf{p}, u) \in L^2(0, T; H(\text{div}, \Omega)) \times H^1(0, T; L^2(\Omega))$ is the unique solution of the dual mixed formulation:

$$\begin{cases} \int_\Omega \mathbf{p}(t).\mathbf{q}\, dx + \int_\Omega u(t)\ \text{div}\,\mathbf{q}\, dx = 0, \quad \forall \mathbf{q} \in H(\text{div}, \Omega), \quad \text{for a.e. } t \in I, \\ \int_\Omega v\ \text{div}\,\mathbf{p}(t) dx = -\int_\Omega (f(t) - u_t(t))\ v\, dx, \quad \forall v \in L^2(\Omega), \quad \text{for a.e. } t \in I, \\ u(0) = g \in \overset{\circ}{H}{}^1(\Omega). \end{cases} \tag{7.2}$$

7.3 The Completely Discretized Problem

Let us consider a family of triangulations $(\mathcal{T}_h)_{h>0}$ on $\overline{\Omega}$. For a triangle K belonging to the triangulation $\mathcal{T}_h$, let us denote by h_K the diameter of K and by ρ_K the interior diameter of K, i.e. the diameter of the largest of disc included in K. As in Theorem 8.4.1.6 p. 392 of [3], we suppose that the family of triangulations $(\mathcal{T}_h)_{h>0}$ has the property that $\max\limits_{K \in \mathcal{T}_h} \frac{h_K}{\rho_K}$ is bounded by a positive constant independent of the parameter h; in that case, one says usually that the family of triangulations is regular (see for example [1] (17.1) p. 131). In accordance with the tradition (see [1] remark 17.1 p. 131) the same letter h may have also another significance: it may denote instead: $h =: \max_{K \in \mathcal{T}_h} h_K$. The true significance of h is always clear from the context.

For the discretization in space, let us recall that we are using the approximant spaces:

$$X_h := \left\{ \mathbf{q}_h \in H(\mathrm{div};\Omega); \forall K \in \mathcal{T}_h \ : \ \mathbf{q}_{h/K} \in RT_0(K) \right\},$$

$$M_h := \left\{ v_h \in L^2(\Omega); \forall K \in \mathcal{T}_h \, , \ v_{h/K} \in P_0\,(K) \right\},$$

where $RT_0(K) := P_0(K)^2 \oplus P_0(K) \binom{x_1}{x_2}$ denotes the real vector space of dimension three of the so called Raviart-Thomas vectorfields of degree 0 on the triangle K ($RT_0(K)$ is denoted $D_1(K)$ in [7, p. 550]), and $P_0\,(K)$ is the real vector space of dimension one of the constant functions on the triangle K. For the complete discretized problem, we use the time subdivision of the interval $\left[0, T\right]$ into N subintervals $\left[t_{n-1}, t_n\right]$ $(N \geq 2)$, such that: $0 = t_0 \leq \cdots \leq t_n \leq \cdots \leq t_N = T$. $\Delta t = t_n - t_{n-1}$ denotes the fixed time step. Let us denote by u_h^n the approximation of the temperature at time $t_n = n\Delta t$ in M_h. For the approximation of $\frac{\partial u}{\partial t}$ at the time t_n, we use the following formula:

$$\bar{\partial} u_h^n = \frac{(u_h^n - u_h^{n-1})}{\Delta t}.$$

7.4 The Crank-Nicolson Scheme

Before giving the complete discretized mixed formulation for the heat diffusion equation by the Crank-Nicolson scheme, we will first define new variables. We set,

$$t_{n-\frac{1}{2}} = \frac{t_n + t_{n-1}}{2}, \ \mathbf{p}_h^{n-\frac{1}{2}} = \frac{\mathbf{p}_h^n + \mathbf{p}_h^{n-1}}{2}, \text{ and } u_h^{n-\frac{1}{2}} = \frac{u_h^n + u_h^{n-1}}{2}. \qquad (7.3)$$

Let us consider the following problem:

$$\begin{cases} \int_\Omega \mathbf{p}_h^{n-\frac{1}{2}} . \mathbf{q}_h \, dx + \int_\Omega u_h^{n-\frac{1}{2}} \ \mathrm{div}\, \mathbf{q}_h \, dx = 0, \quad \forall \mathbf{q}_h \in X_h, \ \forall n \geq 1, \\ \int_\Omega v_h \ \mathrm{div}\, \mathbf{p}_h^{n-\frac{1}{2}} dx = - \int_\Omega (f(t_{n-\frac{1}{2}}) - \bar{\partial} u_h^n) \, v_h \, dx, \quad \forall v_h \in M_h, \ \forall n \geq 1, \\ u_h^0 \, , \text{ given.} \end{cases} \qquad (7.4)$$

Let us note that in (7.4) appear $\mathbf{p}_h^{n-\frac{1}{2}}, u_h^{n-\frac{1}{2}}, u_h^n$ and u_h^{n-1}. Thus, we can choose as unknowns $\mathbf{p}_h^{n-\frac{1}{2}}$ and u_h^n for $n \geq 1$. Another alternative is to consider that the unknowns are the $\mathbf{p}_h^n$ for $n \geq 0$ and u_h^n for $n \geq 1$. u_h^0 denotes the initial condition that is known and $\mathbf{p}_h^0$ is exceptionally defined by the equation:

$$\int_\Omega \mathbf{p}_h^0 . \mathbf{q}_h \, dx + \int_\Omega u_h^0 \operatorname{div} \mathbf{q}_h \, dx = 0, \quad \forall \mathbf{q}_h \in X_h. \tag{7.5}$$

This last choice of unknowns presents the following advantages:

- Traditional unknowns.
- Symmetry of "**p**" and "u" in the problem.

Proposition 7.1 *Problem* (7.4) *admits one and only one solution* $(\mathbf{p}_h^n, u_h^n)_{n \in IN}$.

Proof Let us start by demonstrating the unicity. Therefore, let us show that if $(\mathbf{p}_h^n, u_h^n) \in X_h \times M_h$ verifies :

$$\begin{cases} \int_\Omega \mathbf{p}_h^{n-\frac{1}{2}} . \mathbf{q}_h \, dx + \int_\Omega u_h^{n-\frac{1}{2}} \operatorname{div} \mathbf{q}_h \, dx = 0, \quad \forall \mathbf{q}_h \in X_h, \ \forall n \geq 1 \\ \int_\Omega v_h \operatorname{div} \mathbf{p}_h^{n-\frac{1}{2}} dx = \int_\Omega \overline{\partial} u_h^n \, v_h \, dx, \quad \forall v_h \in M_h, \ \forall n \geq 1 \text{ and } u_h^0 = 0, \end{cases} \tag{7.6}$$

then $(\mathbf{p}_h^n, u_h^n) = 0$. By taking $\mathbf{q}_h = \mathbf{p}_h^{n-\frac{1}{2}}$ in the first equation of (7.6) and $v_h = u_h^{n-\frac{1}{2}}$ in the second equation of (7.6), we obtain:

$$\int_\Omega \left| \mathbf{p}_h^{n-\frac{1}{2}} \right|^2 dx + \int_\Omega \overline{\partial} u_h^n \, u_h^{n-\frac{1}{2}} \, dx = 0. \tag{7.7}$$

On the other hand we have:

$$\overline{\partial} u_h^n \, u_h^{n-\frac{1}{2}} = \frac{1}{2\Delta t} \left(u_h^n - u_h^{n-1} \right) \left(u_h^n + u_h^{n-1} \right) = \frac{1}{2\Delta t} \left(u_h^n \right)^2, \tag{7.8}$$

on the condition of having already demonstrated that $u_h^{n-1} = 0$. So let us take $n = 1$ in (7.8) and since $u_h^0 = 0$ (initial condition), consequently $u_h^1 = 0$, and $\mathbf{p}_h^{\frac{1}{2}} = 0$ by Eq. (7.7) the second term in the left-hand side of Eq. (7.7) being also a priori non-negative due to [6].

According to (7.5), $u_h^0 = 0$ implies that $\mathbf{p}_h^0 = 0$. Knowing already that $\mathbf{p}_h^{\frac{1}{2}} = 0$, it follows that $\mathbf{p}_h^1 = 0$. Thus $\mathbf{p}_h^2 = 0$ and $u_h^2 = 0$ by (7.7) and (7.8) with $n = 2$ and so on for every $n \geq 1$.

For the existence, it is well known by the Riesz Representation Theorem applied to Eq. (7.5) that $\mathbf{p}_h^0$ exists (see Remark 1.5.2 [5, p. 31]). According to the system

(7.4) with $n = 1$, and with the aim of constructing $\mathbf{p}_h^1$ and u_h^1, u_h^0 and $\mathbf{p}_h^0$ being known, we have

$$\begin{cases} \int_\Omega \mathbf{p}_h^1 . \mathbf{q}_h \, dx + \int_\Omega u_h^1 \operatorname{div} \mathbf{q}_h \, dx = \int_\Omega \mathbf{p}_h^0 . \mathbf{q}_h \, dx + \int_\Omega u_h^0 \operatorname{div} \mathbf{q}_h \, dx, \forall \mathbf{q}_h \in X_h, \\ \int_\Omega v_h \operatorname{div} \mathbf{p}_h^1 dx - \frac{2}{\Delta t} \int_\Omega u_h^1 \, v_h \, dx = - \int_\Omega v_h \operatorname{div} \mathbf{p}_h^0 dx - \frac{2}{\Delta t} \int_\Omega u_h^0 \, v_h \, dx \\ -2 \int_\Omega f\left(t_{1/2}\right) \, v_h \, dx, \; \forall v_h \in M_h. \end{cases} \tag{7.9}$$

Let Φ_h: $X_h \times M_h \longmapsto X_h' \times M_h'$, be the application defined by:

$$\left(\mathbf{p}_h^1, u_h^1\right) \longmapsto \left(\mathbf{q}_h \longmapsto \int_\Omega \mathbf{p}_h^1 . \mathbf{q}_h \, dx + \int_\Omega u_h^1 \, \operatorname{div} \mathbf{q}_h \, dx, \right.$$
$$\left. v_h \longmapsto \int_\Omega v_h \, \operatorname{div} \mathbf{p}_h^1 dx - \frac{2}{\Delta t} \int_\Omega u_h^1 \, v_h \, dx \right).$$

Let us prove that it is an isomorphism. Since Φ_h is linear of $X_h \times M_h$ in its dual, and since the two spaces $X_h \times M_h$ and $X_h' \times M_h'$ have the same dimension, it suffices to show that Φ_h is injective to prouve its bijectivity. Let $\left(\mathbf{p}_h^1, u_h^1\right)$ such that:

$$\int_\Omega \mathbf{p}_h^1 . \mathbf{q}_h \, dx + \int_\Omega u_h^1 \, \operatorname{div} \mathbf{q}_h \, dx = 0, \; \forall \mathbf{q}_h \in X_h, \tag{7.10}$$

$$\int_\Omega v_h \, \operatorname{div} \mathbf{p}_h^1 dx - \frac{2}{\Delta t} \int_\Omega u_h^1 \, v_h \, dx = 0, \; \forall v_h \in M_h. \tag{7.11}$$

In (7.10), let us take $\mathbf{q}_h = \mathbf{p}_h^1$ and $v_h = u_h^1$ in (7.11), it follows that :

$$\int_\Omega \left|\mathbf{p}_h^1\right|^2 \, dx + \frac{2}{\Delta t} \int_\Omega \left|u_h^1\right|^2 \, dx = 0. \tag{7.12}$$

We obtain $\mathbf{p}_h^1 = 0$ and $u_h^1 = 0$. It follows from that the injectivity and thus the bijectivity of Φ_h. So it suffices to apply Φ_h^{-1} to the couple of linear forms defined by the two members of the right hand side of the system of Eqs. (7.9). In a similar way, we construct $\left(\mathbf{p}_h^2, u_h^2\right)$ by considering the system (7.4), with $n = 2$ and so on. □

7.4.1 Stability

Now we will give the stability result of the Crank-Nicolson scheme.

Theorem 7.1 *Suppose $\Delta t \leq \frac{1}{2}$. There exists a constant $c > 0$ independent of h such that:*

$$\left\| u_h^N \right\|_{0,\Omega}^2 \lesssim \left\| u_h^0 \right\|_{0,\Omega}^2 + \Delta t \left\| \mathbf{p}_h^0 \right\|_{0,\Omega}^2 + \sum_{n=1}^{N} \Delta t \left\| f(t_{n-\frac{1}{2}}) \right\|_{0,\Omega}^2, \tag{7.13}$$

and

$$\left\| \mathbf{p}_h^N \right\|_{0,\Omega} \leq \left\| \mathbf{p}_h^0 \right\|_{0,\Omega} + \sqrt{\frac{T}{2}} \max_{n=1,\ldots,N} \| f(t_n) \|_{0,\Omega}. \tag{7.14}$$

Proof For the proof of (7.13), we use in particular the discrete inequality of Gronwall [6, p. VI-9], with:

$$\begin{cases} \varphi_n = \left\| u_h^n \right\|_{0,\Omega}^2 + \Delta t \left\| \mathbf{p}_h^n \right\|_{0,\Omega}^2 \\ m_0 = 2 \\ m_1 = \cdots = m_{N-1} = 2\Delta t \\ C = 2 \left\| u_h^0 \right\|_{0,\Omega}^2 + \Delta t \left\| \mathbf{p}_h^0 \right\|_{0,\Omega}^2 + 2\Delta t \sum_{n=1}^{N} \left\| f(t_{n-\frac{1}{2}}) \right\|_{0,\Omega}^2 \end{cases}$$

For the complete demonstration of (7.13) see the proof of Theorem 1.5.15 [5, pp. 52–54]. For the proof of (7.14), see the proof of Proposition 1.5.16 [5, p. 55]. □

7.4.2 Error Estimates on the Temperature and on the Heat Flux Density Vector

In order to demonstrate the results related to the error estimate, proceeding similarly as in the semi-discrete case [2], we decompose the error: $u(t_n) - u_h^n$ into the sum of $u(t_n) - \tilde{u}_h(t_n)$ and $\tilde{u}_h(t_n) - u_h^n$, where $(\tilde{\mathbf{p}}_h(t_n), \tilde{u}_h(t_n)) \in X_h \times M_h$ is the solution of the "elliptic projection problem" at time t_n : find $(\tilde{\mathbf{p}}_h(t_n), \tilde{u}_h(t_n)) \in X_h \times M_h$ solution of,

$$\begin{cases} \int_\Omega \tilde{\mathbf{p}}_h(t_n).\mathbf{q}_h \, dx + \int_\Omega \tilde{u}_h(t_n) \ \operatorname{div} \mathbf{q}_h \, dx = 0, \quad \forall \mathbf{q}_h \in X_h, \\ \int_\Omega v_h \ \operatorname{div} \ \tilde{\mathbf{p}}_h(t_n) dx = - \int_\Omega (f(t_n) - u_t(t_n)) \, v_h \, dx, \quad \forall v_h \in M_h. \end{cases} \tag{7.15}$$

The elliptic problem (7.15) being true for n and $n - 1$, making the sum, we obtain:

$$\begin{cases} \int_\Omega \frac{\tilde{\mathbf{p}}_h(t_n)+\tilde{\mathbf{p}}_h(t_{n-1})}{2}.\mathbf{q}_h \, dx + \int_\Omega \frac{\tilde{u}_h(t_n)+\tilde{u}_h(t_{n-1})}{2} \operatorname{div} \mathbf{q}_h \, dx = 0, \quad \forall \mathbf{q}_h \in X_h, \\ \int_\Omega v_h \operatorname{div} \frac{\tilde{\mathbf{p}}_h(t_n)+\tilde{\mathbf{p}}_h(t_{n-1})}{2} dx = - \int_\Omega \left(\frac{f(t_n)+f(t_{n-1})}{2} - \frac{u_t(t_n)+u_t(t_{n-1})}{2} \right) v_h \, dx, \forall v_h \in M_h. \end{cases}$$

We may rewrite the Crank-Nicolson scheme, with u_h^0 , given by:

$$\begin{cases} \int_\Omega \mathbf{p}_h^{n-\frac{1}{2}}.\mathbf{q}_h \, dx + \int_\Omega u_h^{n-\frac{1}{2}} \operatorname{div} \mathbf{q}_h \, dx = 0, \quad \forall \mathbf{q}_h \in X_h, \forall n \geq 1 \\ \int_\Omega v_h \operatorname{div} \mathbf{p}_h^{n-\frac{1}{2}} dx = - \int_\Omega (f(t_{n-\frac{1}{2}}) - \overline{\partial} u_h^n) \, v_h \, dx, \quad \forall v_h \in M_h, \ \forall n \geq 1 \end{cases}$$

Let us set $\theta_h^n := u_h^n - \tilde{u}_h(t_n)$ and $\boldsymbol{\varepsilon}_h^n := \mathbf{p}_h^n - \tilde{\mathbf{p}}_h(t_n)$. Then, by subtraction we obtain the following system:

$$\begin{cases} \int_\Omega \frac{\boldsymbol{\varepsilon}_h^n + \boldsymbol{\varepsilon}_h^{n-1}}{2}.\mathbf{q}_h \, dx + \int_\Omega \frac{\theta_h^n + \theta_h^{n-1}}{2} \operatorname{div} \mathbf{q}_h \, dx = 0, \quad \forall \mathbf{q}_h \in X_h, \\ \int_\Omega v_h \operatorname{div} \frac{\boldsymbol{\varepsilon}_h^n + \boldsymbol{\varepsilon}_h^{n-1}}{2} \, dx \\ \quad = - \int_\Omega (f(t_{n-\frac{1}{2}}) - \frac{f(t_n)+f(t_{n-1})}{2} - (\overline{\partial} u_h^n - \frac{u_t(t_n)+u_t(t_{n-1})}{2})) \, v_h dx. \ \forall v_h \in M_h, \end{cases} \tag{7.16}$$

Proposition 7.2 *Assume* $f, \frac{df}{dt} \in H^1(0,T; L^2(\Omega))$ *and* $\Delta g + f(0) \in \mathring{H}^1(\Omega)$ *as well as* $\Delta\left(\Delta g + f(0)\right) + \frac{df}{dt}(0) \in \mathring{H}^1(\Omega)$ *. Then*

$$u_{ttt} \in L^2(0,T; L^2(\Omega)).$$

Proof Let $w \in H^1(0,T; L^2(\Omega)) \cap L^2(0,T; L^2(\Omega))$ be the solution of the heat diffusion equation

$$\begin{cases} \frac{dw}{dt}(t) = \Delta w(t) + \frac{d^2 f}{dt^2}(t), \ \forall' t \in [0,T] \\ w(0) = \Delta\left(\Delta g + f(0)\right) + \frac{df}{dt}(0). \end{cases}$$

Set $v(t) = \int_0^t w(s)\, ds + \Delta g + f(0)$, $\frac{dv}{dt}(t) = w(t)$ and $v(0) = \Delta g + f(0) \in \mathring{H}^1(\Omega)$.

By integrating the equation $\frac{dw}{dt}(s) = \Delta w(s) + \frac{d^2 f}{dt^2}(s)$, $\forall' s \in [0,T]$ from 0 to t, we obtain:

$$w(t) - w(0) = \Delta\left(v(t) - \Delta g - f(0)\right) + \frac{df}{dt}(t) - \frac{df}{dt}(0)$$

$$i.e. \ \frac{dv}{dt}(t) - \Delta\left(\Delta g + f(0)\right) - \frac{df}{dt}(0) = \Delta v(t) - \Delta\left(\Delta g + f(0)\right) + \frac{df}{dt}(t) - \frac{df}{dt}(0).$$

Then, by simplifying the 2 members we obtain that v is the solution of the following Cauchy problem:

$$\begin{cases} \frac{dv}{dt}(t) = \Delta v(t) + \frac{df}{dt}(t) \\ v(0) = \Delta g + f(0) \in \overset{\circ}{H}{}^1(\Omega). \end{cases}$$

According to the proof of Proposition 8 in [2], we have $v = \frac{du}{dt}$. Thus $\frac{d^2u}{dt^2} = \frac{dv}{dt} = w \in H^1(0, T; L^2(\Omega))$. And consequently $u_{ttt} \in L^2(0, T; L^2(\Omega))$. □

Theorem 7.2 *Under the hypotheses of Proposition 7.2, there exists a constant $c > 0$ independent of h and of k such that:*

$$\left\|u_h^n - \tilde{u}_h(t_n)\right\|_{0,\Omega} \leq ch\left(\|u_0\|_{H^{2,\alpha}(\Omega)} + \int_0^{t_n} \left\|u_t(s)\right\|_{H^{2,\alpha}(\Omega)} ds\right) + 2\Delta t^2 \times$$
$$\left(\int_0^{t_n} \left\|u_{ttt}(s)\right\|_{0,\Omega} ds + \int_0^{t_n} \left\|f_{tt}(s)\right\|_{0,\Omega} ds\right). \tag{7.17}$$

Proof The first step of this demonstration is to bound $\|\theta_h^n\|$ in terms of $\left\|\theta_h^{n-1}\right\|$ and $\|\omega^n\|$. For this, let's take $v_h = \theta_h^n + \theta_h^{n-1}$ in the second equation of the system (7.16) and $\mathbf{q}_h = \boldsymbol{\varepsilon}_h^n + \boldsymbol{\varepsilon}_h^{n-1}$ in the first. We have:

$$\int_\Omega \frac{\left|\boldsymbol{\varepsilon}_h^n + \boldsymbol{\varepsilon}_h^{n-1}\right|^2}{2} dx = \int_\Omega \left(f(t_{n-\frac{1}{2}}) - \frac{f(t_n) + f(t_{n-1})}{2}\right. \tag{7.18}$$
$$\left. - \left(\overline{\partial} u_h^n - \frac{u_t(t_n) + u_t(t_{n-1})}{2}\right)\right) \times \left(\theta_h^n + \theta_h^{n-1}\right) dx.$$

And as we have :

$$\begin{aligned} \overline{\partial} u_h^n - \frac{u_t(t_n) + u_t(t_{n-1})}{2} &= \overline{\partial}\theta_h^n + \tilde{u}(t_n) - \tilde{u}(t_{n-1}) - \frac{u_t(t_n) + u_t(t_{n-1})}{2} \\ &= \overline{\partial}\theta_h^n + (R_h - I)\,\overline{\partial} u(t_n) + \overline{\partial} u(t_n) \\ &\quad - \frac{u_t(t_n) + u_t(t_{n-1})}{2}, \end{aligned}$$

where, with analogy to the book of Vidar Thomée [8], "$R_h()$" denotes here the component in M_h of the couple of $X_h \times M_h$ "elliptic projection of" (cf definition 6 of [2]). Thus,

$$\begin{aligned}
&\overline{\partial} u_h^n - \frac{u_t(t_n) + u_t(t_{n-1})}{2} + \frac{f(t_n) + f(t_{n-1})}{2} - f(t_{n-\frac{1}{2}}) \\
&= \overline{\partial}\theta_h^n + (R_h - I)\,\overline{\partial} u(t_n) + \left(\overline{\partial} u(t_n) - u_t(t_{n-\frac{1}{2}})\right) + u_t(t_{n-\frac{1}{2}}) \\
&-\frac{u_t(t_n) + u_t(t_{n-1})}{2} + \frac{f(t_n) + f(t_{n-1})}{2} - f(t_{n-\frac{1}{2}}) \qquad (7.19)\\
&= \overline{\partial}\theta_h^n + (R_h - I)\,\overline{\partial} u(t_n) + \left(\overline{\partial} u(t_n) - u_t(t_{n-\frac{1}{2}})\right) \\
&+\Delta\left[u(t_{n-\frac{1}{2}}) - \frac{1}{2}\left(u(t_n) + u(t_{n-1})\right)\right],
\end{aligned}$$

because $\Delta u(t) + f(t) = u_t(t)$, for every $t > 0$.

Let us set $\widetilde{\omega}^n := \widetilde{\omega}_1^n + \widetilde{\omega}_2^n + \widetilde{\omega}_3^n$, with:

$$\begin{aligned}
\widetilde{\omega}_1^n &:= (R_h - I)\,\overline{\partial} u(t_n), \\
\widetilde{\omega}_2^n &:= \left(\overline{\partial} u(t_n) - u_t(t_{n-\frac{1}{2}})\right), \\
\widetilde{\omega}_3^n &:= \Delta\left[u(t_{n-\frac{1}{2}}) - \frac{1}{2}\left(u(t_n) + u(t_{n-1})\right)\right].
\end{aligned}$$

Then, it follows from (7.19) and (7.18), that,

$$\int_\Omega \frac{\left|\boldsymbol{\varepsilon}_h^n + \boldsymbol{\varepsilon}_h^{n-1}\right|^2}{2}\,dx = -\int_\Omega (\overline{\partial}\theta_h^n + \widetilde{\omega}^n)\left(\theta_h^n + \theta_h^{n-1}\right)\,dx. \qquad (7.20)$$

Thus, we obtain:

$$\int_\Omega \overline{\partial}\theta_h^n\left(\theta_h^n + \theta_h^{n-1}\right)\,dx \leq -\frac{1}{2}\left\|\boldsymbol{\varepsilon}_h^n + \boldsymbol{\varepsilon}_h^{n-1}\right\|^2 + \|\widetilde{\omega}^n\|\left\|\theta_h^n + \theta_h^{n-1}\right\|.$$

As

$$\int_\Omega \overline{\partial}\theta_h^n\left(\theta_h^n + \theta_h^{n-1}\right)\,dx = \frac{\|\theta_h^n\|^2 - \left\|\theta_h^{n-1}\right\|^2}{\Delta t}, \qquad (7.21)$$

we deduce that :

$$\left\|\theta_h^n\right\|_{0,\Omega}^2 - \left\|\theta_h^{n-1}\right\|_{0,\Omega}^2 \leq \Delta t \left(-\frac{1}{2}\left\|\boldsymbol{\varepsilon}_h^n + \boldsymbol{\varepsilon}_h^{n-1}\right\|_{0,\Omega}^2 + \left\|\widetilde{\omega}^n\right\|_{0,\Omega}\left\|\theta_h^n + \theta_h^{n-1}\right\|_{0,\Omega}\right). \tag{7.22}$$

And so, a fortiori we get

$$\left\|\theta_h^n\right\|^2 - \left\|\theta_h^{n-1}\right\|^2 \leq \Delta t \left\|\widetilde{\omega}^n\right\|_{0,\Omega}\left(\left\|\theta_h^n\right\|_{0,\Omega} + \left\|\theta_h^{n-1}\right\|_{0,\Omega}\right). \tag{7.23}$$

Thus,

$$\left\|\theta_h^n\right\|_{0,\Omega} \leq \left\|\theta_h^{n-1}\right\|_{0,\Omega} + \Delta t \left\|\widetilde{\omega}^n\right\|_{0,\Omega}, \tag{7.24}$$

so that it suffices to bound $\widetilde{\omega}^n$. Let us start with $\widetilde{\omega}_2^n$. By definition we have

$$\begin{aligned}
\Delta t \left\|\widetilde{\omega}_2^n\right\|_{0,\Omega} &= \Delta t \left\|\overline{\partial} u(t_n) - u_t(t_{n-\frac{1}{2}})\right\|_{0,\Omega} \\
&= \Delta t \left\|\frac{u(t_n) - u(t_{n-1})}{\Delta t} - u_t(t_{n-\frac{1}{2}})\right\|_{0,\Omega} \\
&= \left\|u(t_n) - u(t_{n-1}) - \Delta t\, u_t(t_{n-\frac{1}{2}})\right\|_{0,\Omega}.
\end{aligned}$$

Using Taylor's formula, we get

$$u(t_n) = u(t_{n-\frac{1}{2}}) + \frac{\Delta t}{2}\, u_t(t_{n-\frac{1}{2}}) + \frac{\Delta t^2}{8}\, u_{tt}(t_{n-\frac{1}{2}}) + \frac{1}{2}\int_{t_{n-\frac{1}{2}}}^{t_n} (t_n - s)^2 u_{ttt}(s)\, ds.$$

Hence, at the time t_{n-1}, we have

$$u(t_{n-1}) = u(t_{n-\frac{1}{2}}) - \frac{\Delta t}{2}\, u_t(t_{n-\frac{1}{2}}) + \frac{\Delta t^2}{8}\, u_{tt}(t_{n-\frac{1}{2}}) + \frac{1}{2}\int_{t_{n-\frac{1}{2}}}^{t_{n-1}} (t_{n-1} - s)^2 u_{ttt}(s)\, ds. \tag{7.25}$$

Let us consider the difference of these two above equalities, we obtain

$$\begin{aligned}
&u(t_n) - u(t_{n-1}) - \Delta t\, u_t(t_{n-\frac{1}{2}}) \\
&= \frac{1}{2}\int_{t_{n-\frac{1}{2}}}^{t_n} (t_n - s)^2 u_{ttt}(s)\, ds - \frac{1}{2}\int_{t_{n-\frac{1}{2}}}^{t_{n-1}} (t_{n-1} - s)^2 u_{ttt}(s)\, ds.
\end{aligned}$$

It follows that

$$\begin{aligned}
&\left\| u(t_n) - u(t_{n-1}) - \Delta t \; u_t(t_{n-\frac{1}{2}}) \right\|_{0,\Omega} \\
&\leq \frac{1}{2} \int_{t_{n-\frac{1}{2}}}^{t_n} (t_n - s)^2 \left\| u_{ttt}(s) \right\|_{0,\Omega} ds + \frac{1}{2} \int_{t_{n-1}}^{t_{n-\frac{1}{2}}} (t_{n-1} - s)^2 \left\| u_{ttt}(s) \right\|_{0,\Omega} ds \\
&\leq \frac{\Delta t^2}{8} \int_{t_{n-\frac{1}{2}}}^{t_n} \left\| u_{ttt}(s) \right\|_{0,\Omega} ds + \frac{\Delta t^2}{8} \int_{t_n}^{t_{n-1}} \left\| u_{ttt}(s) \right\|_{0,\Omega} ds \\
&= \frac{\Delta t^2}{8} \int_{t_{n-1}}^{t_n} \left\| u_{ttt}(s) \right\|_{0,\Omega} ds.
\end{aligned}$$

Thus we have demonstrated that

$$\Delta t \; \left\| \widetilde{\omega}_2^n \right\|_{0,\Omega} \leq \frac{\Delta t^2}{8} \int_{t_{n-1}}^{t_n} \left\| u_{ttt}(s) \right\|_{0,\Omega} ds. \tag{7.26}$$

Now, let us try to bound $\left\| \widetilde{\omega}_3^n \right\|_{0,\Omega}$. The Taylor formula gives:

$$\begin{aligned}
\frac{1}{2} u(t_n) &= \frac{1}{2} u(t_{n-\frac{1}{2}}) + \frac{\Delta t}{4} \; u_t(t_{n-\frac{1}{2}}) + \frac{1}{2} \int_{t_{n-\frac{1}{2}}}^{t_n} (t_n - s) u_{tt}(s) ds, \\
\frac{1}{2} u(t_{n-1}) &= \frac{1}{2} u(t_{n-\frac{1}{2}}) - \frac{\Delta t}{4} \; u_t(t_{n-\frac{1}{2}}) + \frac{1}{2} \int_{t_{n-\frac{1}{2}}}^{t_{n-1}} (t_n - s) u_{tt}(s) ds.
\end{aligned}$$

By summing these two above equalities we obtain:

$$u(t_{n-\frac{1}{2}}) - \frac{1}{2} \left(u(t_n) + u(t_{n-1}) \right) = -\frac{1}{2} \int_{t_{n-\frac{1}{2}}}^{t_n} (t_n - s) u_{tt}(s) ds - \frac{1}{2} \int_{t_{n-\frac{1}{2}}}^{t_{n-1}} (t_{n-1} - s) u_{tt}(s) ds.$$

So, by applying the Laplace operator Δ, we obtain

$$\begin{aligned}
&\left\| \Delta \left[u(t_{n-\frac{1}{2}}) - \frac{1}{2} \left(u(t_n) + u(t_{n-1}) \right) \right] \right\|_{0,\Omega} \\
&\leq \frac{1}{2} \int_{t_{n-\frac{1}{2}}}^{t_n} (t_n - s) \left\| \Delta u_{tt}(s) \right\|_{0,\Omega} ds + \frac{1}{2} \int_{t_{n-1}}^{t_{n-\frac{1}{2}}} \left| t_{n-1} - s \right| \left\| \Delta u_{tt}(s) \right\|_{0,\Omega} ds \\
&\leq \frac{\Delta t}{4} \int_{t_{n-1}}^{t_n} \left\| \Delta u_{tt}(s) \right\|_{0,\Omega} ds.
\end{aligned}$$

Consequently,

$$\Delta t \, \left\| \widetilde{\omega}_3^n \right\|_{0,\Omega} \leq \frac{\Delta t^2}{4} \int_{t_{n-1}}^{t_n} \left\| \Delta u_{tt}(s) \right\|_{0,\Omega} ds. \tag{7.27}$$

We still have to bound $\left\| \widetilde{\omega}_1^n \right\|_{0,\Omega}$. Let us recall that

$$\widetilde{\omega}_1^n := (R_h - I) \, \overline{\partial} u(t_n) = (R_h - I) \, \frac{u(t_n) - u(t_{n-1})}{\Delta t}.$$

Thus, by using Proposition 12 in [2, p. 252], there exists a constant $c > 0$ independent of h such that:

$$\begin{aligned}
\Delta t \, \left\| \widetilde{\omega}_1^n \right\|_{0,\Omega} &\leq ch \left\| u(t_n) - u(t_{n-1}) \right\|_{H^{2,\alpha}(\Omega)} \\
&= ch \left\| \int_{t_n-1}^{t_n} u_t(s) ds \right\|_{H^{2,\alpha}(\Omega)}.
\end{aligned}$$

Consequently

$$\Delta t \, \left\| \widetilde{\omega}_1^n \right\|_{0,\Omega} \leq ch \int_{t_n-1}^{t_n} \left\| u_t(s) \right\|_{H^{2,\alpha}(\Omega)} ds. \tag{7.28}$$

According to the inequality (7.24), we have that:

$$\begin{aligned}
\left\| \theta_h^n \right\|_{0,\Omega} &\leq \left\| \theta_h^{n-1} \right\|_{0,\Omega} + \Delta t \, \left\| \widetilde{\omega}^n \right\|_{0,\Omega} \\
&\leq \left\| \theta_h^{n-2} \right\|_{0,\Omega} + \Delta t \left(\left\| \widetilde{\omega}^{n-1} \right\|_{0,\Omega} + \left\| \widetilde{\omega}^n \right\|_{0,\Omega} \right) \\
&\leq \left\| \theta_h^{n-3} \right\|_{0,\Omega} + \Delta t \left(\left\| \widetilde{\omega}^{n-2} \right\|_{0,\Omega} + \left\| \widetilde{\omega}^{n-1} \right\|_{0,\Omega} + \left\| \widetilde{\omega}^n \right\|_{0,\Omega} \right) \\
&\;\;\vdots \\
&\leq \left\| \theta_h^0 \right\|_{0,\Omega} + \Delta t \sum_{i=1}^{n} \left\| \widetilde{\omega}^i \right\|_{0,\Omega},
\end{aligned}$$

recalling that $\theta_h^0 = u_h^0 - \tilde{u}_h(0) = 0$. By using inequalities (7.26), (7.27) and (7.28), we get

$$\Delta t \, \left\| \widetilde{\omega}_2^n \right\|_{0,\Omega} \leq \frac{(\Delta t)^2}{8} \int_{t_{n-1}}^{t_n} \left\| u_{ttt}(s) \right\|_{0,\Omega} ds, \tag{7.29}$$

$$\Delta t\ \left\|\widetilde{\omega}_3^n\right\|_{0,\Omega} \le \frac{(\Delta t)^2}{4}\int_{t_{n-1}}^{t_n} \left\|\Delta u_{tt}(s)\right\|_{0,\Omega} ds, \tag{7.30}$$

$$\Delta t\ \left\|\widetilde{\omega}_1^n\right\|_{0,\Omega} \le ch\int_{t_{n-1}}^{t_n} \left\|u_t(s)\right\|_{H^{2,\alpha}(\Omega)} ds. \tag{7.31}$$

Consequently recalling that $\widetilde{\omega}^n = \widetilde{\omega}_1^n + \widetilde{\omega}_2^n + \widetilde{\omega}_3^n$, we obtain

$$\left\|\theta_h^n\right\|_{0,\Omega} = \left\|u_h^n - \tilde{u}_h(t_n)\right\|_{0,\Omega} \le ch\int_{t_0}^{t_n} \left\|u_t(s)\right\|_{H^{2,\alpha}(\Omega)} ds +$$
$$(\Delta t)^2\left(\int_{t_0}^{t_n} \left\|u_{ttt}(s)\right\|_{0,\Omega} ds + \int_{t_0}^{t_n} \left\|\Delta u_{tt}(s)\right\|_{0,\Omega} ds\right).$$

And since $\Delta u_{tt}(s) = \frac{d^2}{dt^2}\Delta u(s) = u_{ttt}(s) - f_{tt}(s)$, we can replace $\Delta u_{tt}(s)$ by $u_{ttt}(s) - f_{tt}(s)$ in the above inequality.

Finally, we get the following inequality:

$$\left\|u_h^n - \tilde{u}_h(t_n)\right\|_{0,\Omega} \le ch\left(\int_0^{t_n} \left\|u_t(s)\right\|_{H^{2,\alpha}(\Omega)} ds\right) +$$
$$2\,\Delta t^2\left(\int_0^{t_n} \left\|u_{ttt}(s)\right\|_{0,\Omega} ds + \int_0^{t_n} \left\|f_{tt}(s)\right\|_{0,\Omega} ds\right).$$

□

Theorem 7.3 *Let $\{\mathcal{T}_h\}$ be a regular family of triangulations on $\overline{\Omega}$, satisfying the properties (i) and (ii) of Proposition 9 of [2, p. 250]. Under the hypotheses of Proposition 7.2, and for $\alpha \in \left]1 - \frac{\pi}{w}, 1\right[$, there exists a constant $c > 0$ independent of h such that for every $n \ge 1$, we have*

$$\left\|u(t_n) - u_h^n\right\|_{0,\Omega}$$
$$\le c\,h\left(|u(t_n)|_{H^1(\Omega)} + |u(t_n)|_{H^{2,\alpha}(\Omega)} + \int_0^{t_n} \left\|u_t(s)\right\|_{H^{2,\alpha}(\Omega)} ds\right)$$
$$+ 2\,(\Delta t)^2\left(\int_0^{t_n} \left\|u_{ttt}(s)\right\|_{0,\Omega} ds + \int_0^{t_n} \left\|f_{tt}(s)\right\|_{0,\Omega} ds\right). \tag{7.32}$$

Proof It suffices to apply the triangular inequality:

$$\left\|u(t_n) - u_h^n\right\|_{0,\Omega} \le \left\|u(t_n) - \tilde{u}_h(t_n)\right\|_{0,\Omega} + \left\|\tilde{u}_h(t_n) - u_h^n\right\|_{0,\Omega}.$$

By the inequality (7.17) and the inequality 5.6 of Proposition 9 in [2], we obtain the result. □

Similarly to the bound obtained in the implicit case, and in order to demonstrate the error estimate on $\mathbf{p}_h^n$, we need an analogous result to Proposition 1-5-10 of [5, p. 42], but for here $\widetilde{\omega}^n := \widetilde{\omega}_1^n + \widetilde{\omega}_2^n + \widetilde{\omega}_3^n$.

Proposition 7.3 *Let us suppose that* $f \in H^1(0, T; L^2(\Omega))$ *and* $\Delta g + f(0) \in \mathring{H}^1(\Omega)$. *Then,*

$$\bar{\partial} \|\boldsymbol{\varepsilon}_h^n\|^2 \leq \|\widetilde{\omega}^n\|^2, \quad \text{with } \boldsymbol{\varepsilon}_h^n := \mathbf{p}_h^n - \tilde{\mathbf{p}}_h(t_n). \tag{7.33}$$

Proof Let us consider the Crank-Nicolson scheme for the mixed method, written in the following form:

$$\begin{cases} \int_\Omega \mathbf{p}_h^n . \mathbf{q}_h \, dx + \int_\Omega u_h^n \operatorname{div} \mathbf{q}_h \, dx = 0, \quad \forall \mathbf{q}_h \in X_h, \quad \forall n \geq 1 \\ \int_\Omega v_h \operatorname{div} \frac{\mathbf{p}_h^n + \mathbf{p}_h^{n-1}}{2} \, dx = -\int_\Omega (f(t_{n-\frac{1}{2}}) - \bar{\partial} u_h^n) \, v_h \, dx, \quad \forall v_h \in M_h, \forall n \geq 1. \end{cases} \tag{7.34}$$

By subtracting member by member from the first equation of (7.34), the first equation defining the elliptic projection:

$$\int_\Omega \tilde{\mathbf{p}}_h(t_n) . \mathbf{q}_h \, dx + \int_\Omega \tilde{u}_h(t_n) \operatorname{div} \mathbf{q}_h \, dx = 0, \quad \forall \mathbf{q}_h \in X_h,$$

we obtain,

$$\int_\Omega \boldsymbol{\varepsilon}_h^n . \mathbf{q}_h \, dx + \int_\Omega \theta_h^n \operatorname{div} \mathbf{q}_h \, dx = 0, \quad \forall \mathbf{q}_h \in X_h, \tag{7.35}$$

where $\boldsymbol{\varepsilon}_h^n := \mathbf{p}_h^n - \tilde{\mathbf{p}}_h(t_n)$ and $\theta_h^n := u_h^n - \tilde{u}_h(t_n)$. Equation (7.35) being true for n and $n-1$, by making the difference member by member and by dividing by the time step, we obtain:

$$\int_\Omega \bar{\partial} \boldsymbol{\varepsilon}_h^n . \mathbf{q}_h \, dx + \int_\Omega \bar{\partial} \theta_h^n \operatorname{div} \mathbf{q}_h \, dx = 0, \quad \forall \mathbf{q}_h \in X_h. \tag{7.36}$$

Taking $\mathbf{q}_h = \boldsymbol{\varepsilon}_h^n + \boldsymbol{\varepsilon}_h^{n-1}$ in Eq. (7.36), we obtain:

$$\|\boldsymbol{\varepsilon}_h^n\|_{0,\Omega}^2 - \left\|\boldsymbol{\varepsilon}_h^{n-1}\right\|_{0,\Omega}^2 = -\Delta t \int_\Omega \operatorname{div} \left(\boldsymbol{\varepsilon}_h^n + \boldsymbol{\varepsilon}_h^{n-1}\right) \bar{\partial} \theta_h^n \, dx. \tag{7.37}$$

By the equalities (7.16) and (7.19), it follows that:

$$\int_\Omega v_h \ \mathrm{div}\, \frac{\boldsymbol{\varepsilon}_h^n + \boldsymbol{\varepsilon}_h^{n-1}}{2} \ dx = \int_\Omega \left(\overline{\partial}\theta_h^n + \widetilde{\omega}^n\right) \ v_h \, dx\,,$$

$\forall v_h \in M_h$. In particular, if we choose $v_h = 1_K$, for any $K \in \mathcal{T}_h$, we obtain

$$\mathrm{div}\, \frac{\boldsymbol{\varepsilon}_h^n + \boldsymbol{\varepsilon}_h^{n-1}}{2} = P_h^0 \left(\overline{\partial}\theta_h^n + \widetilde{\omega}^n\right) = \overline{\partial}\theta_h^n + P_h^0\, \widetilde{\omega}^n\,, \tag{7.38}$$

since $\overline{\partial}\theta_h^n = \frac{\theta_h^n - \theta_h^{n-1}}{\Delta t} \in M_h$. From (7.37) and (7.38), follows that

$$\left\|\boldsymbol{\varepsilon}_h^n\right\|_{0,\Omega}^2 - \left\|\boldsymbol{\varepsilon}_h^{n-1}\right\|_{0,\Omega}^2 = -2\,\Delta t \left\|\overline{\partial}\theta_h^n\right\|^2 - 2\,\Delta t \int_\Omega P_h^0\, \widetilde{\omega}^n \ \overline{\partial}\theta_h^n \, dx \tag{7.39}$$

$$\leq \Delta t \left\|P_h^0\, \widetilde{\omega}^n\right\|_{0,\Omega}^2 + \Delta t \left\|\overline{\partial}\theta_h^n\right\|_{0,\Omega}^2 - 2\,\Delta t \left\|\overline{\partial}\theta_h^n\right\|^2 \tag{7.40}$$

$$\leq \Delta t \left\|\widetilde{\omega}^n\right\|_{0,\Omega}^2 . \tag{7.41}$$

By dividing both sides of the above inequality (7.41) by the time step Δt, we obtain: $\overline{\partial}\|\boldsymbol{\varepsilon}_h^n\|^2 \leq \|\widetilde{\omega}^n\|^2$. □

Corollary 7.1 *Under the hypotheses of Proposition 7.2, there exists a constant $c > 0$ independent of h and Δt such that:*

$$\|\boldsymbol{\varepsilon}_h^n\|^2 \leq ch^2 \int_0^{t_n} \|u_t(s)\|_{H^{2,\alpha}(\Omega)}^2 \, ds$$
$$+ c\,(\Delta t)^4 \left(\int_0^{t_n} \|u_{ttt}(s)\|_{0,\Omega}^2 \, ds + \int_0^{t_n} \|\Delta u_{tt}(s)\|_{0,\Omega}^2 \, ds\right). \tag{7.42}$$

Proof By inequality (7.31), we have:

$$\Delta t \left\|\widetilde{\omega}_1^j\right\|_{0,\Omega} \leq ch \int_{t_j - 1}^{t_j} \|u_t(s)\|_{H^{2,\alpha}(\Omega)} \, ds. \tag{7.43}$$

Thus:

$$\Delta t \sum_{j=1}^{j=n} \left\|\widetilde{\omega}_1^j\right\|^2 \leq c \frac{h^2}{\Delta t} \sum_{j=1}^{j=n} \left(\int_{t_{j-1}}^{t_j} \|u_t(s)\|_{H^{2,\alpha}(\Omega)} \, ds\right)^2$$

$$\leq ch^2 \sum_{j=1}^{j=n} \int_{t_{j-1}}^{t_j} \|u_t(s)\|_{H^{2,\alpha}(\Omega)}^2 \, ds$$

$$= h^2 \int_{t_0}^{t_n} \|u_t(s)\|_{H^{2,\alpha}(\Omega)}^2 \, ds\ . \tag{7.44}$$

By (7.29), we have $\Delta t \left\|\widetilde{\omega}_2^j\right\|_{0,\Omega} \leq \frac{\Delta t^2}{8} \int_{t_{j-1}}^{t_j} \left\|u_{ttt}(s)\right\|_{0,\Omega} ds.$

It follows from a similar calculus using the inequality of Cauchy-Schwartz:

$$\Delta t \sum_{j=1}^{j=n} \left\|\widetilde{\omega}_2^j\right\|^2 \leq c\Delta t^4 \int_{t_0}^{t_n} \left\|u_{ttt}(s)\right\|_{0,\Omega}^2 ds. \tag{7.45}$$

And also for (7.30), we obtain:

$$\Delta t \sum_{j=1}^{j=n} \left\|\widetilde{\omega}_3^j\right\|^2 \leq c\Delta t^4 \int_{t_0}^{t_n} \left\|\Delta u_{tt}(s)\right\|_{0,\Omega}^2 ds. \tag{7.46}$$

Now, by using Proposition 7.3, we get

$$\begin{cases} \left\|\boldsymbol{\varepsilon}_h^1\right\|^2 - \left\|\boldsymbol{\varepsilon}_h^0\right\|^2 \leq \Delta t \left\|\widetilde{\omega}^1\right\|^2, \\ \left\|\boldsymbol{\varepsilon}_h^2\right\|^2 - \left\|\boldsymbol{\varepsilon}_h^1\right\|^2 \leq \Delta t \left\|\widetilde{\omega}^2\right\|^2, \\ \quad \vdots \qquad\qquad\qquad \vdots \\ \left\|\boldsymbol{\varepsilon}_h^n\right\|^2 - \left\|\boldsymbol{\varepsilon}_h^{n-1}\right\|^2 \leq \Delta t \left\|\widetilde{\omega}^n\right\|^2. \end{cases} \tag{7.47}$$

It follows that by summing up these inequalities, by (7.5) and $u_h^0 = \tilde{u}_h(0)$:

$$\|\boldsymbol{\varepsilon}_h^n\|^2 \leq \Delta t \sum_{j=1}^{j=n} \left\|\widetilde{\omega}^j\right\|^2. \tag{7.48}$$

Since $\widetilde{\omega}^n = \widetilde{\omega}_1^n + \widetilde{\omega}_2^n + \widetilde{\omega}_3^n$, it follows that

$$\|\boldsymbol{\varepsilon}_h^n\|^2 \leq 3\Delta t \sum_{j=1}^{j=n} \left\|\widetilde{\omega}_1^j\right\|^2 + 3\Delta t \sum_{j=1}^{j=n} \left\|\widetilde{\omega}_2^j\right\|^2 + 3\Delta t \sum_{j=1}^{j=n} \left\|\widetilde{\omega}_2^j\right\|^2. \tag{7.49}$$

From inequalities (7.46), (7.45) and (7.44) the assertion follows. □

In conclusion, we get the following result:

Theorem 7.4 *Under the hypotheses of Proposition 7.2, let $\{\mathcal{T}_h\}$ be a regular family of triangulations on $\overline{\Omega}$, satisfying the properties (i) and (ii) of Proposition 9 of [2,* *p. 250]. For* $\alpha \in \left]1 - \frac{\pi}{w}, 1\right[$, *there exists a constant* $c > 0$ *independent of* h *such that for every* $n \geq 1$, *we have*

$$\begin{aligned}&\left\|\mathbf{p}(t_n) - \mathbf{p}_h^n\right\|_{0,\Omega} \\ &\lesssim h\,\left(|u(t_n)|_{H^{2,\alpha}(\Omega)} + \|u_t\|_{L^2(0,t_n;H^{2,\alpha}(\Omega))}\right) \\ &\quad + 2\,(\Delta t)^2\left(\sqrt{\int_0^{t_n}\left\|u_{ttt}(s)\right\|_{0,\Omega}^2 ds} + \sqrt{\int_0^{t_n}\left\|\Delta u_{tt}(s)\right\|_{0,\Omega}^2 ds}\right).\end{aligned}$$

Proof By the triangular inequality:

$$\begin{aligned}\left\|\mathbf{p}(t_n) - \mathbf{p}_h^n\right\|_{0,\Omega} &\leq \left\|\mathbf{p}(t_n) - \tilde{\mathbf{p}}_h(t_n)\right\|_{0,\Omega} + \left\|\tilde{\mathbf{p}}_h(t_n) - \mathbf{p}_h^n\right\|_{0,\Omega}, \\ &= \left\|\mathbf{p}(t_n) - \tilde{\mathbf{p}}_h(t_n)\right\|_{0,\Omega} + \left\|\boldsymbol{\varepsilon}_h^n\right\|_{0,\Omega}.\end{aligned}$$

By Farhloul et al. [2, (5.5) p. 250]:

$$\left\|\mathbf{p}(t_n) - \tilde{\mathbf{p}}_h(t_n)\right\|_{0,\Omega} \lesssim h\,|u(t_n)|_{H^{2,\alpha}(\Omega)}.$$

From this estimate and the preceding corollary, the result follows. □

7.5 Conclusion

In this paper we have demonstrated that the completely discretized problem of the mixed formulation for the heat equation using the Crank-Nicolson scheme for time discretization admits one and only one solution. By refining the meshings according to Raugel's rules near the reentrant corners [9], we have established optimal order of convergence for this completely discretized dual mixed method for the heat diffusion equation in a polygonal domain. And this by using for the spatial discretization, Raviart–Thomas vectorfields of degree 0 for the heat flux density vector, locally constant functions for the scalar field of temperatures, and the Crank-Nicolson scheme for the discretization in time.

References

1. Ciarlet, P.G.: Basic error estimates for elliptic problems. In: Ciarlet, P.G., Lions, J.L. (eds.) Handbook of Numerical Analysis, vol. II, pp. 17–351. North-Holland, Amsterdam (1991)
2. Farhloul, M., Korikache, R., Paquet, L.: The dual mixed finite element method for the heat diffusion equation in a polygonal domain I. In: Amann, H., Arendt, W., Hieber, M., Neubrander, F., Nicaise, S., von Below, J. (eds.) Functional Analysis and Evolution Equations. The Günter Lumer Volume, pp. 239–256. Birkhäuser, Basel (2007)

3. Grisvard, P.: Elliptic Problems in Nonsmooth Domains. Monographs and Studies in Mathematics, vol. 24. Pitman, Boston (1985)
4. Grisvard, P.: Singularities in Boundary Value Problems. Recherches en Mathématiques Appliquées [Research Notes in Applied Mathematics], vol. 22. Masson, Paris; Springer, Berlin (1992)
5. Korikache, R.: Mixed finite element methods for parabolic equations. PhD thesis, Polytechnic University Hauts-De-France (2007)
6. Lions, J.-L.: Cours d'Analyse numérique, Cours de l'Ecole Polytechnique, Promotion 1972. Hermann, Paris (1974)
7. Roberts, J.E., Thomas, J.-M.: Mixed and hybrid methods. In: Ciarlet, P.G., Lions, J.L. (eds.) Handbook of Numerical Analysis, vol. II. Finite Element Methods (Part 1), pp. 523–639. Elsevier, North-Holland, Amsterdam (1991)
8. Thomée, V.: Galerkin Finite Element Methods for Parabolic Problems. Springer Series in Computational Mathematics, vol. 25. Springer, Berlin (2006)
9. Raugel, G.: Résolution numérique par une méthode d'éléments finis du problème de Dirichlet pour le laplacien dans un polygone. C.R.A.S. Paris **286**, 791–794 (1978)
10. Taine, J., Petit, J.-P.: Heat Transfer. Prentice Hall, Englewood Cliffs (1993)

Chapter 8
Economic Statistical Splicing Data Using Smoothing Quadratic Splines

Rim Akhrif, Elvira Delgado-Márquez, Abdelouahed Kouibia, and Miguel Pasadas

Abstract The main interest of this paper is to state a new method which allows to adjust the statistical difficulty when the statistical series are spliced. Hence, we study the scope of different splicing methods in the literature. We present an approximation method for statistical splicing of economic data by using smoothing quadratic splines. Finally, we show the effectiveness of our method by presenting a complete data of Gross Domestic Product for Venezuela by productive economic activity from 1950 to 2005, expressed at prices of the base year of 1997, also by showing the results of some data of Morocco for different economics activities such as the Gross Domestic Product, the agriculture, the trade and the electricity generation from petroleum sources of Morocco between 1971 and 2015.

Keywords Quadratic spline · Splicing · Smoothing spline.

8.1 Introduction

The notion of economic development has become very important in recent years. Recently, internal and external factors have contributed to Venezuela and Morocco's relatively impressive economic growth. The better macroeconomic management, a relatively more stable political climate and strong domestic demand are among the internal factors that have supported economic growth in Venezuela and Morocco. The purpose of this work is to conduct a study of long-term economic growth and

R. Akhrif (✉)
University Abdelmalek Essaadi, F.S.J.E.S. de Tétouan, Tétouan, Morocco

E. Delgado-Márquez
Department of Economics and Statistics, University of León, León, Spain
e-mail: elvira.delgado@unileon.es

A. Kouibia · M. Pasadas
Department of Applied Mathematics, University of Granada, Granada, Spain
e-mail: kouibia@ugr.es; mpasadas@ugr.es

D. Barrera et al. (eds.), *Mathematical and Computational Methods for Modelling, Approximation and Simulation*, SEMA SIMAI Springer Series 29,
https://doi.org/10.1007/978-3-030-94339-4_8

development to solve economic problems when the data are spliced. We introduce an approximation method for statistical splicing of economic data by smoothing quadratic splines, a similar theory developed with many details can be consulted in the references [4], [5] and [6].

To ensure the best possible quality for the spliced data, the information was obtained from official sources, specifically from databases published by the Central Bank of Venezuela and from the National Accounts data files of the Organization for Economic Cooperation and Development (OECD) of Morocco [3, 8, 9].

The remainder of the paper is organized as follows. Section 8.2 presents the splicing data methods, on the one hand the variation method and on the other hand the linear interpolation method [3], Sect. 8.3 describes splicing data by smoothing quadratic spline method [5], Sect. 8.4 we compute the resulting function, Sect. 8.5.1 shows some applications of our method. Finally, we give some calculation of the relative error for different values of economics activities to show the validity and the effectiveness of our method. This paper is finished by given a conclusion.

8.2 Methodology of Statistical Splicing Data

Statistics have developed a technique known as chaining or splicing to make two data comparable to solve the problem of comparing those data when there is a change of base year of an economic data. Generally there are two widely used methods are recognized in the literature, on the one hand the method of variation and on the other hand the method of linear interpolation. This work presents a new method of smoothing quadratic splines, which takes into account not only the nominal adjustments but also the structural adjustments.

8.2.1 Splicing by Variation

The method of variation represents the most used method, it is known for its simplicity, its ease of calculation and interpretation. It consists in reprogramming at constant prices the previous data by repeating the relative variations observed in the data expressed with respect to the previous base. Moreover, the only data required to apply this methodology are the data at constant prices of the different bases. However, in the face of base year changes or structural changes, this method is unable to collect such data over the long term, especially when relatively long data is constructed, that is,

$$R_t^{Basis\ t^1} = R_{t-1}^{Basis\ t^1} + d\,R_t^{Basis\ t^0},$$

where $R_t^{Basis\ t^1}$ is the value in real terms of the new base t^1 and $dR_t^{basis\ t^0}$ is the first relative difference or percentage variation of the data in terms of the previous base t^0.

8.2.2 *Splicing by Linear Interpolation*

The linear interpolation method is a splicing method that allows linear adjustment of nominal values between different base years. This method consists in defining the measurement error detected during the base change.

$$\eta = \frac{N_t^{Basis\ t^1}}{N_t^{Basis\ t^0}},$$

where $N_t^{Basis\ t^k}$ corresponds to the nominal value in the period t estimated by the established estimation method in the base year t^k. Subsequently, this discrepancy is geometrically distributed as

$$\gamma_t = \sqrt[n]{\eta^{t-t^0}},$$

where n is the number of periods observed between the different base years $t^1 - t^0$. Interpolation of the nominal values for each period t is performed as follows

$$N_t^{Basis\ t^1} = N_t^{Basis\ t^0} \gamma_t.$$

8.3 Splicing by Smoothing Quadratic Splines

In order to define the problem, suppose we have $n + 1$ time bases $t^0, \ldots, t^n$ and, for $i = 0, \ldots, n-1$, let $N_t^{Basis\ t^j}$ be the nominal value in the period t estimated by the established estimation method in the base year t^j, for $j = i, i+1$ and $t^i \leq t \leq t^{i+1}$. Denote by $d^t = N_{t+1}^{Basis\ t^i} - N_t^{Basis\ t^i}$, for any $t^i \leq t < t^{i+1}$ and $i = 0, \ldots, n-1$, and let $\Delta_n = \{t^0, t^1, \ldots, t^n\}$.

Let $S_2(\Delta_n)$ be the space of C^1 quadratic splines constructed from the partition T_n and $\{B_1, \ldots, B_{n+2}\}$ be the B-spline basis of $S_2(\Delta_n)$. It is well-known that that $\dim S_2(\Delta_n) = n + 2$.

Let $\varepsilon > 0$ and $J : S_2(\Delta_n) \to \mathbb{R}$ defined by

$$J(v) = \sum_{t=t^0}^{t^n - 1} (v'(t) - d^t)^2 + \varepsilon \int_{t^0}^{t^n} v''(t)^2\, dt.$$

The first term of $J(v)$ indicates how well v' approaches the data d^t. The second one represents the classical smoothness measure, while the parameter ε weights the importance given to such smoothness in order to avoid any oscillation.

We define $H = \{v \in S_2(\Delta_n) : v(t^i) = N_{t^i}^{Basis\ t^i},\ i = 0, \ldots, n\}$ and we consider the following problem:

Problem 1 *Find $S \in H$ such that minimizes the functional J.*

Theorem 8.1 *The problem* 1 *has a unique solution which is also the unique solution of the following variational problem: Find $S \in H$ and $\tau_0, \ldots, \tau_n \in \mathbb{R}$ such that*

$$\sum_{t=t^0}^{t^n-1} S'(t)v'(t) + \varepsilon \int S''(t)v''(t)dt + \sum_{i=0}^{n} \tau_i S(t^i) = \sum_{t=t^0}^{t^n-1} d^t v'(t),$$

for any $v \in S_2(\Delta_n)$.
Finally, we define the statistical splicing of given economic data as

$$N_t^{Basis\ t^{i+1}} = S(t),\ t^i \leq t < t^{i+1},\ i = 0, \ldots, n-1.$$

Proof We consider the following norm:

$$v \longrightarrow [[v]] = \sum_{i=1}^{n} v(t^i) + \varepsilon \int v''(t)^2 dt$$

is equivalent in $v \in S_2(\Delta_n)$ to the norm $\|.\|$, one easily checks that the symmetric bilinear form:
$\tilde{a} : S_2(\Delta_n) \times S_2(\Delta_n) \longrightarrow \mathbb{R}$ given by:

$$\tilde{a}(u, v) = \sum_{i=1}^{n} u(t^i)v(t^i) + \varepsilon \int_{t^0}^{t^n} u''v''dt$$

is continuous and $S_2(\Delta_n)$ elliptic.

Likewise, the linear form

$$\varphi : v \in S_2(\Delta_n) \longrightarrow \varphi\subsetneq v) = \sum_{i=1}^{n} \mathrm{N}_{t^i}^{Basis\ t^i}\ v(t^i)$$

is continuous.

The result is then a consequence of the Lax-Milgram Lemma (see [2]). □

8.4 Computation

Let $S \in H$ be the unique solution of Problem 1. Then, there exist coefficients α_i such that $S = \sum_{i=0}^{n+2} \alpha_i B_i$.

By linearity, and applying Theorem 1, we obtain that $\alpha := (\alpha_1, \ldots, \alpha_{n+2})^t$ and $\tau := (\tau_0, \ldots, \tau_n)^t$ are the solution of the following linear system

$$\left(\begin{array}{c|c} \mathcal{A}\mathcal{A}^t + \mathcal{R} & \mathcal{D}^t \\ \hline \mathcal{D} & 0 \end{array}\right) \left(\begin{array}{c} \boldsymbol{\alpha} \\ \hline \tau \end{array}\right) = \left(\begin{array}{c} \mathbf{d} \\ \hline \mathbf{0} \end{array}\right),$$

where

$$\mathcal{A} := \left(B_i^{\prime j}\right)_{\substack{1\leq i\leq n+2 \\ 0\leq j\leq n}},$$

$$\mathcal{D} := \left(B_i\left(t^j\right)\right)_{\substack{1\leq i\leq n+2 \\ 0\leq j\leq n}},$$

$$\mathcal{R} := \left(\int_{t^0}^{t^n} B_i'' B_j''\right)_{1\leq i,j\leq n+2},$$

$$d = \left(\sum_{t=t^0}^{t^n-1} d^t B_i'(t)\right)_{1\leq i\leq n+1}.$$

Proposition *The matrix* $C := \left(B_i'(t^j) + \int_{t^0}^{t^n} B_i'' B_j'' dt\right)_{\substack{1\leq i\leq n+2 \\ 0\leq j\leq n}}$ *is symmetric, positive definite and of band type.*

Proof Obviously, C is symmetric.

Let $p = (p_1, \ldots, p_{n+2}) \in \mathbb{R}^{n+2}$ and $\omega = \sum_{i=0}^{n+2} p_i B_i$, and if we suppose that $p.C.p^T = 0$, then we obtain that $[[\omega]]^2 = 0$ where $[[.]]$ designs the norm defined in the proof of Theorem 1. It means that $\omega = 0$.

Moreover, for the independent linearity of the family $(B_i)_{0\leq i\leq n+2}$, because it is basis, we obtain that $p = 0$ and C is positive.

Finally, the matrix C is of band type because for each $i = 0, \ldots.., n+2$ the function B_i has a local support. □

8.5 Case Studies

8.5.1 Splicing Data of Economics Activities for Venezuela Between 1950 and 2005

The purpose of this investigation is to present a complete data of GDP from 1950 to 2005 expressed at 1997 prices broken down into the largest amount of economic activities.

After setting up the system of national accounts, Venezuela has witnessed four base year changes: 1957, 1968, 1984 and 1997, which on average have had an extension of approximately fifteen years.

The first estimate data at constant prices starts in the fifties and it covers the period 1950–1968. They used the recommendations of the United Nations Development Program in their periodic revisions of the Manual of National Accounts as a basis for calculation, and therefore were based on constant price accounting around 1957 as a reference base. Because relative price stability prevailed during this period, estimates at constant prices during this period of time were not significantly different from those estimated at prices currents (Antiveros, 1992 see [1]). Likewise, Palacios et al. [7] points out that although the Gross Domestic Product estimates did not directly include the activity estimate of financial services, they did consider them in aggregate manner within the sector accounts of the rest of the economy. the recommendations as a basis for calculation.

Figure 8.1 shows the data at constant prices of GDP (Gross Domestic Product) for Venezuela between 1971–2015.

Figure 8.2 presents the original data and splicing data by using smoothing quadratic splines during the period 1950–2005.

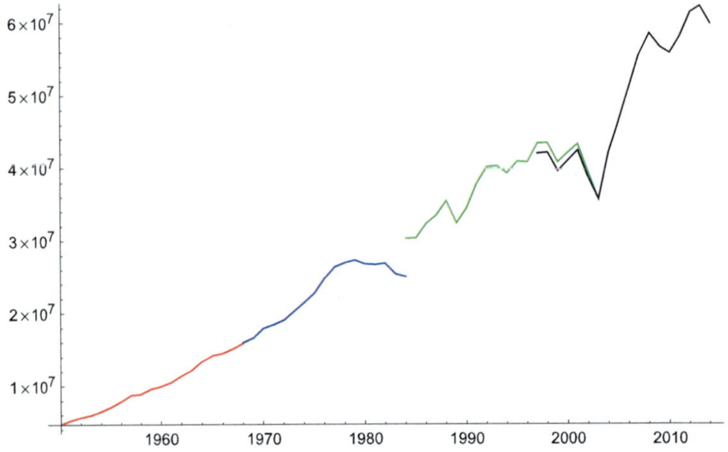

Fig. 8.1 Data at constant prices during the period 1950–2005

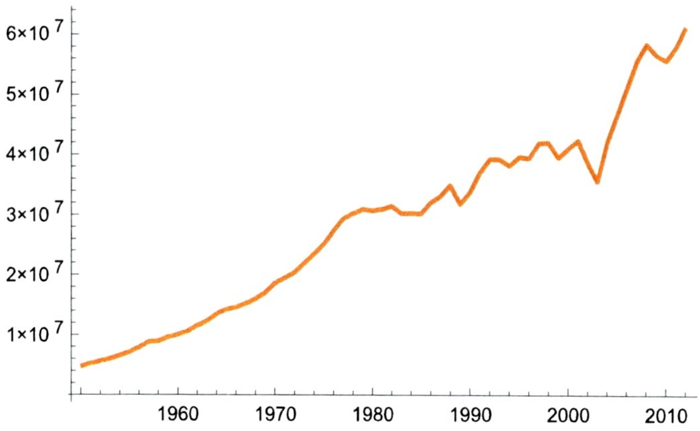

Fig. 8.2 Splicing data at constant prices during the period 1950–2005 by linear interpolation

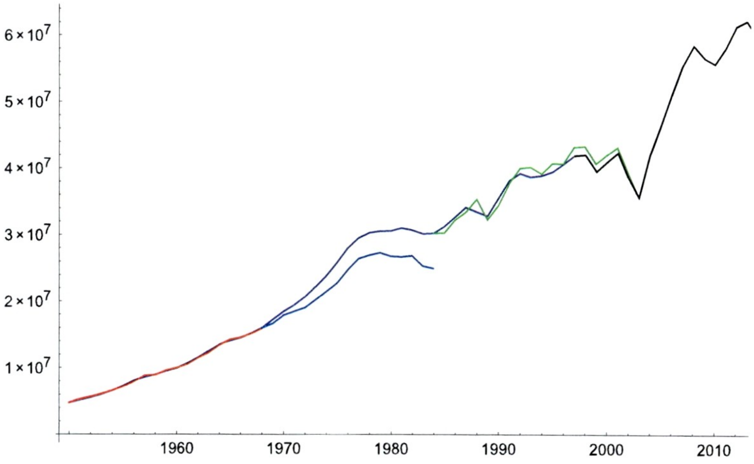

Fig. 8.3 Data at constant prices during the period 1950–2005

Figure 8.3 shows the original series and the splicing by linear interpolation method.

Figure 8.4 presents the splicing data between 1950–2005 by using the smoothing quadratic splines method.

8.5.2 Approximating Some Data of Economics Activities of Morocco Between 1971 and 2015

On the African economic front, Morocco has experienced rapid economic growth over the long period 1971–2015. This period was marked by its rapid socio-economic transformations resulting from the convergence of several factors: the rise

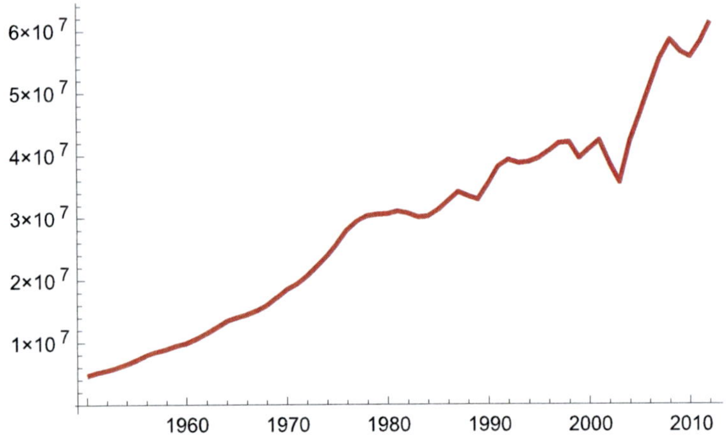

Fig. 8.4 Splicing data at constant prices during the period 1950–2005 by smoothing quadratic splines

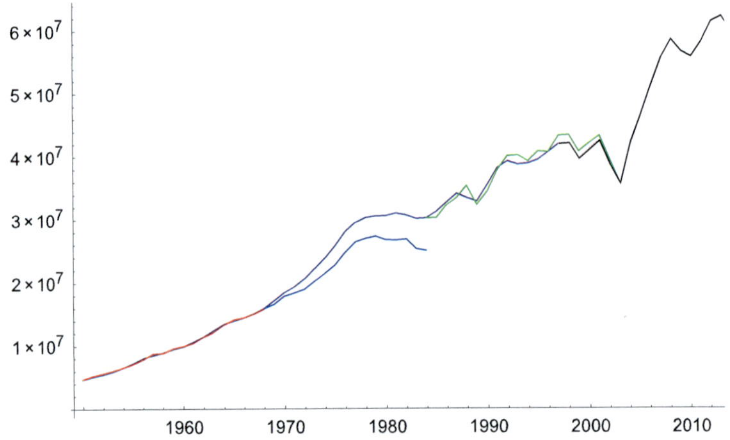

Fig. 8.5 Original series and the splicing by smoothing quadratic splines

in phosphate prices, the sectoralization of investment codes, the moroccanization and the anarchic growth of public enterprises.

Figure 8.5 shows the original series and the splicing by smoothing quadratic splines.

8.5.2.1 Data of Gross Domestic Product for Morocco Between 1971 and 2015

Economic growth is a long-term phenomenon requiring structural policies, the effects of which usually only appear after several years. Between 1971 and 2015

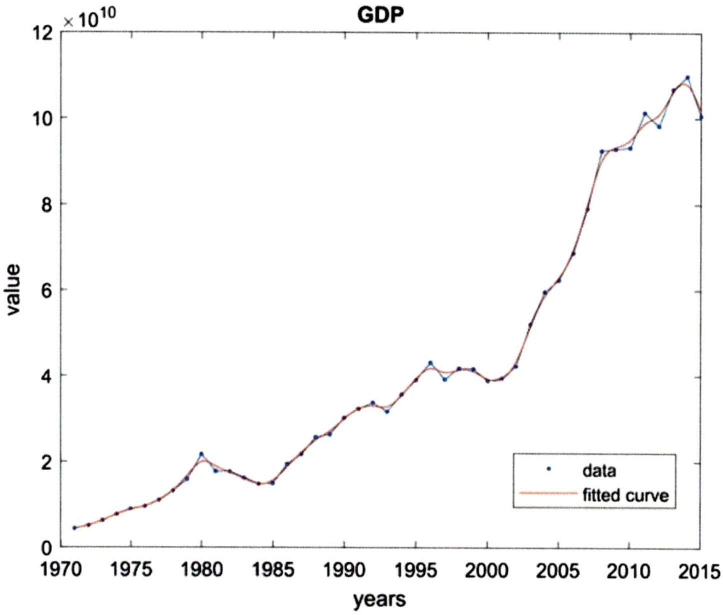

Fig. 8.6 Data of GDP during the period 1971–2015 and its approximating curve defined by a smoothing quadratic spline

the Moroccan economy has progressed in real terms the Gross Domestic Product which is distributed in several economic activities (agriculture, trade, industry ...)

Figure 8.6 shows the data and its approximating curve defined by a smoothing quadratic spline for the variable GDP (Gross Domestic Product) of Morocco during the period 1971–2015.

8.5.2.2 Data of Agriculture for Morocco Between 1971 and 2015

The agricultural sector plays an important role at the social level through its exchanges with other sectors, such as the agri-food industry, through the acquisition of capital goods and sales of agricultural products.

The economic growth in the agriculture sector has more effective effects on the increase of the income of the poorest populations.

Figure 8.7 shows the data and its approximating curve defined by a smoothing quadratic spline for the agriculture sector of Morocco between 1971–2015.

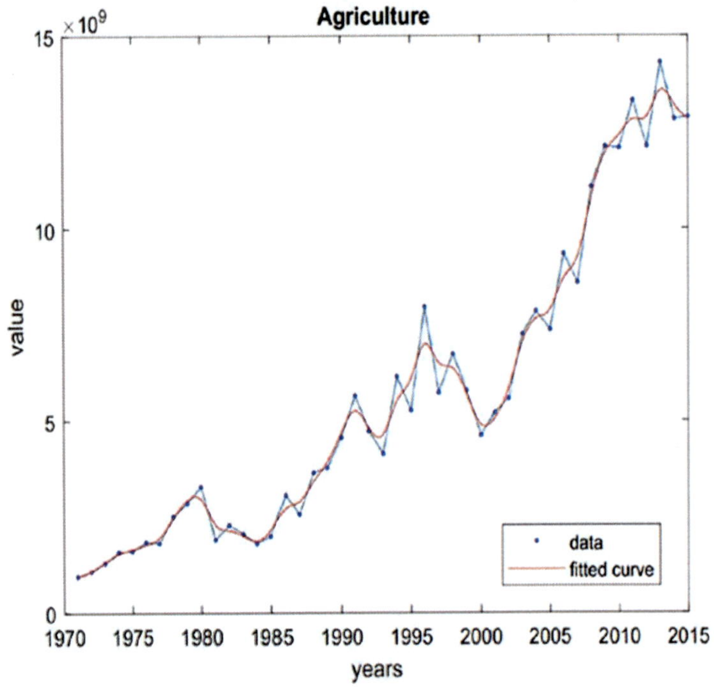

Fig. 8.7 Data of agriculture during the period 1971–2015 and its approximating curve defined by a smoothing quadratic spline

8.5.2.3 Data of Electricity for Morocco Between 1971 and 2015

In recent years, the energy sector has undergone major structural changes, as part of a process of liberalization and gradual opening up to the private sector, as well as far-reaching reforms affecting all of its components: electricity, oil and natural gas.

Figure 8.8 shows the data and its approximating curve defined by a smoothing quadratic spline for the electricity generation from petroleum sources of Morocco during the period 1971–2015.

8.5.2.4 Data of Trade for Morocco Between 1971 and 2015

Trade is a set of commercial activities required to produce, ship and sell goods and services. It encompasses a set of intricate and complex processes including a large number of actors whose roles are not always well optimized but whose impact can lengthen the process time and significantly reduce the cost of operations.

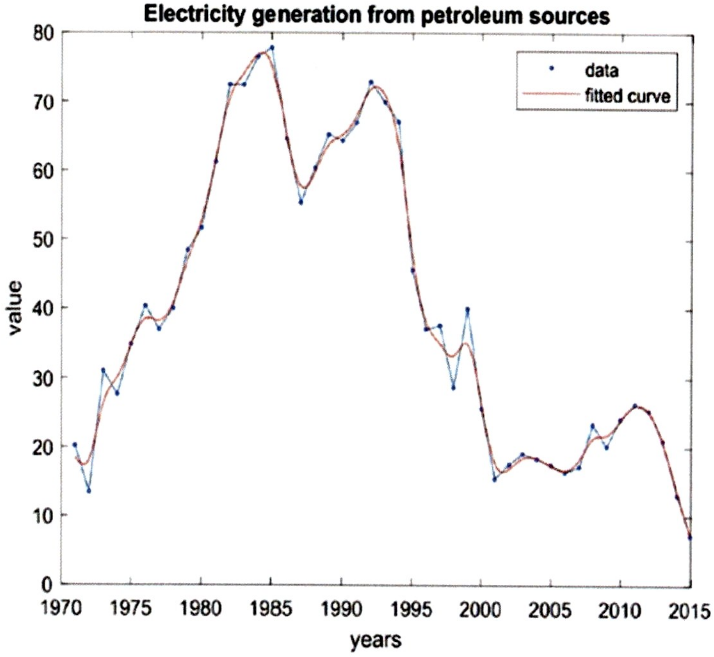

Fig. 8.8 Data of agriculture during the period 1971–2015 and its approximating curve defined by a smoothing quadratic spline

Figure 8.9 shows the data its approximating curve defined by a smoothing quadratic spline for the trade of Morocco during the period 1971–2015.

8.6 Error Estimate

In this section, we present estimates of the relative error for different values of economics activities such as: the GDP (Gross Domestic Product), the agriculture, the trade and the electricity generation from petroleum sources of Morocco between 1971 and 2015 defined by this expression

$$R.Error = \sqrt{\sum_{j=0}^{100} \frac{|S(a_j) - N_{a_j}^{Basis\ t^j})|^2}{|N_{a_j}^{Basis\ t^j}|^2}}$$

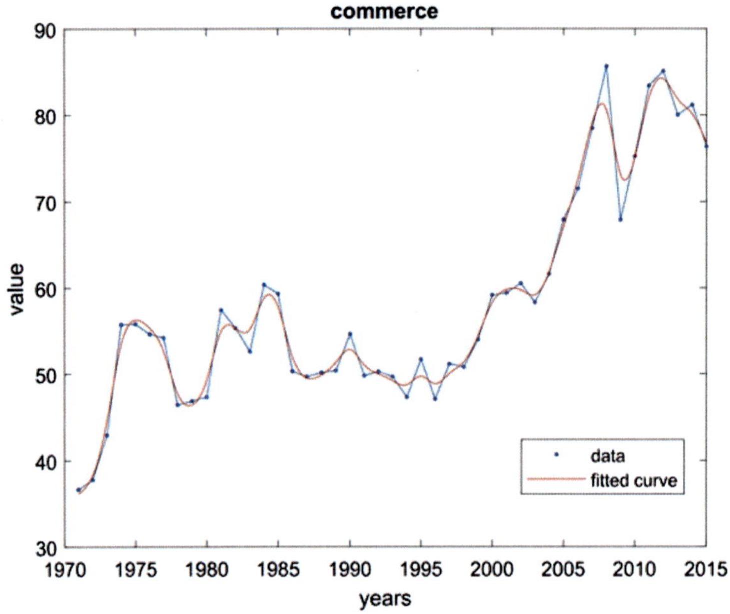

Fig. 8.9 Data of trade during the period 1971–2015 and its approximating curve defined by a smoothing quadratic spline

with $N_{a_j}^{Basis\ t^j}$ is the nominal value in the random period a_j, for j= 0,..,100, estimated by the established estimation method in the base year t^j.

By applying the smoothing quadratic spline method described in the Sect. 12.4, we have computed the following Tables with $e = 10$.

From Tables 8.1 and 8.2, one can observe that the order of the relative error reaches 10^{-3} for GDP and Agriculture during the period 1971–2015, meanwhile Table 8.3 shows that such order reaches the 10^{-4} for the electricity during the same period. Also for trade, the error reaches the order 10^{-3} (see Table 8.4). These results clearly support the usefulness and effectiveness of our method.

8.7 Conclusion

The main interest of this research has been to introduce the interpolation methodology structural to splice data. For this reason, the results are presented by three different methods (variation, linear interpolation and smoothing quadratic splines) of the Gross Domestic Product by economic activity. In this paper, we present the GDP for Venezuela from 1950 to 2005, expressed at base year 1997 prices, also we present the data of Morocco for different economics activities such as the GDP

Table 8.1 Relative error for the variable GDP during the period 1971–2015

Years	GDP	Relative error	Years	GDP	R. error
1971	4356633663.00	1.0733033e−02	1994	35604137423.00	4.7453702e−03
1972	5074117545.00	1.1913534e−02	1995	39030285468.00	6.5676721e−03
1973	6242177798.00	6.7856879e−03	1996	43161452678.00	3.6538642e−02
1974	7675408486.00	2.9927668e−03	1997	39147844526.00	4.0087454e−02
1975	8984824183.00	1.8849485e−02	1998	41806219379.00	1.2313121e−02
1976	9584323309.00	1.0511380e−02	1999	41632027600.00	1.5369068e−02
1977	11049896742.00	1.1547941e−03	2000	38857251336.00	1.0158166e−02
1978	13236854105.00	5.5876230e−03	2001	39459581217.00	2.3622727e−03
1979	15912133569.00	5.0597796e−02	2002	42236836821.00	2.5275881e−02
1980	21728770055.00	8.9897580e−02	2003	52064058834.00	1.1581181e−02
1981	17788171722.00	5.7907228e−02	2004	59626020162.00	1.8346499e−02
1982	17692341358.00	1.2273575e−02	2005	62343022651.00	8.9463713e−03
1983	16251460689.00	1.2059341e−02	2006	68640825481.00	1.0191109e−02
1984	14824728528.00	4.4819000e−03	2007	79041294874.00	1.0365441e−02
1985	14991283216.00	4.1005428e−02	2008	92507257784.00	2.6302769e−02
1986	19462175322.00	3.1804024e−02	2009	92897320376.00	4.7846649e−03
1987	21765261042.00	1.5908469e−02	2010	93216746662.00	1.8767542e−02
1988	25705296184.00	2.6464541e−02	2011	101370000000.00	2.5870214e−02
1989	26314220188.00	2.7820520e−02	2012	98266306615.00	2.6711477e−02
1990	30180108562.00	8.0556302e−03	2013	106826000000.00	5.4807297e−03
1991	32285388165.00	1.3116159e−03	2014	109881000000.00	1.7629283e−02
1992	33711069431.00	2.3216181e−02	2015	100593000000.00	1.1762913e−02
1993	31655473664.00	3.4230743e−02			

(Gross Domestic Product), the agriculture, the trade and the electricity generation from petroleum sources of Morocco between 1971 and 2015. The method used is smoothing quadratic splines as it constitutes an appropriate methodology when the periods between updates of the base year are relatively long, as is the case of the Venezuelan economy.

Table 8.2 Relative error for the variable Agriculture during the period 1971–2015

Years	Agriculture	Relative error	Years	Agriculture	R. error
1971	952673267.3	1.0827532e−02	1994	6145098848	1.1835831e−01
1972	1078508086	1.6204135e-02	1995	5266219754	1.3220398e-01
1973	1300007305	1.1605037e−02	1996	7958963285	1.3943000e−01
1974	1585427251	3.2730558e−02	1997	5728966086	1.1871503e−01
1975	1608143122	2.4691441e−02	1998	6736022250	6.0068405e−02
1976	1845314869	4.4513205e−02	1999	5772506711	1.4886370e−02
1977	1810450114	7.0643301e−02	2000	4615184397	5.8644048e−02
1978	2504379965	1.9360673e−02	2001	5205492931	3.0853167e−02
1979	2850914314	2.3757303e−02	2002	5575482824	4.1225745e−02
1980	3285855815	1.1734124e−01	2003	7248934064	3.1202748e−02
1981	1900638594	1.7120260e−01	2004	7847323602	2.6140359e−02
1982	2285063922	7.2733561e−02	2005	7366470220	6.5123028e−02
1983	2051484257	2.6658099e−02	2006	9336855132	7.3715734e−02
1984	1797227172	3.6269878e−02	2007	8586355480	7.1570153e−02
1985	1992481367	7.2401422e−02	2008	11063055624	2.6962298e−02
1986	3056556171	1.2940991e−01	2009	12103238138	1.4485261e−02
1987	2560577567	1.1378525e−01	2010	12065532481	2.6373602e−02
1988	3651031173	6.7100110e−02	2011	13299793570	3.8122152e−02
1989	3776591032	3.7614010e−02	2012	12115803625	5.8697809e−02
1990	4555740327	2.6974832e−02	2013	14303491761	5.4611235e−02
1991	5649761078	7.6404456e−02	2014	12823953463	2.7533265e−02
1992	4733963633	1.3838541e−02	2015	12871890458	4.7243397e−03
1993	4137182826	1.0647724e−01			

In short, we can conclude that we have presented a simple method to approximate some data, but such method is very useful and interesting to solve some serious problems for the System of National Accounts that have arisen in Economy.

We have taken the data from Venezuela and Morocco because they are countries of increasing economy in recent years, although the method is valid to be applied to any economy of any country.

Table 8.3 Relative error for the variable electricity during the period 1971–2015

Years	Electricity	Relative error	Years	Electricity	R. error
1971	20.21834061	9.2915403e−02	1994	67.06190061	5.3697053e−02
1972	13.46595256	2.5909006e−01	1995	45.59212703	5.3827582e−02
1973	31.02608696	1.8163670e−01	1996	37.02344872	2.9676105e−02
1974	27.64015645	8.0262243e−02	1997	37.5708085	7.5255623e−02
1975	34.81186907	2.9852514e−04	1998	28.67828501	1.3592016e−01
1976	40.33823122	4.9318889e−02	1999	40.03101978	1.4836584e−01
1977	36.91962057	3.3529108e−02	2000	25.63943093	3.4113619e−02
1978	39.99544211	1.9037434e−02	2001	15.48584341	1.2745601e−01
1979	48.41450777	3.3477664e−02	2002	17.55169019	3.8538483e−02
1980	51.64856108	2.3446761e−02	2003	19.10966341	3.4020714e−02
1981	61.24301676	8.1905433e−03	2004	18.37334086	2.8606033e−03
1982	72.46541904	2.8900845e−02	2005	17.47537584	7.6606536e−03
1983	72.40502543	2.2794995e−02	2006	16.45690697	1.2382964e−02
1984	76.520734	3.9431669e−03	2007	17.20359853	5.6286978e−02
1985	77.80803268	3.5140649e−02	2008	23.36575875	9.6704980e−02
1986	64.64451314	9.7601532e−03	2009	20.18917002	7.2973606e−02
1987	55.34308211	4.1941399e−02	2010	24.14667117	7.1303609e−03
1988	60.39626002	1.1302664e−02	2011	26.30568664	1.0367151e−02
1989	65.25930851	2.3354810e−02	2012	25.31187123	6.6852162e−03
1990	64.35396759	1.2734507e−02	2013	20.95657192	1.6667070e−02
1991	66.95274307	1.4515605e−02	2014	13.10443192	4.2458818e−02
1992	72.91902459	1.5808752e−02	2015	7.173913043	1.8275347e−02
1993	69.94954591	1.4928514e−02			

Table 8.4 Relative error for the variable trade during the period 1971–2015

Years	Trade	Relative error	Years	Trade	R. error
1971	36.67924185	1.4948489e−02	1994	47.31355373	2.8461930e−02
1972	37.79695866	1.3586843e−02	1995	51.71502538	4.0275353e−02
1973	42.94273678	3.6666332e−02	1996	47.09554345	3.5764066e−02
1974	55.74925462	4.3531671e−02	1997	51.15015465	2.2781114e−02
1975	55.81884595	7.8960701e−03	1998	50.79768215	1.1188231e−02
1976	54.64444235	1.2524589e−02	1999	53.98669265	8.2302680e−03
1977	54.21715802	2.9334447e−02	2000	59.16182679	1.4895363e−02
1978	46.43924285	2.8155636e−02	2001	59.4180039	6.3387211e−03
1979	46.87474816	1.0097258e−02	2002	60.5340864	1.2908555e−02
1980	47.34206783	3.8067022e−02	2003	58.32774268	1.4619264e−02
1981	57.4360558	4.6673680e−02	2004	61.59653347	4.1516439e−03
1982	55.34027036	1.7247737e−04	2005	67.91485449	1.2324514e−02
1983	52.61409502	4.5964093e−02	2006	71.49628678	1.5790551e−02
1984	60.36267952	2.7222253e−02	2007	78.48717434	8.0065653e−03
1985	59.31507463	2.5172482e−02	2008	85.6728209	6.3878468e−02
1986	50.28060358	3.4602387e−02	2009	67.91510295	7.0561398e−02
1987	49.65750551	2.8298379e−03	2010	75.24763454	1.5500302e−03
1988	50.13760317	6.2541264e−03	2011	83.42680017	1.6813706e−02
1989	50.35084475	1.8063834e−02	2012	85.12491729	1.1084358e−02
1990	54.62668742	3.4013878e−02	2013	80.02055856	2.1614393e−02
1991	49.77929548	2.4856511e−02	2014	81.17703227	1.3724499e−02
1992	50.26331618	5.5552048e−03	2015	76.37919291	4.7391907e−03
1993	49.66845661	9.6442035e−03			

References

1. Antiveros, I.: Series Estadisticas de Venezuela de los últimos cincuenta años. Tomo I y II. Cuentas Nacionales (Capítulos I-II-III-IV-V). Recopilador: Mireya de Cabre, Caracas (1992).
2. Brezis, H.: Analyse fonctionnelle. Théorie et Applications. Masson, Paris (1983)
3. Correa, V., Escanlón, A.S., Venegas, A.L.: Empalme PIB: Series Anuales y Trimestrales 1986–1995, Base 1996. Documento Metodológico. Banco Central de Chile, vol. 179 (2002)
4. Kouibia, A., Pasadas, M.: Approximation by discrete variational splines. J. Comput. Appl. Math. **116**(1), 145–156 (2000)
5. Kouibia, A., Pasadas, M.: Approximation by shape preserving interpolation splines. Appl. Numer. Math. **37**(3), 271–288 (2001)
6. Kouibia, A., Pasadas, M., Torrens, J.J.: Construction of surfaces with parallelism conditions. Numer. Algorithms **33**, 331–342 (2003)
7. Palacios, L.C., Puente, A.A., Frank Gómez, F.: Venezuela, Crecimiento y Petróleo, Mimeografía, Banco Central de Venezuela, Caracas (2005)
8. Paracare, E.: Empalme de las series de PIB 1984–1996, base 1997. Banco Central de Venezuela (2005)
9. Rodríguez, F.: The Anarchy of Numbers: Attempting to Understand Venezuelan Economic Performance. Oficina de Asesoria Económica y Financiera de la Asamblea Nacional de Venezuela (2005)

Chapter 9
Some Properties of Convexity Structure and Applications in b-Menger Spaces

Abderrahim Mbarki and Rachid Oubrahim

Abstract We discuss, in Menger spaces, the notion of convexity using the convex structure introduced by Takahashi (Kodai Math Sem Rep 22:142–149, 1970), then we develop some geometric and topological properties. Furthermore, we introduce the notion of strong convex structure and we compare it with the Takahashi convex structure. At the end, we prove the existence and uniqueness of a solution for a Volterra type integral equation.

Keywords Takahashi convex structure · Probabilistic strong convex structure · Volterra type integral equation

9.1 Introduction and Preliminaries

In the paper [1] authors introduced the concept of probabilistic b-metric space (b-Menger space), which the probabilistic b-metric mapping F is not necessarily continuous, and which generalizes the concept of probabilistic metric space (Menger space [8, 9]) and b-metric space. They discussed its topological and geometrical properties and they showed the fixed point and common fixed point property for nonlinear contractions in these spaces [5]. Furthermore, they defined the notion of fully convex structure and established [4] in fully convex b-Menger spaces the existence of common fixed point for nonexpansive mapping by using the normal structure property. Also, they showed a fixed point theorem in b-Menger spaces using B-contraction with cyclical conditions (See [2, 3] and [6]).

A. Mbarki (✉)
ANO Laboratory, National School of Applied Sciences, Mohammed First University, Oujda, Morocco

R. Oubrahim
ANO Laboratory, Faculty of Sciences, Mohammed First University, Oujda, Morocco

D. Barrera et al. (eds.), *Mathematical and Computational Methods for Modelling, Approximation and Simulation*, SEMA SIMAI Springer Series 29,
https://doi.org/10.1007/978-3-030-94339-4_9

Definition 9.1 ([1]) A b-Menger space is a quadruple (X, F, T, s) where X is a nonempty set, F is a function from $X \times X$ into Δ^+, T is a t-norm, $s \geq 1$ is a real number, and the following conditions are satisfied:

For all $p, r, q \in X$ and $x, y > 0$,

1. $F_{pp} = H$,
2. $F_{pr} = H \Rightarrow p = r$,
3. $F_{pr} = F_{rp}$,
4. $F_{pr}(s(x + y)) \geq T(F_{pq}(x), F_{qr}(y))$.

It should be noted that a Menger space is a b-Menger space with $s = 1$.

Definition 9.2 Let (X, F) be a probabilistic semimetric space (i.e., (1), (2) and (3) of Definition 9.1 are satisfied). For p in X and $t > 0$, the strong t-neighborhood of p is the set

$$N_p(t) = \{q \in X : F_{pq}(t) > 1 - t\}.$$

The strong neighborhood system at p is the collection

$$\wp_p = \{N_p(t) : t > 0\}$$

and the strong neighborhood system for X is the union

$$\wp = \bigcup_{p \in X} \wp_p.$$

In probabilistic semimetric space, the convergence of sequence is defined as follow.

Definition 9.3 Let $\{x_n\}$ be a sequence in a probabilistic semimetric space (X, F). Then

1. The sequence $\{x_n\}$ is said to be convergent to $x \in X$, if for every $\epsilon > 0$, there exists a positive integer $N(\epsilon)$ such that $F_{x_n x}(\epsilon) > 1 - \epsilon$ whenever $n \geq N(\epsilon)$.
2. The sequence $\{x_n\}$ is called a Cauchy sequence, if for every $\epsilon > 0$ there exists a positive integer $N(\epsilon)$ such that $n, m \geq N(\epsilon) \Rightarrow F_{x_n x_m}(\epsilon) > 1 - \epsilon$.
3. (X, F) is said to be complete if every Cauchy sequence has a limit.

Schweizer and Sklar proved that if (X, F, T) is a Menger space with T is continuous, then the family $\Im$ consisting of $\emptyset$ and all unions of elements of this strong neighborhood system for X determines a Hausdorff topology for X (see [11] and [10]).

Lemma 9.1 ([11]) *Let (X, F, T) be a Menger space with T is continuous, X be endowed with the topology $\Im$ and $X \times X$ with the corresponding product topology. Then F is a uniformly continuous mapping from $X \times X$ into Δ^+.*

However, Mbarki et al. [1] showed that in b-Menger space (X, F, T, s), the probabilistic b-metric F is not continuous in general even though T is continuous.

Example 9.1 ([1]) Let $X = \mathbb{N} \cup \{\infty\}$, $0 < a \leq 1$. Define $F^a : X \times X \to \Delta^+$ as follow:

$$F^a_{bc}(t) = \begin{cases} H(t) & \text{if } b = c, \\ H(t-7) & \text{if } b \text{ and } c \text{ are odd and } b \neq c, \\ H(t - |\frac{a}{b} - \frac{a}{c}|) & \text{if } b \text{ and } c \text{ are even or } bc = \infty, \\ H(t-3) & \text{otherwise.} \end{cases}$$

It easy to show that $(X, F^a, T_M, 4)$ is a b-Menger space with T_M is continuous. In the sequel, we take $a = 1$. Consider the sequence $x_n = 2n$, $n \in \mathbb{N}$. Then $F_{2n\infty}(t) = H(t - \frac{1}{2n})$. Therefore $x_n \to \infty$, but $F_{2n1}(t) = H(t-3) \neq H(t-1) = F_{1\infty}(t)$. Hence F is not continuous at ∞.

In this work we give two notions of structures convex, first one is the probabilistic extension of the Takahashi convex structure defined in ordinary metric spaces, second one is the strong convex structure more general than the probabilistic Takahashi convex structure and we study a properties and relationship of the theory of betweenness in Menger spaces using those convex structures. We finish this work by showing the existence and uniqueness of a solution for a Volterra type integral equation.

9.2 Probabilistic Takahashi Convex Structure

In 1970, Takahashi [12] introduced the concept of convexity in a metric space. We give the probabilistic version of this convex structure.

Definition 9.4 Let (X, F, T) be a Menger space, and let I be the closed unit interval $[0, 1]$. A probabilistic Takahashi convex structure (PTCS) on X is a function $W : X \times X \times I \to X$ which has the property that for every $x, y \in X$ and $\lambda \in I$ we have

$$F_{zW(x,y,\lambda)}(\lambda s + (1-\lambda)t) \geq T(F_{zx}(s), F_{zy}(t))$$

for all $z \in X$, and $s, t > 0$. If (X, F, T) is equipped with PTCS, we call X a convex Menger space.

In the sequel of this section we suppose that $ImF \subset \mathcal{D}^+$.

Definition 9.5 Let W be a PTCS on a Menger space (X, F, T). We say that W is a strict PTCS if it has a property that whenever $w \in X$ and there is $(x, y, \lambda) \in X \times X \times (0, 1)$ for which

$$F_{zw}(2t) \geq T(F_{zx}(\frac{t}{\lambda}), F_{zy}(\frac{t}{1-\lambda})) \quad \textit{for every } z \in X$$

then $w = W(x, y, \lambda)$.

Proposition 9.1 *Let W be a strict PTCS on the Menger space (X, F, T). Then for all $x, y \in X$ and $\lambda \in I$ we have*

$$W(x, y, \lambda) = W(y, x, 1-\lambda).$$

Proof The equality is true for $\lambda = 0$ and $\lambda = 1$. Let $\lambda \in (0, 1)$, we have

$$\begin{aligned} F_{zW(y,x,1-\lambda)}(2t) &\geq T(F_{zy}(\frac{t}{1-\lambda}), F_{zx}(\frac{t}{\lambda})) \\ &= T(F_{zx}(\frac{t}{\lambda}), F_{zy}(\frac{t}{1-\lambda})), \end{aligned}$$

for all $z \in X$.

By strictness we get

$$W(x, y, \lambda) = W(y, x, 1-\lambda).$$

□

Theorem 9.1 *Let W be a strict PTCS on the Menger space (X, F, T_M) under the t-norm T_M. Then for every $x, y \in X$ and $\alpha, \beta \in [0, \frac{1}{2})$, we have*

$$W(W(x, y, \alpha), y, \beta) = W(x, y, 2\alpha\beta).$$

Proof Let $x, y \in X$. The assertion is true for $\alpha = 0$ or $\beta = 0$. Let $\alpha, \beta \in (0, \frac{1}{2})$. For all $z \in X$ we have

$$\begin{aligned} F_{zW(W(x,y,\alpha),y,\beta)}(2t) &\geq \min\left(F_{zW(x,y,\alpha)}\left(\frac{t}{\beta}\right), F_{zy}\left(\frac{t}{1-\beta}\right)\right) \\ &\geq \min\left(\min(F_{zx}\left(\frac{t}{2\beta\alpha}\right), F_{zy}\left(\frac{t}{2\beta(1-\alpha)}\right),\right. \\ &\quad \left.F_{zy}\left(\frac{t}{1-\beta}\right)\right) \\ &= \min\left(F_{zx}\left(\frac{t}{2\beta\alpha}\right), \min\left(F_{zy}\left(\frac{t}{2\beta(1-\alpha)}\right),\right.\right. \\ &\quad \left.\left.F_{zy}\left(\frac{t}{1-\beta}\right)\right)\right) \\ &\geq \min\left(F_{zx}\left(\frac{t}{2\beta\alpha}\right), F_{zy}\left(\frac{t}{1-2\beta\alpha}\right)\right). \end{aligned}$$

Because of $\frac{1}{2\beta(1-\alpha)} \geq \frac{1}{1-2\beta\alpha}$ and $\frac{1}{1-\beta} \geq \frac{1}{1-2\beta\alpha}$, it holds

$$W(W(x, y, \alpha), y, \beta) = W(x, y, 2\alpha\beta).$$

□

Definition 9.6 A convex Menger space (X, F, T) with a probabilistic Takahashi convex structure W will be called strictly convex if, for arbitraries $x, y \in X$ and $\lambda \in (0, 1)$ the element $W(x, y, \lambda)$ is the unique element which satisfies

$$F_{xy}(\frac{t}{\lambda}) = F_{W(x, y, \lambda)y}(t), \quad F_{xy}(\frac{t}{1-\lambda}) = F_{W(x, y, \lambda)x}(t),$$

for all $t > 0$.

Theorem 9.2 *Let W be a strict PTCS on a strictly convex Menger space (X, F, T). Then for every $x, y \in X$ with $x \neq y$ the mapping $\lambda \mapsto W(x, y, \lambda)$ is an injective from $[0, \frac{1}{2})$ into X.*

Proof Let $\alpha, \beta \in [0, \frac{1}{2})$ such that $\alpha \neq \beta$ and assume, without loss of generality, that $\alpha < \beta$. Let $x, y \in X$ such that $x \neq y$. We have

$$\begin{aligned} F_{W(x, y, \alpha)W(x, y, \beta)}(t) &= F_{W((W(x, y, \beta), y, \frac{\alpha}{2\beta})W(x, y, \beta)}(t) \\ &= F_{W(x, y, \beta)y}(\frac{t}{1-\frac{\alpha}{2\beta}}) \\ &= F_{xy}(\frac{t}{\beta(1-\frac{\alpha}{2\beta})}) \end{aligned}$$

for all $t > 0$.

Since $x \neq y$, then $F_{xy} \neq H$, hence there exists $t_0 > 0$ such that $F_{xy}(t_0) < 1$, for $\frac{t}{\beta(1-\frac{\alpha}{2\beta})} < t_0$ we have $F_{W(x, y, \alpha)W(x, y, \beta)}(t) < 1$. So $F_{W(x, y, \alpha)W(x, y, \beta)} \neq H$ and therefore $W(x, y, \alpha) \neq W(x, y \beta)$. □

Theorem 9.3 *Let W be a strict PTCS on a compact Menger space (X, F, T). Then for each $\lambda \in (0, 1)$, $W_\lambda : (x, y) \mapsto W_\lambda(x, y) = W(x, y, \lambda)$ is continuous as a mapping from $X \times X$ into X.*

Proof Given $\lambda \in (0, 1)$ and let $\{(x_n, y_n)\}_{n=1}^{\infty}$ be a sequence in $X \times X$ which converges to (x, y) and let w be a cluster point of the sequence $\{W(x_n, y_n, \lambda)\}_{n=1}^{\infty}$. By using Theorems 2.2-2.5 of [7], select a subsequence $\{W(x_{n_k}, y_{n_k}, \lambda)\}_{n=1}^{\infty}$ which converges to w. Then for any $z \in X$, we have

$$F_{zW(x_{n_k}, y_{n_k}, \lambda)}(2t) \geq T(F_{zx_{n_k}}(\frac{t}{\lambda}), F_{zy_{n_k}}(\frac{t}{1-\lambda})) \quad for \quad k = 1, 2, \ldots.$$

Strictness and using the fact that the set of the points of discontinuity of F is countable now guarantees that

$$w = W(x, y, \lambda).$$

It follows that $W(x, y, \lambda)$ is the only cluster point of the sequence $\{W(x_n, y_n, \lambda)\}_{n=1}^{\infty}$. Therefore, in view of Theorem 2.4 of [7], $\{W(x_n, y_n, \lambda)\}_{n=1}^{\infty}$ must converges to $W(x, y, \lambda)$ which complete the proof. □

9.3 Probabilistic Strong Convex Structure

In this section, we give a relationship between $W(W(x, y, s_1), z, s_2)$ and $W(W(x, z, t_1), y, t_2)$ for $s_1, s_2, t_1, t_2 \in [0, 1]$.

Definition 9.7 Let (X, F, T) be a Menger space, and let $P = \{(\alpha, \beta, \gamma) \in I \times I \times I : \alpha + \beta + \gamma = 1\}$. A probabilistic strong convex structure (PSCS) on X is a continuous function $K : X \times X \times X \times P \to X$ with the property that for each $(x, y, z; (\alpha, \beta, \gamma)) \in X \times X \times X \times P$, $K(x, y, z, (\alpha, \beta, \gamma))$ is the unique point of X which satisfies

$$F_{wK(x, y, z, (\alpha, \beta, \gamma))}(\alpha s + \beta t + \gamma r) \geq T(T(F_{wx}(s), F_{wy}(t)), F_{wz}(r)) \quad (9.1)$$

for every $w \in X$ and for all $s, t, r > 0$.

Remark 9.1 The uniqueness assumption in last definition guarantees that if p is a permutation of $\{1, 2, 3\}$, then, for $(x_1, x_2, x_3, (\alpha_1, \alpha_2, \alpha_3)) \in X \times X \times X \times P$, we have

$$K(x_1, x_2, x_3, (\alpha_1, \alpha_2, \alpha_3)) = K(x_{p(1)}, x_{p(2)}, x_{p(3)}, (\alpha_{p(1)}, \alpha_{p(2)}, \alpha_{p(3)})).$$

Proposition 9.2 *Let (X, F, T) be a strong convex Menger space with $ImF \subset \mathcal{D}^+$ and K its PSCS. Define $W_K : X \times X \times I \to X$ by $W_K(x, y, \lambda) = K(x, y, x, (\lambda, 1-\lambda, 0))$. Then W_K is a PTCS on X.*

Proof Let $s, t > 0, x, y \in X$ and $\lambda \in I$. We have

$$\begin{aligned} F_{wW_K(x, y, \lambda)}(\lambda s + (1-\lambda)t) &= F_{wK(x, y, x, (\lambda, 1-\lambda;0))}(\lambda s + (1-\lambda)t) \\ &\geq T(T(F_{wx}(s), F_{wy}(t)), F_{wz}(r)), \end{aligned}$$

for all $w \in X$ and $r > 0$.

Setting $r \to \infty$ we get

$$F_{wW_K(x, y, \lambda)}(\lambda s + (1-\lambda)t) \geq T(F_{wx}(s), F_{wy}(t)),$$

for all $w \in X$.

Then W_K is a PTCS on X. □

Theorem 9.4 *For any three points x, y, z in a strong convex Menger space (X, F, T), if $\beta \leq \frac{1}{2}$ and $\alpha \in I$. Then,*

$$W(W(x, y, \alpha), z, \beta) = K(x, y, z, (\frac{4\beta\alpha}{3}, \frac{4\beta(1-\alpha)}{3}, \frac{3-4\beta}{3})).$$

Proof Let $\beta \leq \frac{1}{2}$, $\alpha \in I$ and $x, y, z \in X$. For all $w \in X$ we have

$$\begin{aligned} F_{wW(W(x, y, \alpha), z, \beta)}(3t) &\geq T(F_{wW(x, y, \alpha)}(\frac{3t}{2\beta}), F_{wz}(\frac{3t}{2(1-\beta)})) \\ &\geq T(T(F_{wx}(\frac{3t}{4\beta\alpha}), F_{wy}(\frac{3t}{4\beta(1-\alpha)})), F_{wz}(\frac{3t}{2(1-\beta)})) \\ &\geq T(T(F_{wx}(\frac{3t}{4\beta\alpha}), F_{wy}(\frac{3t}{4\beta(1-\alpha)})), F_{wz}(\frac{3t}{3-4\beta})), \end{aligned}$$

because $\frac{3}{2(1-\beta)} \geq \frac{3}{3-4\beta}$.

Since $\frac{4\beta\alpha}{3} + \frac{4\beta(1-\alpha)}{3} + \frac{3-4\beta}{3} = 1$, then by uniqueness

$$W(W(x, y, \alpha), z, \beta) = K(x, y, z, (\frac{4\beta\alpha}{3}, \frac{4\beta(1-\alpha)}{3}, \frac{3-4\beta}{3})).$$

□

Corollary 9.1 *For any three points x, y, z in a strong convex Menger space (X, F, T), if $\frac{1}{4} \leq \beta \leq \frac{1}{2}$ and $\alpha \leq \frac{4\beta-1}{4\beta}$, then*

$$W(W(x, y, \alpha), z, \beta) = W(W(x, z, \beta\alpha[\frac{3-4\beta(1-\alpha)}{4}]^{-1}), y, \frac{3-4\beta(1-\alpha)}{4}).$$

Proof The conditions $\frac{1}{4} \leq \beta \leq \frac{1}{2}$ and $\alpha \leq \frac{4\beta-1}{4\beta}$ imply that $\frac{3-4\beta(1-\alpha)}{4} \leq \frac{1}{2}$ and $\gamma = \beta\alpha[\frac{3-4\beta(1-\alpha)}{4}]^{-1} \in (0, 1)$. We apply the Theorem 9.4 we get

$$W(W(x, z, \gamma), y, \frac{3-4\beta(1-\alpha)}{4}) = K(x, z, y, (\frac{4\beta\alpha}{3}, \frac{3-4\beta}{3}, \frac{4\beta(1-\alpha)}{3})),$$

and by permutation we obtain

$$W(W(x, y, \alpha), z, \beta) = W(W(x, z, \beta\alpha[\frac{3-4\beta(1-\alpha)}{4}]^{-1}), y, \frac{3-4\beta(1-\alpha)}{4}).$$

□

9.4 Application to An Integral Equation

As an application of the Theorem 4.1 in [1], we will consider the following Volterra type integral equation:

$$x(t) = g(t) + \int_0^t \Omega(t, \alpha, x(\alpha))d\alpha, \tag{9.2}$$

for all $t \in [0, k]$, where $k > 0$.

Theorem 9.5 *Let $\Omega \in C([0, k] \times [0, k] \times \mathbb{R}, \mathbb{R})$ be an operator satisfying the following conditions:*

1. $\|\Omega\|_\infty = sup_{t, \alpha\in[0, k], x\in C([0, k], \mathbb{R})}|\Omega(t, \alpha, x(\alpha))| < \infty$.
2. *There exists $L > 0$ such that for all $t, \alpha \in [0, k]$ and $x, y \in C([0, k], \mathbb{R})$ we obtain*

$$|\Omega(t, \alpha, fx(\alpha)) - \Omega(t, \alpha, fy(\alpha))| \leq \frac{L}{\sqrt{2}}|x(\alpha) - y(\alpha)|,$$

where $f : C([0, k], \mathbb{R}) \to C([0, k], \mathbb{R})$ is defined by

$$fx(t) = g(t) + \int_0^t \Omega(t, \alpha, fx(\alpha))d\alpha, \qquad g \in C([0, k], \mathbb{R}).$$

Then the Volterra type integral equation (9.2) has a unique solution $x^* \in C([0, k], \mathbb{R})$.

Proof We define the mapping $F : C([0, k], \mathbb{R}) \times C([0, k], \mathbb{R}) \to \mathcal{D}^+$ by

$$F_{xy}(t) = H(t - \max_{t\in[0, k]}(|x(t) - y(t)|^2 e^{-2Lt})), \quad t > 0, \ x, \ y \in C([0, k], \mathbb{R}).$$

From Lemma 3.1 of [1], $(C([0, k], \mathbb{R}), F, T_M, 2)$ is a complete b-Menger space with coefficient $s = 2$. Therefore, for all $x, y \in C([0, k], \mathbb{R})$, we get

$$\begin{aligned}
&F_{fxfy}(r) \\
&= \mathcal{H}(r - \max_{t\in[0, k]}(|fx(t) - fy(t)|^2 e^{-2Lt})) \\
&= \mathcal{H}(r - \max_{t\in[0, k]}(|\int_0^t (\Omega(t, \alpha, fx(\alpha)) - \Omega(t, \alpha, fx(\alpha)))d\alpha|^2 e^{-2Lt})) \\
&= \mathcal{H}(r - \max_{t\in[0, k]}(|\int_0^t (\Omega(t, \alpha, fx(\alpha)) - \Omega(t, \alpha, fx(\alpha)))e^{-L\alpha}e^{L(\alpha-t)}d\alpha|^2)) \\
&\geq \mathcal{H}(r - \frac{L^2}{2}\max_{t\in[0, k]}(|x(t) - y(t)|^2 e^{-2Lt})\max_{t\in[0, k]}(\int_0^t e^{L(\alpha-t)}d\alpha)^2)
\end{aligned}$$

$$= \mathcal{H}(r - \frac{1}{2}(1 - e^{-Lk})^2 \max_{t\in[0,\ k]}(|x(t) - y(t)|^2 e^{-2Lt}))$$
$$= \mathcal{H}(r - \frac{c}{2}\max_{t\in[0,k]}(|x(t) - y(t)|^2 e^{-2Lt}))$$
$$= \mathcal{H}(\frac{2r}{c} - \max_{t\in[0,k]}(|x(t) - y(t)|^2 e^{-2Lt}))$$
$$= F_{xy}(\frac{2r}{c}),$$

where $c = (1 - e^{-Lk})^2$.

Therefore, in view of Theorem 4.1 in [1] with $\varphi(r) = cr,\ c \in [0,\ 1]$, we deduce that the operator f has a unique fixed point $x^* \in C([0,\ k],\ \mathbb{R})$, which is the unique solution of the integral equation (9.2). □

References

1. Mbarki, A., Oubrahim, R.: Probabilistic b-metric spaces and nonlinear contractions. Fixed Point Theory Appl. **2017**, 29 (2017)
2. Mbarki, A., Oubrahim, R.: Cyclical contractive conditions in probabilistic metric spaces. Adv. Sci. Technol. Eng. Syst. J. **5**(2), 100–103 (2017)
3. Mbarki, A., Oubrahim, R.: Fixed point theorems with cyclical contractive conditions in b-Menger spaces. Results Fixed Point Theory Appl. **2018**, 1–7 (2018)
4. Mbarki, A., Oubrahim, R.: Common fixed point theorem in b-Menger spaces with a fully convex structure. Int. J. Appl. Math. **32**(2), 219–238 (2019)
5. Mbarki, A., Oubrahim, R.: Common fixed point theorems in b-Menger spaces. In: Recent Advances in Intuitionistic Fuzzy Logic Systems, pp. 283–289. Springer, Berlin (2019)
6. Mbarki, A., Oubrahim, R.: Fixed point theorem satisfying cyclical conditions in b-Menger spaces. Moroccan J. Pure Appl. Anal. **1**(5), 31–36 (2019)
7. Mbarki, A., Ouahab, A., Naciri, R.: On compactness of probabilistic metric space. Appl. Math. Sci. **8**(35), 1703–1710 (2014)
8. Menger, K.: Statistical metrics. Proc. Nat. Acad. Sci. **28**, 535–537 (1942)
9. Menger, K.: Géométrie générale. Mémor. Sci. Math. **124**, 80+i (1954). Gauthier-Villars, Paris
10. Schweizer, B., Sklar, A.: Statistical metric spaces. Pac. J. Math. **10**, 313–334 (1960)
11. Schweizer, B., Sklar, A.: Probabilistic Metric Spaces. North-Holland Series in Probability and Applied Mathematics, vol. 5. North-Holland, New York (1983)
12. Takahashi, W.: A convexity in metric spaces and nonexpansive mappings, I. Kodai Math. Sem. Rep. **22**, 142–149 (1970)

Chapter 10
A Super-Superconvergent Cubic Spline Quasi-Interpolant

Afaf Rahouti, Abdelhafid Serghini, Ahmed Tijini, and Ahmed Zidna

Abstract In this paper we use the developed B-spline representation to construct a cubic super-superconvergent quasi-interpolant with an optimal approximation order which improves the efficiency and accuracy over traditional methods.

Keywords Hermite interpolation · Finite element · Splines · Quasi-interpolation · Superconvergence

10.1 Introduction

Spline quasi-interpolants are very useful approximants in practice, they have a direct construction without solving any system of equations and with the minimum possible computation time. In general, a spline quasi-interpolant for a given function f is obtained as a linear combination of some elements of a suitable set of basis functions. In order to achieve stability and local control, these functions are required to be positive and to have small local supports. The coefficients of the linear combination are the values of linear functionals, depending on f and (or) its derivatives or integrals. Many works concerning the construction of quasi-interpolants are developed in the literature (see [1, 2, 7, 8], for instance).

Superconvergence is a useful tool in numerical analysis, it is a phenomenon where the order of convergence of the approximation error at certain special points is higher than the one over the definition's domain (see [5, 9]). Very recently, the

A. Rahouti (✉) · A. Tijini
FSO, University Mohammed First, Oujda, Morocco

A. Serghini
ESTO, University Mohammed First, Oujda, Morocco
e-mail: a.serghini@ump.ma

A. Zidna
LITA, Université de Lorraine, Metz, France
e-mail: zidna@univ-metz.fr

D. Barrera et al. (eds.), *Mathematical and Computational Methods for Modelling, Approximation and Simulation*, SEMA SIMAI Springer Series 29,
https://doi.org/10.1007/978-3-030-94339-4_10

autors introduced in [4] an efficient method to build superconvergent discrete quasi-interpolants of a function u which are of order $m+1$, $m \geq 3$ at the knots of the initial partition τ. This property is not only true for approximating function values but also for approximating first derivatives and second derivatives (m and $m-1$, respectively). To improve the numerical results given by this operator, we introduce a new concept, called the super-superconvergence phenomenon, which allows us to increase the approximation order of the superconvergent discrete quasi-interpolant. This new approach can be considered as an interesting approximation tool and it can be applied for solving some numerical analysis problems, such as differential and integral equations.

We develop our results in the following manner. In Sect. 10.2, we recall from [4] some results concerning the construction and the properties of cubic Hermite B-splines. In Sect. 10.3, we give how to construct super-superconvergent discrete quasi-interpolants. Finally in Sect. 10.4, we illustrate the theoretical results by some numerical tests.

10.2 Normalized Basis

10.2.1 Finite Element of Class C^2 and Degree 3

Let

$$\tau := (a = x_0 < x_1 < \cdots < x_n = b),$$

be the uniform partition of the interval $I := [a, b]$ and let $x_i = a + 3ih$ where $h := \frac{b-a}{3n}$. We suppose that the values of a function u and its first and second derivatives at the knots x_i, $i = 0, \ldots, n$, are available.

To construct a cubic finite element of class C^2 over the partition τ, we search its local expression in each subinterval $[x_i, x_{i+1}]$, in terms of the function and its first and second derivative values at knots x_i and x_{i+1}. As we have more data than it is needed to determine a cubic polynomial on $[x_i, x_{i+1}]$, the local expression of our finite element must be a cubic piecewise polynomial. For this, we consider a new refinement τ_1 of τ obtained by adding two arbitrary knots $x_{i,1}$ and $x_{i,2}$ in $]x_i, x_{i+1}[$ and by imposing the C^2 smoothness at these new knots. In this case the construction of a cubic C^2 finite element is possible.

For $i = 0, \ldots, n-1$, let $\tau_{i,1} := (x_i < x_{i,1} < x_{i,2} < x_{i+1})$, be a subdivision of $[x_i, x_{i+1}]$ into three subintervals $[x_i, x_{i,1}]$, $[x_{i,1}, x_{i,2}]$ and $[x_{i,2}, x_{i+1}]$, and $\mathcal{P}_i$ be a spline of degree 3 and class C^2 defined on $[x_i, x_{i+1}]$. Denote by

$$\mathcal{P}_i|_{[x_i,x_{i,1}]} = \mathcal{P}_i^1, \quad \mathcal{P}_i|_{[x_{i,1},x_{i,2}]} = \mathcal{P}_i^2 \quad \text{and} \quad \mathcal{P}_i|_{[x_{i,2},x_{i+1}]} = \mathcal{P}_i^3,$$

the restrictions of the spline $\mathcal{P}_i$ in each subinterval of $[x_i, x_{i+1}]$. The polynomials $\mathcal{P}_i^1$, $\mathcal{P}_i^2$ and $\mathcal{P}_i^3$ are written in the Bernstein basis.

The unknown coefficients are determined by the conditions of interpolation at knots x_i and x_{i+1} and the C^2 smoothness at the knots $x_{i,1}$ and $x_{i,2}$.

Define $\tau_1 := \cup_{i=0}^{n-1} \tau_{i,1}$ as a refinement of τ. The space of C^2 cubic splines on the interval I endowed with τ_1 is defined by

$$\mathbb{P}_3^2(I, \tau_1) := \{\mathcal{P} \in C^2(I) : \mathcal{P}_i^j \in \mathbb{P}_3(\mathbb{R}), \quad i = 0, \ldots, n-1, \ j = 0, 1, 2\},$$

where $\mathbb{P}_3(\mathbb{R})$ is the space of polynomials of degree three.

In [10] it has been proved that for a given data $u(x_i)^{(j)}$, $j = 0, 1, 2$, $i = 0, \ldots, n$, there exists a unique spline $\mathcal{P} \in \mathbb{P}_3^2(I, \tau_1)$, solution of the following Hermite interpolation problem

$$\mathcal{P}^{(j)}(x_i) = u^{(j)}(x_i), \ \ i = 0, \ldots, n, \quad j = 0, 1, 2. \tag{10.1}$$

Therefore, the dimension of the space $\mathbb{P}_3^2(I, \tau_1)$ equals $3(n + 1)$.

10.2.2 Hermite B-Splines

Let φ_i, ψ_i and ξ_i, $i = 0, \ldots, n$, be solution functions of the problem (10.1) in $\mathbb{P}_3^2(I, \tau_1)$ which satisfy the following interpolation conditions

$$\begin{array}{lll} \varphi_i(x_j) = \delta_{ij}, & \varphi_i'(x_j) = 0, & \varphi_i''(x_j) = 0, \\ \psi_i(x_j) = 0, & \psi_i'(x_j) = \delta_{ij}, & \psi_i''(x_j) = 0, \\ \xi_i(x_j) = 0, & \xi_i'(x_j) = 0, & \xi_i''(x_j) = \delta_{ij}, \end{array}$$

where δ_{ij}, $i, j = 0, \ldots, n$, stands for the Kronecker symbol.
The built functions φ_i, ψ_i and ξ_i, $i = 0, \ldots, n$, constitute the Hermite basis of the space $\mathbb{P}_3^2(I, \tau_1)$ and $\operatorname{supp}\varphi_i = \operatorname{supp}\psi_i = \operatorname{supp}\xi_i = \left[x_{i-1}, x_{i+1}\right]$.

The Hermite B-splines $H_{i,s}$ of $\mathbb{P}_3^2(I, \tau_1)$ are constructed as follows

$$H_{i,s}(x) = \alpha_{i,s}\varphi_i(x)+\beta_{i,s}\psi_i(x)+\gamma_{i,s}\xi_i(x), \quad s = 0, 1, 2, \ i = 0, \ldots, n. \tag{10.2}$$

They form a basis of the space $\mathbb{P}_3^2(I, \tau_1)$, are non-negative, have a local support $[x_{i-1}, x_{i+1}]$ and form a partition of unity, i.e., for each $x \in I$

$$H_{i,s} \geq 0, \quad \sum_{i=0}^{n} \sum_{s=0}^{2} H_{i,s} = 1.$$

The necessary conditions so that the B-splines $H_{i,s}$ satisfy the partition of the unity [4] are given by

$$\begin{cases} \alpha_{i,0} + \alpha_{i,1} + \alpha_{i,2} = 1 \\ \beta_{i,0} + \beta_{i,1} + \beta_{i,2} = 0 \\ \gamma_{i,0} + \gamma_{i,1} + \gamma_{i,2} = 0, \end{cases}$$

where

$$H_{i,s}(x_i) = \alpha_{i,s}; \quad H'_{i,s}(x_i) = \beta_{i,s}; \quad H''_{i,s}(x_i) = \gamma_{i,s}.$$

All details are reported in [4]. The following theorem gives the non-negativity condition.

Theorem 10.1 ([4]) *The Hermite B-splines $H_{i,s}$ are non-negative if for each $i = 1, \ldots, n-1$, we have*

$$\alpha_{i,s} \geq \frac{h}{3}|\beta_{i,s}|, \quad \alpha_{i,s} + \frac{h^2}{3}\gamma_{i,s} \geq h|\beta_{i,s}|, \quad \text{and} \quad \frac{3}{2}\alpha_{i,s} + \frac{h^2}{4}\gamma_{i,s} \geq h|\beta_{i,s}|,$$

for $i = 0$:

$$\alpha_{i,s} \geq \left(\frac{-h}{3}\beta_{i,s}\right)_+, \quad \alpha_{i,s} + \frac{h^2}{3}\gamma_{i,s} \geq -h\beta_{i,s}, \quad \text{and} \quad \frac{3}{2}\alpha_{i,s} + \frac{h^2}{4}\gamma_{i,s} \geq -h\beta_{i,s} \tag{10.3}$$

and for $i = n$:

$$\alpha_{i,s} \geq \left(\frac{h}{3}\beta_{i,s}\right)_+, \quad \alpha_{i,s} + \frac{h^2}{3}\gamma_{i,s} \geq h\beta_{i,s}, \quad \text{and} \quad \frac{3}{2}\alpha_{i,s} + \frac{h^2}{4}\gamma_{i,s} \geq h\beta_{i,s},$$

where $(x)_+ = \max(x, 0)$.

Let the choice of B-splines used in the Subsection 3.4 of [4] i.e for $i = 1, \ldots, n-1$

$$\begin{array}{lll} \alpha_{i,0} = \dfrac{1}{6}, & \alpha_{i,1} = \dfrac{2}{3}, & \alpha_{i,2} = \dfrac{1}{6}, \\ \beta_{i,0} = \dfrac{-m}{3(m-1)h}, & \beta_{i,1} = 0, & \beta_{i,2} = \dfrac{m}{3(m-1)h}, \\ \gamma_{i,0} = \dfrac{2m}{3(m-1)h^2}, & \gamma_{i,1} = \dfrac{-4m}{3(m-1)h^2}, & \gamma_{i,2} = \dfrac{2m}{3(m-1)h^2}, \end{array} \tag{10.4}$$

with the particular values of $\alpha_{0,s}$, $\beta_{0,s}$, $\gamma_{0,s}$, $\alpha_{n,s}$, $\beta_{n,s}$ and $\gamma_{n,s}$ at the extremities x_0 and x_n, as follows

$$\left\{ \begin{array}{lll} \alpha_{0,0} = 1, & \alpha_{0,1} = 0, & \alpha_{0,2} = 0, \\ \beta_{0,0} = \dfrac{-2m}{h(m-1)}, & \beta_{0,1} = \dfrac{2m}{h(m-1)}, & \beta_{0,2} = 0, \\ \gamma_{0,0} = \dfrac{4m}{h^2(m-1)}, & \gamma_{0,1} = \dfrac{-6m}{h^2(m-1)}, & \gamma_{0,2} = \dfrac{2m}{h^2(m-1)}, \\ \text{and} & & \\ \alpha_{n,0} = 0, & \alpha_{n,1} = 0, & \alpha_{n,2} = 1, \\ \beta_{n,0} = 0, & \beta_{n,1} = \dfrac{-2m}{h(m-1)}, & \beta_{n,2} = \dfrac{2m}{h(m-1)}, \\ \gamma_{n,0} = \dfrac{2m}{h^2(m-1)}, & \gamma_{n,1} = \dfrac{-6m}{h^2(m-1)}, & \gamma_{n,2} = \dfrac{4m}{h^2(m-1)}. \end{array} \right. \tag{10.5}$$

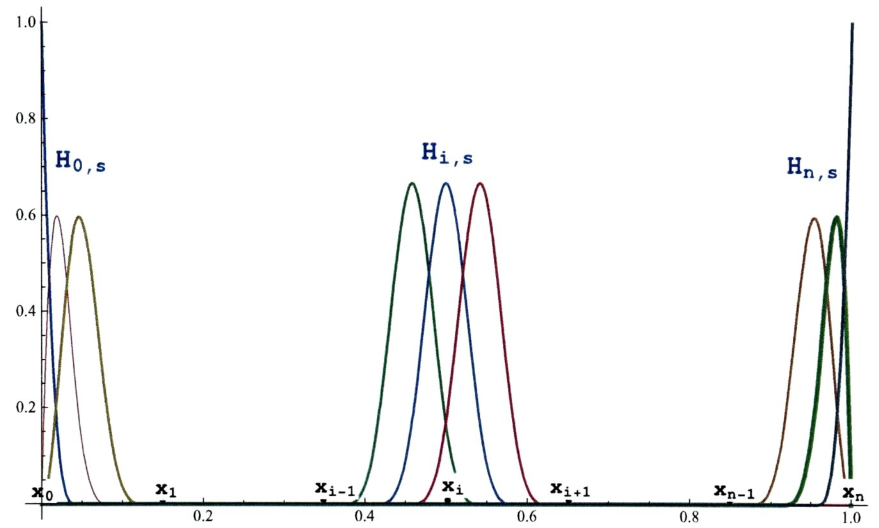

Fig. 10.1 Normalized B-splines on the interval [0,1]

These values satisfy the necessary conditions of non-negativity and of partition of the unity.

By the choices (10.4) and (10.5) for $m = 3$, we can display the cubic Hermite B-splines as shown in Fig. 10.1.

10.2.3 Marsden's Identity in Polar Form

Let us denote by $x_i^{(m)}$ the expression $\underbrace{x_i, x_i, \ldots, x_i}_{m\ times}$. The polar form of a polynomial is a transformation that reduces its complexity by adding new variables while having a certain symmetry property i.e, the polar form $\hat{p}$ of p, is a function of m variables satisfying the multi-affine, symmetry and diagonal properties (see [3, 6]).

At each knot x_i, $i = 0, \ldots, n$, we assume that the vectors $(\alpha_{i,s}, \beta_{i,s}, \gamma_{i,s})^\top$, $s = 0, 1, 2$ are given such that $H_{i,s}$ are non-negatives and linearly independent. Let us denote e_l the monomial of degree l, $l \in \mathbb{N}$, $l \leq m$ with m an integer greater than or equal to 3. In the following theorem we give the Hermite interpolant $\mathcal{H}e_l$ of any monomial $e_l \in \mathbb{P}_m$.

Theorem 10.2 **([4])** *For any monomial e_l of degree $l \leq m$, there exist two points $y_{i,s,m,1}$ and $y_{i,s,m,2}$ such that*

$$\mathcal{H}e_l(x) = \sum_{i=0}^{n} \sum_{s=0}^{2} \hat{e}_l(y_{i,s,m,1}, y_{i,s,m,2}, x_i^{(m-2)}) H_{i,s}(x), \quad \forall x \in I. \qquad (10.6)$$

In the following proposition we give the expression of the Hermite interpolant of any polynomial $p \in \mathbb{P}_m$ in terms of its polar form and the Hermite B-splines basis.

Proposition 10.1 ([4]) *For each $p \in \mathbb{P}_m(\mathbb{R})$, the Hermite interpolant $\mathcal{H}p$ of p in the space $\mathbb{P}_3^2(I, \tau_1)$ is expressed by*

$$\mathcal{H}p(x) = \sum_{i=0}^{n} \sum_{s=0}^{2} \widehat{p}(y_{i,s,m,1}, y_{i,s,m,2}, x_i^{(m-2)}) H_{i,s}(x) \quad \forall x \in I. \tag{10.7}$$

Corollary 10.1 ([4]) *(Marsden's Identity) For $p \in \mathbb{P}_3$, we have*

$$p(x) = \sum_{i=0}^{n} \sum_{s=0}^{2} \widehat{p}(y_{i,s,3,1}, y_{i,s,3,2}, x_i) H_{i,s}(x), \quad \forall x \in I. \tag{10.8}$$

10.3 Spline Quasi-Interpolant in $\mathbb{P}_3^2(I, \tau_1)$

10.3.1 Construction of a Superconvergent Discrete Quasi-Interpolant

We are interested in the discrete quasi-interpolants built in [4] of the form

$$Q_m u = \sum_{i=0}^{n} \sum_{s=0}^{2} \mu_{i,s}(u) H_{i,s}, \tag{10.9}$$

where, $\mu_{i,s}$ $i = 0, 1, \ldots, n$, $s = 0, 1, 2$, are linear functionals defined using values of u at some points in the neighbourhood of supports of the B-splines $H_{i,s}$. Supported by these values, we construct in a neighbourhood of supp $H_{i,s}$ a local linear polynomial operator $\mathcal{I}_{i,s}$ that reproduces the space of polynomials of degree at most m. Then we have the following result.

Theorem 10.3 ([4]) *Let u be a function defined on I such that the values of u are given at some discrete points in the neighbourhood of the support of $H_{i,s}$, $i = 0, \ldots, n$, $s = 0, 1, 2$. If we denote $\mathcal{I}_{i,s}(u)$ by $p_{i,s}$, then the quasi-interpolant defined by (10.9) with*

$$\mu_{i,s}(u) = \widehat{p}_{i,s}\left(y_{i,s,m,1}, y_{i,s,m,2}, x_i^{(m-2)}\right) \tag{10.10}$$

satisfies

$$Q_m p = \mathcal{H}p, \quad \forall p \in \mathbb{P}_m.$$

To build a superconvergent discrete spline quasi-interpolant, it suffices to take $m+1$ distinct interpolation points in the neighbourhood of the support of $H_{i,s}$ for $i=0,\dots,n$ and $s=0,1,2$. Let $t_{i,s,k}$, $k=0,\dots,m$ be these points and let $p_{i,s}$ be the Lagrange interpolant, i.e.,

$$p_{i,s}=\sum_{k=0}^{m}u(t_{i,s,k})L_{i,s,k}, \tag{10.11}$$

where $L_{i,s,k}$ are the Lagrange basis functions of $\mathbb{P}_m$ associated with the points $t_{i,s,k}$. Then, the quasi-interpolant defined by (10.9) and (10.10) satisfies

$$Q_m p=\mathcal{H}p,\ \forall p\in\mathbb{P}_m.$$

Let $\eta_{i,s}\in\mathbb{R}$ such that $x_i=\eta_{i,s}y_{i,s,m,1}+(1-\eta_{i,s})y_{i,s,m,2}$. In the following theorem, we give an explicit formula of the coefficients $\mu_{i,s}(u)$ in terms of the data values $u(t_{i,s,k})$ for $k=0,\dots,m$.

Theorem 10.4 ([4]) *Let $t_{i,s,k}=\varepsilon_{i,s,k}y_{i,s,m,1}+(1-\varepsilon_{i,s,k})y_{i,s,m,2}$ for $k=0,\dots,m$, be $m+1$ distinct points in the neighbourhood of the support of $H_{i,s}$. The quasi-interpolant defined by (10.9) with*

$$\mu_{i,s}(u)=\sum_{k=0}^{m}q_{i,s,k}u(t_{i,s,k}) \tag{10.12}$$

satisfies

$$Q_m p=\mathcal{H}p,\quad\forall\ p\in\mathbb{P}_m$$

if and only if

$$q_{i,s,k}=\frac{1}{m(m-1)}\,\frac{\displaystyle\sum_{\alpha,\theta=0,\alpha,\theta\neq k,\alpha\neq\theta}^{m}(\varepsilon_{i,s,\alpha}-1)\varepsilon_{i,s,\theta}\prod_{\gamma=0,\ \gamma\neq\alpha,\theta,k}^{m}(\eta_{i,s}-\varepsilon_{i,s,\gamma})}{\displaystyle\prod_{\alpha=0,\ \alpha\neq k}^{m}(\varepsilon_{i,s,k}-\varepsilon_{i,s,\alpha})} \tag{10.13}$$

The cubic spline quasi-interpolant provides an improvement of the approximation order at knots by the superconvergence phenomenon. In the following theorem, we give the error estimates associated with Q_m and its first and second derivatives at the knots x_i.

Theorem 10.5 ([4]) *For any $u \in C^{m+1}(I)$ the error estimates associated with Q_m and its first and second derivatives at the knots are*

$$| Q_m^{(k)} u(x_i) - u^{(k)}(x_i) |= O(h^{m+1-k}), \quad i = 0, \dots, n, \quad k = 0, 1, 2.$$

Proof Let $u \in C^{m+1}(I)$. The Taylor expansion of u around x_i, $i = 0, \dots, n$, is given by

$$u(x) = \sum_{j=0}^{m} \frac{u^{(j)}(x_i)}{j!}(x - x_i)^j + O((x - x_i)^{m+1}).$$

Denote by R_m the polynomial part of the Taylor expansion. Then for each point x in the support of $H_{i,s}$, we have

$$u(x) = R_m(x) + O((x - x_i)^{m+1}).$$

From Theorem 10.3, we have $Q_m R_m = \mathcal{H} R_m$, and using the fact that $\mathcal{H} R_m(x_i) = R_m(x_i)$, we get

$$|Q_m u(x_i) - u(x_i)| = |Q_m u(x_i) - R_m(x_i)| = |Q_m(u - R_m)(x_i)|.$$

From (10.12) and using the fact that $q_{i,s,k}$ are bounded by a constant K, we obtain

$$|\mu_{i,s}(u - R_m)| = |\sum_{k=0}^{m} q_{i,s,k}(u(t_{i,s,k}) - R_m(t_{i,s,k}))| \le K \sum_{k=0}^{m} |(u(t_{i,s,k}) - R_m(t_{i,s,k}))|.$$

Then,

$$|\mu_{i,s}(u - R_m)| = O((t_{i,s,k} - x_i)^{m+1})$$

and therefore

$$|Q_m(u - R_m)(x_i)| = O((t_{i,s,k} - x_i)^{m+1}).$$

Hence

$$| Q_m u(x_i) - u(x_i) |= O(h^{m+1}).$$

In a similar way, we prove that

$$| Q'_m u(x_i) - u'(x_i) |= O(h^m) \quad \text{and} \quad | Q''_m u(x_i) - u''(x_i) |= O(h^{m-1}),$$

which completes the proof. □

10.3.2 Super-Superconvergence Phenomenon

From a numerical observation, we found that the approximation's order is $O\left(h^{m+2}\right)$ and $O\left(h^m\right)$ (in approximating function values and second derivative values, respectively) at knots when the degree m of the local polynomials is even, which leads to improve the properties of approximation. Unfortunately, this phenomenon does not happen for an arbitrary data sites, whence the necessity to choose quasi-interpolation points in order to achieve the super-superconvergence phenomenon.

For each $i = 1, \ldots, n-1$, let $t_{i,s,\frac{m}{2}}$ be the midpoint of $[y_{i,s,m,1}, y_{i,s,m,2}]$.

Theorem 10.6 *For every $u \in C^{m+2}(I)$ such that m is even, the quasi-interpolant Q_m is super-superconvergent at x_i, i.e,*

$$\left|Q_m^{(r)}u(x_i) - u^{(r)}(x_i)\right| = O\left(h^{m+2-r}\right), \quad i = 1, \ldots, n-1, \quad r = 0, 2, \tag{10.14}$$

if

1. *the set of the local interpolation points corresponding to $\mu_{i,0}(u)$ is symmetric to the one corresponding to $\mu_{i,2}(u)$ with respect to x_i*
2. *the local interpolation points corresponding to $\mu_{i,1}(u)$ are symmetric between them with respect to x_i;*
3. *the point $t_{i,0,\frac{m}{2}}$ is symmetric to the point $t_{i,2,\frac{m}{2}}$ with respect to x_i, $i = 1, \ldots, n-1$;*
4. *$\alpha_{i,0} = \alpha_{i,2} = \alpha_i$ and $\gamma_{i,0} = \gamma_{i,2} = \gamma_i$.*

Proof We will proof the theorem by a similar way as in [4, 5]. Let $u \in C^{m+2}(I)$. The Taylor expansion of u around x_i for $i = 1, \ldots, n-1$ is given by

$$u(x) = \sum_{j=0}^{m+1} \frac{u^{(j)}(x_i)}{j!}(x - x_i)^j + O((x - x_i)^{m+2}).$$

Denote by R_m the polynomial part of the Taylor expansion and by f_{m+1} its last term

$$f_{m+1}(x) = \frac{u^{(m+1)}(x_i)}{(m+1)!}(x - x_i)^{m+1}.$$

Using a similar way as in the proof of Theorem 10.5, we get for $r = 0, 2$

$$\begin{aligned}|Q_m^{(r)}u(x_i) - u^{(r)}(x_i)| &= |Q_m^{(r)}u(x_i) - Q_m^{(r)}R_m(x_i) + Q_m^{(r)}R_m(x_i) - u^{(r)}(x_i)| \\ &\le |Q_m^{(r)}(u - R_m)(x_i)| + |Q_m^{(r)} f_{m+1}(x_i)|.\end{aligned} \tag{10.15}$$

From (10.12) and using the fact that $q_{i,s,k}$ are bounded by a constant C, we get

$$|\mu_{i,s}(u - R_m)| \leq C \sum_{k=0}^{m} |(u(t_{i,s,k}) - R_m(t_{i,s,k}))|.$$

This implies that

$$|\mu_{i,s}(u - R_m)| = O((t_{i,s,k} - x_i)^{m+2}),$$

and therefore

$$|Q_m^{(r)}(u - R_m)(x_i)| = O((t_{i,s,k} - x_i)^{m+2-r}), \quad r = 0, 2. \tag{10.16}$$

For any function $f \in C^2(I)$, and for $r = 0, 2$ we have

$$Q_m^{(r)} f(x_i) = \sum_{i=0}^{n} \sum_{s=0}^{2} \mu_{i,s}(f) H_{i,s}^{(r)}(x_i).$$

First of all, we show that the coefficients $q_{i,s,k}$, $s = 0, 2$ are the same for two symmetrical interpolation points with respect to x_i. i.e., $t_{i,0,k} + t_{i,2,k} = 2x_i$.

As

$$t_{i,s,k} = \varepsilon_{i,s,k} y_{i,s,m,1} + (1 - \varepsilon_{i,s,k}) y_{i,s,m,2}, \quad k = 0, \ldots, m, \tag{10.17}$$

we have

$$\varepsilon_{i,0,k} y_{i,0,m,1} + (1 - \varepsilon_{i,0,k}) y_{i,0,m,2} + \varepsilon_{i,2,k} y_{i,2,m,1} + (1 - \varepsilon_{i,2,k}) y_{i,2,m,2} = 2x_i.$$

Also, as $t_{i,0,\frac{m}{2}}$ and $t_{i,2,\frac{m}{2}}$ are symmetric with respect to x_i, we have

$$\begin{aligned} &\varepsilon_{i,0,k} y_{i,0,m,1} + (1 - \varepsilon_{i,0,k}) y_{i,0,m,2} + \varepsilon_{i,2,k} y_{i,2,m,1} + (1 - \varepsilon_{i,2,k}) y_{i,2,m,2} \\ &\quad = 2x_i = t_{i,0,\frac{m}{2}} + t_{i,2,\frac{m}{2}}. \end{aligned}$$

The fact that $t_{i,s,\frac{m}{2}} = \frac{1}{2}(y_{i,s,m,1} + y_{i,s,m,2})$ gives

$$\begin{aligned} &\varepsilon_{i,0,k} y_{i,0,m,1} + (1 - \varepsilon_{i,0,k}) y_{i,0,m,2} + \varepsilon_{i,2,k} y_{i,2,m,1} + (1 - \varepsilon_{i,2,k}) y_{i,2,m,2} \\ &\quad = \frac{1}{2}(y_{i,0,m,1} + y_{i,0,m,2} + y_{i,2,m,1} + y_{i,2,m,2}), \end{aligned}$$

then

$$\varepsilon_{i,0,k} = \varepsilon_{i,2,k} = \frac{1}{2}.$$

By considering that $\eta_{i,2} = 1 - \eta_{i,0}$ and from Theorem 10.4 we easily obtain

$$q_{i,0,k} = q_{i,2,k} = q_{i,k}, \quad k = 0, \dots, m. \tag{10.18}$$

Now, we will show that $q_{i,1,k} = q_{i,1,m-k}$ for $k = 0, \dots, m$. As the local interpolation points corresponding to $\mu_{i,1}$ are symmetric between them with respect to x_i, we get

$$t_{i,1,k} + t_{i,1,m-k} = 2x_i.$$

Then $x_i = t_{i,1,\frac{m}{2}}$ and

$$t_{i,1,k} + t_{i,1,m-k} = 2t_{i,1,\frac{m}{2}} = y_{i,1,m,1} + y_{i,1,m,2}$$

and by using (10.17) we get

$$\begin{aligned}
&\varepsilon_{i,1,k} y_{i,1,m,1} + (1 - \varepsilon_{i,1,k}) y_{i,1,m,2} + \varepsilon_{i,1,m-k} y_{i,1,m,1} + (1 - \varepsilon_{i,1,m-k}) y_{i,2,m,2} \\
&\quad = y_{i,1,m,1} + y_{i,1,m,2},
\end{aligned}$$

which gives

$$\varepsilon_{i,1,m-k} = 1 - \varepsilon_{i,1,k}. \tag{10.19}$$

Then,

$$\begin{aligned}
q_{i,1,m-k} &= \frac{\displaystyle\sum_{\alpha,\theta=0,\alpha\neq k,\alpha\neq\theta}^{m} (\varepsilon_{i,1,m-\alpha} - 1)\varepsilon_{i,1,m-\theta} \prod_{\gamma=0,\ \gamma\neq\alpha,\theta,k}^{m} (\eta_{i,1} - \varepsilon_{i,1,m-\gamma})}{\displaystyle m(m-1) \prod_{\alpha=0,\ \alpha\neq k}^{m} (\varepsilon_{i,1,m-k} - \varepsilon_{i,1,m-\alpha})} \\
&= \frac{\displaystyle\sum_{\alpha,\theta=0,\alpha\neq k,\alpha\neq\theta}^{m} \varepsilon_{i,1,\alpha}(\varepsilon_{i,1,\theta} - 1) \prod_{\gamma=0,\ \gamma\neq\alpha,\theta,k}^{m} (1 - \eta_{i,1} - \varepsilon_{i,1,\gamma})}{\displaystyle m(m-1) \prod_{\alpha=0,\ \alpha\neq k}^{m} (\varepsilon_{i,1,k} - \varepsilon_{i,1,\alpha})}.
\end{aligned}$$

By using (10.19), we get

$$t_{i,1,\frac{m}{2}} = \frac{1}{2} y_{i,1,m,1} + \frac{1}{2} y_{i,1,m,2}.$$

Then

$$\eta_{i,1} = \frac{1}{2}.$$

Thus from Theorem 10.4

$$q_{i,1,k} = q_{i,1,m-k}. \tag{10.20}$$

By hypothesis (4) and from (10.18) we get

$$Q_m f(x_i) = \alpha_i \sum_{k=0}^{m} q_{i,k} \left(f(t_{i,0,k}) + f(t_{i,2,k}) \right) + \alpha_{i,1} \sum_{k=0}^{m} q_{i,1,k} f(t_{i,1,k})$$

and

$$Q''_m f(x_i) = \gamma_i \sum_{k=0}^{m} q_{i,k} \left(f(t_{i,0,k}) + f(t_{i,2,k}) \right) + \gamma_{i,1} \sum_{k=0}^{m} q_{i,1,k} f(t_{i,1,k}).$$

Particularly, for $f = f_{m+1}$ we obtain

$$f_{m+1}(t_{i,0,k}) + f_{m+1}(t_{i,2,k}) = \left((t_{i,0,k} - x_i)^{m+1} + (t_{i,2,k} - x_i)^{m+1} \right) \frac{u^{(m+1)}(x_i)}{(m+1)!}.$$

Knowing that m is even and $t_{i,s,k}$ for $s = 0, 2$ are symmetric with respect to x_i, then

$$f_{m+1}(t_{i,0,k}) + f_{m+1}(t_{i,2,k}) = 0, \quad \forall i = 1, \ldots, n-1.$$

According to (10.20), we have

$$\sum_{k=0}^{m} q_{i,1,k} f_{m+1}(t_{i,1,k}) = \sum_{k=0}^{m/2-1} q_{i,1,k} \left(f_{m+1}(t_{i,1,k}) + f_{m+1}(t_{i,1,m-k}) \right) + q_{i,1,\frac{m}{2}} f_{m+1}(t_{i,1,\frac{m}{2}}).$$

As $t_{i,1,k}$ and $t_{i,1,m-k}$ are symmetric with respect to x_i, and using the fact that $t_{i,1,\frac{m}{2}} = x_i$, we obtain for $k = 0, \ldots, m/2$

$$f_{m+1}(t_{i,1,k}) + f_{m+1}(t_{i,1,m-k}) = 0 \text{ and } f_{m+1}(t_{i,1,\frac{m}{2}}) = 0, \quad \forall i = 1, \ldots, n-1.$$

Therefore, for $i = 1, \ldots, n-1$,

$$Q_m f_{m+1}(x_i) = 0 \quad \text{and} \quad Q''_m f_{m+1}(x_i) = 0. \tag{10.21}$$

Using (10.15), (10.16), and (10.21), we get

$$|Q_m^{(r)}u(x_i) - u^{(r)}(x_i)| = O(h^{m+2-r}), \quad r = 0, 2,$$

which completes the proof. □

10.4 Numerical Examples

For $m = 3, 4, 5$, we present the numerical experiments of the same choice of B-splines (10.4) and (10.5). We consider the following test functions defined on $I = [0, 1]$ by

$$u_1(x) = \exp(-3x)\sin\left(\frac{\pi}{2}x\right) \text{ and } u_2(x) = \frac{x}{4} + \exp\left(\frac{-x}{2}\right)\sin(-3x\pi).$$

We define the local error between a function g and the quasi-interpolant $Q_m g$ at the knots of τ by the following relation:

$$E_{m,n}^{(k)}(g) := \max_{0\leqslant i\leqslant n}\left|Q_m^{(k)}g(x_i) - g^{(k)}(x_i)\right|, \quad k = 0, 1, 2,$$

and the numerical convergence order by

$$NCO_m^{(k)} := \frac{\log\left(\frac{E_{m,n_1}^{(k)}(g)}{E_{m,n_2}^{(k)}(g)}\right)}{\log\left(\frac{n_2}{n_1}\right)},$$

for $m = 3, 4, 5$.

To illustrate numerically the superconvergence characteristics of the supercovergent quasi-interpolants Q_m, we choose arbitary the interpolation points in the interval $[y_{i,s,m,1}, y_{i,s,m,2}]$ such that $\varepsilon_{i,s,k} \in [0, 1]$, $k = 0, \dots, m$, shown in Table 10.1, with

$$\begin{cases} \eta_{i,0} = -1, \quad \eta_{i,2} = 2, \quad \forall i = 0, \dots, n, \\ \eta_{i,1} = \frac{1}{2}, \quad \forall i = 1, \dots, n-1, \\ \eta_{0,1} = 1, \text{ and } \eta_{n,1} = 0. \end{cases}$$

We choose, to illustrate the super-superconvergence phenomenon, the sets of interpolation points corresponding to $\mu_{i,0}$ and $\mu_{i,2}$ such that they are symmetric with respect to x_i and the local interpolation points corresponding to $\mu_{i,1}$ are symmetric between them with respect to x_i. We take $\varepsilon_{i,0,k} = \varepsilon_{i,1,k} = \varepsilon_{i,2,k} \in [0, 1]$, $k = 0, \dots, m$, as it is shown in Table 10.2.

Table 10.1 The values of $\varepsilon_{i,s,k}$ for $m = 3, 4, 5$

m	$\varepsilon_{i,0,k}$	$\varepsilon_{i,1,k}$	$\varepsilon_{i,2,k}$
3	$\frac{2}{10}, \frac{3}{10}, \frac{6}{10}, \frac{8}{10}$	$\frac{1}{10}, \frac{3}{10}, \frac{6}{10}, \frac{9}{10}$	$\frac{1}{10}, \frac{4}{10}, \frac{6}{10}, \frac{9}{10}$
4	$\frac{2}{10}, \frac{3}{10}, \frac{5}{10}, \frac{7}{10}, \frac{8}{10}$	$\frac{1}{10}, \frac{3}{10}, \frac{4}{10}, \frac{7}{10}, \frac{9}{10}$	$\frac{1}{10}, \frac{3}{10}, \frac{5}{10}, \frac{7}{10}, \frac{9}{10}$
5	$\frac{2}{10}, \frac{3}{10}, \frac{4}{10}, \frac{6}{10}, \frac{7}{10}, \frac{8}{10}$	$\frac{1}{10}, \frac{2}{10}, \frac{5}{10}, \frac{6}{10}, \frac{7}{10}, \frac{9}{10}$	$\frac{1}{10}, \frac{3}{10}, \frac{4}{10}, \frac{6}{10}, \frac{7}{10}, \frac{9}{10}$

Table 10.2 The values of $\varepsilon_{i,s,k}$ for $m = 3, 4, 5$

m	$\varepsilon_{i,s,k}$
3	$\frac{1}{10}, \frac{3}{10}, \frac{6}{10}, \frac{9}{10}$
4	$\frac{1}{10}, \frac{3}{10}, \frac{5}{10}, \frac{7}{10}, \frac{9}{10}$
5	$\frac{1}{10}, \frac{3}{10}, \frac{4}{10}, \frac{6}{10}, \frac{7}{10}, \frac{9}{10}$

10.4.1 Superconvergence

10.4.1.1 Approximating Function Values

We give in Table 10.3 (resp. Table 10.4), for different values of n, the maximum absolute errors $E^{(0)}_{m,n}(u_1)$ (resp. $E^{(0)}_{m,n}(u_2)$) at knots associated with the superconvergent quasi-interpolants Q_m for $m = 3, 4, 5$ and the numerical convergence orders $NCO^{(0)}_m$.

Through these examples we remark that, when we increase n or m, we get a quasi-interpolant with small errors and the numerical convergence order is in good agreement with the theoretical one.

Table 10.3 The maximum absolute errors $E^{(0)}_{m,n}(u_1)$ and the numerical convergence orders $NCO^{(0)}_m$

n	$E^{(0)}_{3,n}(u_1)$	$NCO^{(0)}_3$	$E^{(0)}_{4,n}(u_1)$	$NCO^{(0)}_4$	$E^{(0)}_{5,n}(u_1)$	$NCO^{(0)}_5$
16	3.2526×10^{-7}	–	2.9322×10^{-9}	–	7.7024×10^{-10}	–
32	2.2001×10^{-8}	3.88593	8.4598×10^{-11}	5.11523	1.2296×10^{-11}	5.96897
64	1.4290×10^{-9}	3.94444	2.5408×10^{-12}	5.05726	1.9301×10^{-13}	5.99342
128	9.1028×10^{-11}	3.97262	7.7883×10^{-14}	5.02783	3.0199×10^{-15}	5.99804
Theoretical value	–	04	–	05	–	06

Table 10.4 The maximum absolute errors $E^{(0)}_{m,n}(u_2)$ and the numerical convergence orders $NCO^{(0)}_m$

n	$E^{(0)}_{3,n}(u_2)$	$NCO^{(0)}_3$	$E^{(0)}_{4,n}(u_2)$	$NCO^{(0)}_4$	$E^{(0)}_{5,n}(u_2)$	$NCO^{(0)}_5$
16	2.2774×10^{-5}	–	1.1865×10^{-6}	–	9.4917×10^{-7}	–
32	1.4161×10^{-6}	4.00735	2.5851×10^{-8}	5.52033	1.7446×10^{-8}	5.76563
64	8.9129×10^{-8}	3.98996	6.7715×10^{-10}	5.25462	2.8308×10^{-10}	5.94557
128	5.5686×10^{-9}	4.00049	1.9670×10^{-11}	5.10540	4.4644×10^{-12}	5.98661
Theoretical value	–	04	–	05	–	06

Table 10.5 The maximum absolute errors $E^{(1)}_{m,n}(u_1)$ and the numerical convergence orders $NCO^{(1)}_m$

n	$E^{(1)}_{3,n}(u_1)$	$NCO^{(1)}_3$	$E^{(1)}_{4,n}(u_1)$	$NCO^{(1)}_4$	$E^{(1)}_{5,n}(u_1)$	$NCO^{(1)}_5$
16	1.3792×10^{-5}	–	3.5624×10^{-7}	–	1.0091×10^{-8}	–
32	1.7416×10^{-6}	2.98529	2.2456×10^{-8}	3.98765	1.7790×10^{-10}	5.82586
64	2.1881×10^{-7}	2.99266	1.4094×10^{-9}	3.99395	4.1873×10^{-12}	5.40892
128	2.7421×10^{-8}	2.99633	8.8271×10^{-11}	3.99701	1.2997×10^{-13}	5.00969
Theoretical value	–	03	–	04	–	05

10.4.1.2 Approximating First Derivative Values

As in above, we illustrate numerically in Tables 10.5 and 10.6 the superconvergence phenomenon when first derivative values are approximated. The same comments given previously are true in this case.

Table 10.6 The maximum absolute errors $E^{(1)}_{m,n}(u_2)$ and the numerical convergence orders $NCO^{(1)}_m$

n	$E^{(1)}_{3,n}(u_2)$	$NCO^{(1)}_3$	$E^{(1)}_{4,n}(u_2)$	$NCO^{(1)}_4$	$E^{(1)}_{5,n}(u_2)$	$NCO^{(1)}_5$
16	2.5917×10^{-4}	–	8.3940×10^{-5}	–	7.8849×10^{-5}	–
32	2.8162×10^{-5}	3.20205	5.3898×10^{-6}	3.96104	1.3308×10^{-6}	5.88864
64	3.2522×10^{-6}	3.11427	3.4084×10^{-7}	3.98303	2.1210×10^{-8}	5.97147
128	3.8967×10^{-7}	3.06109	2.1419×10^{-8}	3.99213	3.4174×10^{-10}	5.95571
Theoretical value	–	03	–	04	–	05

Table 10.7 The maximum absolute errors $E^{(2)}_{m,n}(u_1)$ and the numerical convergence orders $NCO^{(2)}_m$

n	$E^{(2)}_{3,n}(u_1)$	$NCO^{(2)}_3$	$E^{(2)}_{4,n}(u_1)$	$NCO^{(2)}_4$	$E^{(2)}_{5,n}(u_1)$	$NCO^{(2)}_5$
16	5.8810×10^{-3}	–	1.7911×10^{-5}	–	1.1748×10^{-5}	–
32	1.4715×10^{-3}	1.99877	1.5738×10^{-6}	3.50853	6.8845×10^{-7}	4.09301
64	3.6801×10^{-4}	1.99948	1.5714×10^{-7}	3.32409	4.1129×10^{-8}	4.06510
128	9.2018×10^{-5}	1.99976	2.1328×10^{-8}	2.88130	2.5033×10^{-9}	4.03825
Theoretical value	–	02	–	03	–	04

10.4.1.3 Approximating Second Derivative Values

We give numerical results in Tables 10.7 and 10.8 of the superconvergence phenomenon when second derivative values are approximated. The same comments given previously are true in this case.

Table 10.8 The maximum absolute errors $E^{(2)}_{m,n}(u_2)$ and the numerical convergence orders $NCO^{(2)}_m$

n	$E^{(2)}_{3,n}(u_2)$	$NCO^{(2)}_3$	$E^{(2)}_{4,n}(u_2)$	$NCO^{(2)}_4$	$E^{(2)}_{5,n}(u_2)$	$NCO^{(2)}_5$
16	8.3292×10^{-2}	–	1.1901×10^{-2}	–	1.6565×10^{-2}	–
32	2.0392×10^{-2}	2.03013	8.1578×10^{-4}	3.86687	7.4544×10^{-4}	4.47396
64	5.1176×10^{-3}	1.99451	6.1828×10^{-5}	3.72184	4.8298×10^{-5}	3.94806
128	1.2775×10^{-3}	2.00209	5.8223×10^{-6}	4.40859	3.0490×10^{-6}	3.98555
Theoretical value	–	02	–	03	–	04

10.4.2 Super-Superconvergence

10.4.2.1 Approximating Function Values

In Table 10.9 (resp. Table 10.10) we illustrate the super-superconvergence phenomenon for different values of n, by giving the maximum absolute errors $E^{(0)}_{m,n}(u_1)$ (resp. $E^{(0)}_{m,n}(u_2)$) at knots associated with the superconvergent quasi-interpolants Q_m for $m = 3, 4, 5$ and the numerical convergence orders $NCO^{(0)}_m$.

Table 10.9 The maximum absolute errors $E^{(0)}_{m,n}(u_1)$ and the numerical convergence orders $NCO^{(0)}_m$

n	$E^{(0)}_{3,n}(u_1)$	$NCO^{(0)}_3$	$E^{(0)}_{4,n}(u_1)$	$NCO^{(0)}_4$	$E^{(0)}_{5,n}(u_1)$	$NCO^{(0)}_5$
16	3.2137×10^{-7}	–	6.6701×10^{-10}	–	7.5543×10^{-10}	–
32	2.1718×10^{-8}	3.88721	1.0506×10^{-11}	5.98833	1.2065×10^{-11}	5.96835
64	1.4100×10^{-9}	3.94510	1.6432×10^{-13}	5.99867	1.8939×10^{-13}	5.99334
128	8.9799×10^{-11}	3.97292	2.5687×10^{-15}	5.99930	2.9633×10^{-15}	5.99923
Theoretical value	–	04	–	06	–	06

Table 10.10 The maximum absolute errors $E^{(0)}_{m,n}(u_2)$ and the numerical convergence orders $NCO^{(0)}_m$

n	$E^{(0)}_{3,n}(u_2)$	$NCO^{(0)}_3$	$E^{(0)}_{4,n}(u_2)$	$NCO^{(0)}_4$	$E^{(0)}_{5,n}(u_2)$	$NCO^{(0)}_5$
16	2.2463×10^{-5}	–	9.5890×10^{-7}	–	9.2945×10^{-7}	–
32	1.3964×10^{-6}	4.00770	1.5442×10^{-8}	5.95638	1.7112×10^{-8}	5.76322
64	8.7915×10^{-8}	3.98955	2.4310×10^{-10}	5.94508	2.7776×10^{-10}	5.94508
128	5.4925×10^{-9}	4.00006	3.8056×10^{-12}	5.98650	4.3808×10^{-12}	5.98650
Theoretical value	–	04	–	06	–	06

10.4.2.2 Approximating Second Derivative Values

We give numerical results in Tables 10.11 and 10.12 of the superconvergence phenomenon when second derivative values are approximated.

A comparison with the previous results allow us to see that when m is even the associated errors of the super-superconvergence phenomenon are smaller than the ones of the superconvergence phenomenon.

Table 10.11 The maximum absolute errors $E^{(2)}_{m,n}(u_1)$ and the numerical convergence orders $NCO^{(2)}_m$

n	$E^{(2)}_{3,n}(u_1)$	$NCO^{(2)}_3$	$E^{(2)}_{4,n}(u_1)$	$NCO^{(2)}_4$	$E^{(2)}_{5,n}(u_1)$	$NCO^{(2)}_5$
16	5.6151×10^{-3}	–	1.0443×10^{-5}	–	1.1723×10^{-5}	–
32	1.4023×10^{-3}	2.00148	6.3043×10^{-7}	4.05017	6.8691×10^{-7}	4.09314
64	3.5037×10^{-4}	2.00086	3.8611×10^{-8}	4.02923	4.1034×10^{-8}	4.06520
128	8.7565×10^{-5}	2.00046	2.3870×10^{-9}	4.01576	2.4974×10^{-9}	4.03832
Theoretical value	–	02	–	04	–	04

Table 10.12 The maximum absolute errors $E^{(2)}_{m,n}(u_2)$ and the numerical convergence orders $NCO^{(2)}_m$

n	$E^{(2)}_{3,n}(u_2)$	$NCO^{(2)}_3$	$E^{(2)}_{4,n}(u_2)$	$NCO^{(2)}_4$	$E^{(2)}_{5,n}(u_2)$	$NCO^{(2)}_5$
16	7.8989×10^{-2}	–	1.1962×10^{-2}	–	1.6544×10^{-2}	–
32	1.9325×10^{-2}	2.03117	7.7026×10^{-4}	3.95700	7.4433×10^{-4}	4.47420
64	4.8490×10^{-3}	1.99471	4.8497×10^{-5}	3.98935	4.7400×10^{-5}	3.97290
128	1.2101×10^{-3}	2.00251	3.0367×10^{-6}	3.99734	1.1121×10^{-6}	3.98550
Theoretical value	–	02	–	04	–	04

10.5 Conclusion

In this paper we have presented a simple method to get a super-superconvergence phenomenon of cubic spline quasi-interpolants. This important property can be a powerful tool to improve the accuracy and efficiency of the traditional methods used for solving some numerical analysis problems. Our aim in future works is to use these super-superconvergent quasi-interpolants for solving differential equation problems.

References

1. Abbadi, A., Barrera, D., Ibáñez, M.J., Sbibih, D.: A general method for constructing quasi-interpolants from B-splines. J. Comput. Appl. Math. **234**, 1324–1337 (2010)
2. Barrera, D., Ibáñez, M. J., Sablonnière, P., Sbibih, D.: On near-best discrete quasi-interpolation on a four-directional mesh. J. Comput. Appl. Math. **233**, 1470–1477 (2010)
3. de Casteljau, P.: Shape Mathematics and CAD. Kogan Page, 120 Pentonville Road, London N1, UK. £13. Comput. Aided Des. **18**, 338 (1986)
4. Rahouti, A., Serghini, A., Tijini, A.: Construction of superconvergent quasi-interpolants using a new normalized C^2 cubic B-splines. Math. Comput. Simul. **178**, 603–624 (2020)
5. Rahouti, A., Serghini, A., Tijini, A.: Construction of a normalized basis of a univariate quadratic C^1 spline space and application to the quasi-interpolation Bol. Soc. Paran. Mat. **40**, 1–21 (2022)
6. Ramshaw, L.: Blossoming are polar forms. Comput. Aided Geom. Des. **6**, 323–358 (1989)
7. Remogna, S.: Constructing good coefficient functionals for bivariate C^1 quadratic spline quasi-interpolants. In: Daehlen, M. et al. (Eds.) Mathematical Methods for Curves and Surfaces. Lecture Notes in Computer Science, vol. 5862, pp. 329–346. Springer, Berlin (2010)
8. Sbibih, D., Serghini, A., Tijini, A.: Polar forms and quadratic splines quasi-interpolants over Powell-Sabin triangulation. Appl. Num. Math. **59**, 938–958 (2009)
9. Sbibih, D., Serghini, A., Tijini, A.: Superconvergent C^1 cubic spline quasi-interpolants on Powell-Sabin partitions. BIT Numer. Math. **55**(3), 797–821 (2015)
10. Schmidt, J.W., He*B*, W.: An always successful method in univariate convex C^2 interpolation. Numer. Math. **71**, 237–252 (1995)

Chapter 11
Calibration Adjustment for Dealing with Nonresponse in the Estimation of Poverty Measures

María Illescas-Manzano, Sergio Martínez-Puertas, María del Mar Rueda, and Antonio Arcos-Rueda

Abstract The analysis of poverty measures has been receiving increased attention in recent years. This paper contributes to the literature by developing percentile ratio estimators when there are missing data. Calibration adjustment is used for treating the non-response bias. Variances of the proposed estimators could be not expressible by simple formulae and resampling techniques are investigated for obtaining variance estimators. A numerical example based on data from the Spanish Household Panel Survey is taken up to illustrate how suggested procedures can perform better than existing ones.

Keywords Missing data · Sampling · Calibration · Poverty

11.1 Introduction

The analysis of poverty measures is a topic of increased interest to society. The official poverty rate and the number of people in poverty are important measures of the country's economic wellbeing. The common characteristic of many poverty measures is their complexity. The literature on survey sampling is usually focused on the goal of estimating linear parameters. However when the variable of interest is a measure of wages or income, the distribution function is a relevant tool because is required to calculate the poverty line, the low income proportion, the poverty gap and other poverty measures.

M. Illescas-Manzano · S. Martínez-Puertas
Department of Economy and Company, University of Almería, Almería, Spain
e-mail: mim434@inlumine.ual.es; spuertas@ual.es

M. del M. Rueda (✉) · A. Arcos-Rueda
Department of Statistics and O. R., University of Granada, Granada, Spain
e-mail: mrueda@ugr.es; aarcos@correo.ugr.es

D. Barrera et al. (eds.), *Mathematical and Computational Methods for Modelling, Approximation and Simulation*, SEMA SIMAI Springer Series 29,
https://doi.org/10.1007/978-3-030-94339-4_11

The lack of response is a growing problem in economic surveys. Although there are many procedures for their treatment, few efficient techniques have been developed for their treatment in the estimation of non-linear parameters. Recently, in [15] various estimators for the distribution function in the presence of missing data have been proposed. Using these estimators, we first propose new estimators for several poverty measures, which efficiently use auxiliary information at the estimation stage. Due to the complexity of the percentile ratios and the complex sampling designs used by the official sample surveys, variances of these complex statistics could be not expressible by simple formulae. Additional techniques for variance estimation are therefore required under this scenario.

This paper is organized as follows. Section 11.2 introduces the estimation of the distribution function when there are missing data. In Sect. 11.3, the proposed percentile ratio estimators are described. In Sect. 11.4 we derive resampling techniques for the problem of the variance estimation of percentile ratio estimators. A simulation study based on data derived from the Spanish Household Panel Survey is presented in Sect. 11.5. This study shows how the proposed estimates of the poverty measures perform in reduction of bias and precision when calibration is used for nonresponse that is not missing at random.

11.2 Calibrating the Distribution Function for Treating the Non-response

Consider a finite population $U = \{1, \ldots, N\}$ consisting of N different and identifiable units. Let us assume a sampling design d defined in U with positive first-order inclusion probabilities π_i $i, \in U$. Let $d_i = \pi_i^{-1}$ denote the sampling design-basic weight for unit $i \in U$ which is known. We assume missing data on the sample s obtained by the sampling design d. Let us denote by s_r, the respondent sample of size r, and s_m the non-respondent sample of size $n - r$.

Let y_i be the value of the character under study. The distribution function $F_y(t)$ can be estimated by the Horvitz-Thompson estimator:

$$\widehat{F}_{HT}(t) = \frac{1}{N} \sum_{k \in s_r} d_k \Delta(t - y_k), \tag{11.1}$$

where

$$\Delta(t - y_k) = \begin{cases} 0 & \text{if } t < y_k, \\ 1 & \text{if } t \geq y_k, \end{cases}$$

and $d_k = 1/\pi_k$, the basic design weights.

This estimator is biased for the distribution function. There are several approach for dealing with nonresponse. The most important method is weighting. We assume the existence of auxiliary information relative to several variables related to the main variable y, $\mathbf{x} = (x_1, x_2, \ldots, x_J)'$. Based on this auxiliary information, calibration weighting is used in [15] to propose three methods to reduce the non-response bias in the estimation of the distribution function:

- The first method is based on the methodology proposed in [14]. We define a pseudo-variable $g_k = \widehat{\beta}' \mathbf{x}_k$ for $k = 1, 2, \ldots N$, where

$$\widehat{\beta} = \left(\sum_{j \in s_r} d_j \mathbf{x}_j \mathbf{x}_j' \right)^{-1} \cdot \sum_{j \in s_r} d_j \mathbf{x}_j y_j.$$

Thus we define a calibrated estimator by imposing that the calibrated weights w_k evaluated in the observed sample give perfect estimates for the distribution function in a set of predetermined points t_j for $j = 1, 2, \ldots, P$ that we choose arbitrarily:

$$\frac{1}{N} \sum_{k \in s_r} w_k \Delta(\mathbf{t} - g_k) = F_g(\mathbf{t}), \tag{11.2}$$

where $F_g(\mathbf{t})$ denotes the finite distribution function of the pseudo-variable g_k evaluated at the point $\mathbf{t} = (t_1, \ldots, t_p)'$ and $\Delta(\mathbf{t} - g_k) = (\Delta(t_1 - g_k), \ldots, \Delta(t_p - g_k))'$.

A common way to compute calibration weights is linearly (using the chi-square distance method) and we obtain an explicit expression of the estimator as:

$$\begin{aligned} \hat{F}_{cal}^{(1)}(t) &= \frac{1}{N} \sum_{k \in s_r} w_k^{(1)} \Delta(t - y_k) \\ &= \widehat{F}_{HT}(t) + \left(F_g(\mathbf{t}) - \frac{1}{N} \sum_{k \in s_r} d_k \Delta(\mathbf{t} - g_k) \right)' \cdot T^{-1} \cdot H \end{aligned} \tag{11.3}$$

where $T = \sum_{k \in s_r} d_k \Delta(\mathbf{t} - g_k) \Delta(\mathbf{t} - g_k)'$ and $H = \sum_{k \in s_r} d_k \Delta(\mathbf{t} - g_k) \Delta(t - y_k)$.

Following [14], if we denote by $k_i = \sum_{k \in s_r} d_k \Delta(t_i - g_k)$ for $i = 1, \ldots, P$ the condition $k_i > k_{i-1}$ for $i = 2, \ldots, P$ guarantees the existence of T^{-1}.

- The second method is based on two-step calibration weighting as in the work in [10]:
  1. The first calibration is designed to remove the non-response bias.
     Consider the M vector of explanatory model variables, $\mathbf{x}_k^*$ which population totals $\sum_U \mathbf{x}_k^*$ are know. The calibration under the restrictions $\sum_{s_r} v_k^{(1)} \mathbf{x}_k^* = \sum_U \mathbf{x}_k^*$ yields the calibrations weights $v_k^{(1)}$, $k = 1, \ldots, s_r$.
  2. The second one to decrease the sampling error in the estimation of the distribution function.

The auxiliary information of the calibration variables $\mathbf{x}$ is incorporated through the calibrated weights $v_k^{(2)}$ obtained with the restrictions $\sum_{s_r} v_k^{(2)} \Delta(\mathbf{t} - g_k) = F_g(\mathbf{t})$. The final estimator is given by

$$\hat{F}_{cal}^{(2)}(t) = \frac{1}{N} \sum_{k \in s_r} w_k^{(2)} \Delta(t - y_k) = \frac{1}{N} \sum_{k \in s_r} v_k^{(2)} v_k^{(1)} \Delta(t - y_k). \tag{11.4}$$

This method allows different variables to be used in each phase (model variables $\mathbf{x}_k^*$ and calibration variables $\mathbf{x}$), since the model for non-response and the predictive model can be very different.

- The last method is based on instrumental variables (see [6] and[11]). The calibration is done in a single stage, but different variables are also used to model the lack of response and for the calibration equation. By assuming that the probability of response can be modeled by: $\theta_k = f(\gamma' \mathbf{x}_k^*)$ for some vector parameter γ, where $h(\cdot) = 1/f(\cdot)$ is a known and everywhere monotonic and twice differentiable function. We denote as $\mathbf{z}_k = \Delta(\mathbf{t} - g_k)$.

  The calibration equation is given by

$$\frac{1}{N} \sum_{k \in s_r} \frac{d_k}{f(\hat{\gamma}' \mathbf{x}_k^*)} \mathbf{z}_k = \frac{1}{N} \sum_{k \in s_r} d_k h(\hat{\gamma}' \mathbf{x}_k^*) \mathbf{z}_k = F_g(\mathbf{t}) \tag{11.5}$$

and the resulting calibrated estimator is:

$$\hat{F}_{cal}^{(3)}(t) = \frac{1}{N} \sum_{k \in s_r} w_k^{(3)} \Delta(t - y_k) = \frac{1}{N} \sum_{k \in s_r} d_k h(\hat{\gamma}' \mathbf{x}_k^*) \Delta(t - y_k), \tag{11.6}$$

where $\hat{\gamma}$ is a consistent estimator of vector γ. Authors use several approximation methods for deriving the solution of the minimization problem. We denote by $\widehat{F}_{Dcal}^{(3)}(t)$ the estimator based in the Deville's approach [6] which needs to meet the condition $M = P$. To consider more calibration restrictions in Eq. (11.5) than M, we consider the estimator $\widehat{F}_{KL1cal}^{(3)}(t)$ and $\widehat{F}_{KL2cal}^{(3)}(t)$ based on [11] where $P > M$.

11.3 Poverty Measures Estimation with Missing Values

Currently, poverty measurement, wage inequality, inequality and life condition are overriding issues for governments and society. Some indices and poverty measures used in the poverty evaluation and income inequality measurement are based on quantile and quantiles ratios. Thereby, Eurostat currently set the poverty line (the population threshold for classification into poor and nonpoor) equal to sixty percent of the equivalized net income median Q_{50}. On the other hand, the percentile ratios

Q_{95}/Q_{20}; Q_{90}/Q_{10} and Q_{80}/Q_{20} [9]; Q_{95}/Q_{50} and Q_{50}/Q_{10} (see [12] and [4]); Q_{50}/Q_5 and Q_{50}/Q_{25} [7] have been considered as measures for wage inequality. We focus on estimating the poverty measures based on percentile ratios.

The population α-quantile of y is defined as follows

$$Q_y(\alpha) = \inf\{t : F_y(t) \geq \alpha\} = F_y^{-1}(\alpha). \tag{11.7}$$

A general procedure to incorporate the auxiliary information in the estimation of $Q_y(\alpha)$ is based on the obtainment of an indirect estimator $\widehat{F}_y(t)$ of $F_y(t)$ that fulfills the distribution function's properties. Under this assumption, the quantile $Q_y(\alpha)$ can be estimated in a following way:

$$\widehat{Q}_y(\alpha) = \inf\{t : \widehat{F}_y(t) \geq \alpha\} = \widehat{F}_y^{-1}(\alpha). \tag{11.8}$$

The distribution function estimators described in the previous section allow us to incorporate the auxiliary information in the estimation of quantiles in the presence of non-response and obtain estimators for percentile ratios. Perhaps, some of these calibrated estimators do not satisfy all the properties of the distribution function and consequently for its application in the estimation of quantiles some modifications are necessary. Specifically, the properties that an estimator $\widehat{F}_y(t)$ of the distribution function $F_y(t)$ must meet are the following:

i. $\widehat{F}_y(t)$ is continuous on the right.
ii. $\widehat{F}_y(t)$ is monotone nondecreasing,
iii. (a) $\lim_{t\to-\infty} \widehat{F}_y(t) = 0$ and (b) $\lim_{t\to+\infty} \widehat{F}_y(t) = 1$.

Firstly, it is easy to see that all estimators satisfy the conditions (i) and iii.(a). Secondly, following [14], the estimator $\widehat{F}_{cal}^{(1)}(t)$ meet the rest of conditions if t_P is sufficiently large (i.e $F_g(t_P) = 1$). On the other hand, it's easy to see that $\widehat{F}_{cal}^{(2)}(t)$ satisfy the condition iii.(b) if t_P is sufficiently large but it is not monotone nondecreasing in general. Thus, we can apply the procedure described in [13]. This procedure, for a general estimator $\widehat{F}_y$, is defined in the following way:

$$\tilde{F}_y(y_{[1]}) = \widehat{F}_y(y_{[1]}), \quad \tilde{F}_y(y_{[i]}) = \max\{\widehat{F}_y(y_{[i]}),\ \tilde{F}_y(y_{[i-1]})\} \quad i = 2, \ldots, r. \tag{11.9}$$

Finally, all estimators based on $\widehat{F}_{cal}^{(3)}(t)$ are nondecreasing if $\theta_k = f(\gamma' \mathbf{x}_k^*) \geq 0$ for all $k \in U$ (response model based on logit, raking and logistic methods) because the calibration weights $\omega_k^{(3)} \geq 0$. Moreover, $\widehat{F}_{Dcali}^{(3)}(t)$ fulfills condition iii.(b) with t_P sufficiently large whereas following [15], $\widehat{F}_{KL1cali}^{(3)}(t)$ and $\widehat{F}_{KL2cali}^{(3)}(t)$ meet condition iii.(b) if in addition to considering t_P sufficiently large, a component of the vector $\mathbf{x}_k^*$ contains all 1's.

Based on the population distribution function $F_y(t)$, given two values $1 > \alpha_1 > \alpha_2 > 0$, the percentile ratio $R(\alpha_1, \alpha_2)$ is define as follow:

$$R(\alpha_1, \alpha_2) = \frac{Q_y(\alpha_1)}{Q_y(\alpha_2)} \tag{11.10}$$

and it can be estimated with a generic quantile estimator $\widehat{Q}_y(\alpha)$ as follows:

$$\widehat{R}(\alpha_1, \alpha_2) = \frac{\widehat{Q}_y(\alpha_1)}{\widehat{Q}_y(\alpha_2)}. \tag{11.11}$$

Thus, the quantile estimator derived from $\widehat{F}_{cal}^{(1)}$; $\widehat{F}_{cal}^{(2)}$ and $\widehat{F}_{cal}^{(3)}$ can be employed in the estimation of $R(\alpha_1, \alpha_2)$.

11.4 Variance Estimation for Percentile Ratio Estimators with Resampling Method

Given the complexity of the proposed percentile ratio estimators, we have considered the use of bootstrap techniques for estimating variance and developing confidence intervals associated with the proposed calibration estimators. In this study, we consider the frameworks proposed in [1], [2], and [3].

First, the bootstrap procedure described in [3] consider the repetition of sample units for creating artificial bootstrap populations. The bootstrap samples are drawing with the original sampling design from artificial populations. Specifically, if the population size $N = n \cdot q + m$ with $0 < m < n$, the artificial population U_B is obtained with q repetitions of s and an additional sample of size m selected by simple random sampling without replacement from s. Given a generic percentile ratio estimator $\widehat{R}(\alpha_1, \alpha_2)$, if we consider M independent artificial populations U_B^j with $j = 1, \ldots, M$ and for each pseudo population U_B^j we select K bootstrap samples $s_1^j, \ldots, s_K^j$ with sample size n, we can compute the bootstrap estimates $\widehat{R}^*(\alpha_1, \alpha_2)_h^j$ with the sample s_h^j for the population U_B^j and following [5], we can compute

$$\widehat{V}_j = \frac{1}{K-1} \sum_{h=1}^{K} (\widehat{R}^*(\alpha_1, \alpha_2)_h^j - \widehat{R}_j^*(\alpha_1, \alpha_2))^2, \tag{11.12}$$

where

$$\widehat{R}_j^*(\alpha_1, \alpha_2) = \frac{1}{K} \sum_{h=1}^{K} \widehat{R}^*(\alpha_1, \alpha_2)_h^j, \tag{11.13}$$

Finally, the variance estimation for the estimator $\widehat{R}(\alpha_1, \alpha_2)$ is given by

$$\widehat{V}(\widehat{R}(\alpha_1, \alpha_2)) = \frac{1}{M}\sum_{j=1}^{M} \widehat{V}_j. \tag{11.14}$$

On the other hand, in [1] and [2] a direct bootstrap method has been proposed, where it is not necessary to obtain an artificial population, since the bootstrap samples are drawn from s under a sampling scheme different from the original sampling design. Both frameworks (see [1] and [2]) can be applied under several sample designs, but particularly, if the sample s is drawing with simple random sampling without replacement, the sampling design proposed by Antal and Tillé [1] select two samples from s, the first one is drawing by simple random sampling without replacement and the second one is drawing with one-one sampling design (a sampling design for resampling). Similarly, under simple random sampling without replacement, the sampling design proposed by [2] draw a first sample with Bernoulli design and a second sample with double half sampling design (another sampling design for resampling). For more details see [1] and [2].
For two frameworks, given a percentile ratio estimator $\widehat{R}(\alpha_1, \alpha_2)$, we draw M bootstrap samples $s_1^*, \ldots, s_M^*$ from s, according to the sampling schemes of [1] and [2] respectively. The bootstrap estimation for variance of the estimator $\widehat{R}(\alpha_1, \alpha_2)$ is given by

$$\widehat{V}(\widehat{R}(\alpha_1, \alpha_2)) = \frac{1}{M}\sum_{j=1}^{M}(\widehat{R}(\alpha_1, \alpha_2)_j^* - \bar{R}(\alpha_1, \alpha_2)^*)^2, \tag{11.15}$$

where $\widehat{R}(\alpha_1, \alpha_2)_j^*$ is the bootstrap estimator computed with the bootstrap sample s_j^* and

$$\bar{R}(\alpha_1, \alpha_2)^* = \frac{1}{M}\sum_{j=1}^{M} \widehat{R}(\alpha_1, \alpha_2)_j^*. \tag{11.16}$$

Finally, based on the variance estimation $\widehat{V}(\widehat{R}(\alpha_1, \alpha_2))$ obtained with a bootstrap method, the $1-\alpha$ level confidence interval based on the approximation by a standard normal distribution is defined as follows:

$$\left[\widehat{R}(\alpha_1, \alpha_2) - z_{1-\alpha/2}\cdot\sqrt{\widehat{V}(\widehat{R}(\alpha_1, \alpha_2))}, \widehat{R}(\alpha_1, \alpha_2) + z_{1-\alpha/2}\cdot\sqrt{\widehat{V}(\widehat{R}(\alpha_1, \alpha_2))}\right], \tag{11.17}$$

where z_α is the α quantile of the standard normal distribution. For all bootstrap methods included in this study, we can compute with this procedure the respective confident interval.

11.5 Simulation Study

To determine the behaviour of the estimators when they are applied to real data we consider data from the region of Andalusia of 2016 Spanish living conditions survey carried out by the Instituto Nacional de Estadística (INE) of Spain. The survey data collected are considered as a population with size $N = 1442$ and samples are selected from it. The study variable y is the equivalised net income and the auxiliary variables included are the following dummy variables b_1 = "Home without mortgage", b_2 = "Four-bedroom home" and b_3 ="Can the home afford to go on vacation away from home, at least one week a year?". We considered the vector of model variables $(x_k^*)^{'} = (1, b_{1k})$ and the vector of calibration variables $(x_k)^{'} = (1, b_{1k}, b_{2k}, b_{3k})$.

We consider four response mechanism where the probability of the k-th individual of responde is given by

$$\theta_k = \frac{1}{exp(A + b_{1k}/B)} \tag{11.18}$$

with different values for A and B.

The ratio estimators considered in this simulation study, based on the respondent sample s_r, are obtained from the Horvitz-Thompson estimator $\widehat{F}_{HT}(t)$. We denoted by $\widehat{R}_D^{(3)}$ the calibration estimator based on [6] and we denoted by $\widehat{R}_{KL1}^{(3)}$ and $\widehat{R}_{KL2}^{(3)}$ the calibration estimators based on [11]. The estimator $\widehat{R}_{cal}^{(1)}$ has been included only with comparative purposes with respect to the rest of proposed estimators because it only considers the respondent sample and it does not deal with nonresponse whereas the rest of the estimators proposed try to deal with the bias produced by nonresponse. Although the real response mechanism considered is based on raking method, for $\widehat{R}_D^{(3)}$, $\widehat{R}_{KL1}^{(3)}$ and $\widehat{R}_{KL2}^{(3)}$ three versions of them are computed based on linear, raking and logit (l; u) response models.

We selected $W = 1000$ samples with several sample sizes, $n = 100$, $n = 125$, $n = 150$ and $n = 200$, under simple random sampling without replacement (SRSWOR) and for each estimator included in the simulation study, we computed estimates of $R(\alpha_1, \alpha_2)$ for 50th/25th, 80th/20th, 90th/10th, 90th/20th, 95th/20th and 95th/50th. The performance of each estimator is measured by the relative bias, (RB), and the relative efficiency (RE), given by

$$\text{RB}(\widehat{R}(\alpha_1, \alpha_2)) = \sum_{w=1}^{W} \frac{\left(\widehat{R}(\alpha_1, \alpha_2)\right)_w - R(\alpha_1, \alpha_2)}{R(\alpha_1, \alpha_2)}, \tag{11.19}$$

$$\text{RE}(\widehat{R}(\alpha_1, \alpha_2)) = \frac{\sum_{w=1}^{W} \left[\left(\widehat{R}(\alpha_1, \alpha_2)\right)_w - R(\alpha_1, \alpha_2)\right]^2}{\sum_{w=1}^{W} \left[\left(\widehat{R}_{HT}(\alpha_1, \alpha_2)\right)_w - R(\alpha_1, \alpha_2)\right]^2}, \tag{11.20}$$

where $\widehat{R}(\alpha_1, \alpha_2)$ is a percentile ratio estimator and $\widehat{R}_{HT}(\alpha_1, \alpha_2)$ is the percentile ratio estimator based in the Horvitz-Thompson $\widehat{F}_{HT}(t)$ estimator .

Tables 11.1, 11.2, 11.3, 11.4, 11.5 and 11.6 provide the values of the relative bias and the relative efficiency for this population for several sample sizes and response mechanism of the estimators compared.

Results from Tables 11.1, 11.2, 11.3, 11.4, 11.5, and 11.6 show an important bias for the estimator $\widehat{R}_{HT}$ in almost percentile ratios. The estimator $\widehat{R}^{(1)}_{cal}$ is not capable of correcting the bias in several situations, giving worse estimates than the HT estimator for some ratios and some response mechanisms. The proposed estimators have better values of RB with slight differences between them, although there is no uniformly better estimator than the rest.

Table 11.1 RB and RE for several sample sizes of the estimators of $R(0.5, 0.25)$. SRSWOR from the 2016 SPANISH LIVING CONDITIONS SURVEY

	RB	RE	RB	RE	RB	RE	RB	RE
Estimator	$n = 100$		$n = 125$		$n = 150$		$n = 200$	
$1/\exp(0.5 + b_1/3)$								
$\widehat{R}_{HT}$	1.4641	1	1.6163	1	1.7126	1	1.932	1
$\widehat{R}^{(1)}_{cal}$	0.0543	0.0049	0.0509	0.0032	0.0481	0.0024	0.0461	0.0014
$\widehat{R}^{(2)}_{cal}$	0.0107	0.0038	0.0056	0.0022	0.0023	0.0016	−0.001	9e−04
$\widehat{R}_{KL1cali}$	0.0116	0.0035	0.0051	0.0021	0.0019	0.0015	−8e−04	8e−04
$\widehat{R}_{KL1calr}$	0.0109	0.0035	0.0048	0.0021	0.0021	0.0015	−0.0011	9e−04
$\widehat{R}_{KL1calo}$	0.011	0.0036	0.0048	0.0021	0.002	0.0015	−0.0012	8e−04
$\widehat{R}_{KL2cali}$	0.0117	0.0035	0.005	0.0021	0.002	0.0015	−8e−04	8e−04
$\widehat{R}_{KL2calr}$	0.0108	0.0035	0.0046	0.0021	0.0019	0.0015	−0.0012	8e−04
$\widehat{R}_{KL2calo}$	0.0108	0.0035	0.0046	0.0021	0.0019	0.0015	−0.0012	8e−04
$\widehat{R}_{Dcali}$	0.0101	0.0036	0.0043	0.0021	0.0011	0.0016	−0.001	9e−04
$\widehat{R}_{Dcalr}$	0.0092	0.0036	0.004	0.0021	0.0011	0.0016	−0.0013	9e−04
$\widehat{R}_{Dcalo}$	0.0092	0.0036	0.004	0.0021	0.0011	0.0016	−0.0013	9e−04
$1/\exp(0.25 + b_1/2)$								
$\widehat{R}_{HT}$	0.5152	1	0.502	1	0.4624	1	0.4441	1
$\widehat{R}^{(1)}_{cal}$	0.046	0.0264	0.041	0.0236	0.0391	0.0249	0.0385	0.0266
$\widehat{R}^{(1)}_{cal}$	0.0074	0.0204	0.0024	0.0176	−0.0018	0.0175	−0.0019	0.0172
$\widehat{R}_{KL1cali}$	0.0072	0.0188	0.003	0.0168	−0.0014	0.0173	−0.0014	0.0169
$\widehat{R}_{KL1calr}$	0.0069	0.0187	0.0026	0.0168	−0.0015	0.0173	−0.0017	0.0169
$\widehat{R}_{KL1calo}$	0.0068	0.0188	0.0026	0.0168	−0.0015	0.0172	−0.0016	0.0169
$\widehat{R}_{KL2cali}$	0.0072	0.0188	0.0029	0.0168	−0.0014	0.0173	−0.0014	0.0169
$\widehat{R}_{KL2calr}$	0.0067	0.0188	0.0026	0.0168	−0.0014	0.0173	−0.0016	0.0169
$\widehat{R}_{KL2calo}$	0.0067	0.0188	0.0026	0.0168	−0.0014	0.0173	−0.0016	0.0169
$\widehat{R}_{Dcali}$	0.0063	0.0196	0.0025	0.0174	−0.0018	0.0175	−0.0016	0.0171
$\widehat{R}_{Dcalr}$	0.0056	0.0196	0.0023	0.0174	−0.0018	0.0175	−0.0019	0.0171
$\widehat{R}_{Dcalo}$	0.0056	0.0196	0.0023	0.0174	−0.0018	0.0175	−0.0019	0.0171

(continued)

Table 11.1 (continued)

	RB	RE	RB	RE	RB	RE	RB	RE
Estimator	$n = 100$		$n = 125$		$n = 150$		$n = 200$	
$1/\exp(0.75 + b_1/8)$								
$\widehat{R}_{HT}$	1.7102	1	1.9192	1	2.1328	1	2.4505	1
$\widehat{R}_{cal}^{(1)}$	0.0601	0.0046	0.0539	0.0028	0.0491	0.002	0.0467	0.0012
$\widehat{R}_{cal}^{(2)}$	0.0111	0.0033	0.0066	0.0021	0.0039	0.0014	0.0015	7e−04
$\widehat{R}_{KL1cali}$	0.0132	0.0031	0.0073	0.002	0.003	0.0013	6e−04	7e−04
$\widehat{R}_{KL1calr}$	0.0121	0.0031	0.0067	0.002	0.0027	0.0013	2e−04	7e−04
$\widehat{R}_{KL1calo}$	0.0121	0.0031	0.0069	0.002	0.0028	0.0013	3e−04	7e−04
$\widehat{R}_{KL2cali}$	0.013	0.0031	0.0072	0.002	0.0028	0.0013	6e−04	7e−04
$\widehat{R}_{KL2calr}$	0.0121	0.0031	0.0069	0.002	0.0028	0.0013	3e−04	7e−04
$\widehat{R}_{KL2calo}$	0.0121	0.0031	0.0069	0.002	0.0028	0.0013	3e−04	7e−04
$\widehat{R}_{Dcali}$	0.0105	0.0031	0.0063	0.002	0.0019	0.0013	1e−04	7e−04
$\widehat{R}_{Dcalr}$	0.0095	0.0031	0.0061	0.002	0.0019	0.0013	−3e−04	7e−04
$\widehat{R}_{Dcalo}$	0.0095	0.0031	0.0061	0.002	0.0019	0.0013	−3e−04	7e−04
$1/\exp(0.125 + b_1/2)$								
$\widehat{R}_{HT}$	0.2424	1	0.2331	1	0.2224	1	0.2312	1
$\widehat{R}_{cal}^{(1)}$	0.039	0.1481	0.0381	0.1505	0.0339	0.1305	0.0341	0.1001
$\widehat{R}_{cal}^{(1)}$	0.0042	0.1224	0.0021	0.1145	−0.0014	0.0951	−0.0013	0.0679
$\widehat{R}_{KL1cali}$	0.0058	0.1182	0.0019	0.1096	−0.0017	0.0917	−0.0011	0.0668
$\widehat{R}_{KL1calr}$	0.0052	0.1159	0.0022	0.1094	−0.0018	0.0913	−0.0013	0.0667
$\widehat{R}_{KL1calo}$	0.0052	0.1158	0.0021	0.1094	−0.0019	0.0912	−0.0013	0.0667
$\widehat{R}_{KL2cali}$	0.0057	0.1178	0.0019	0.1096	−0.0017	0.0917	−0.0011	0.0667
$\widehat{R}_{KL2calr}$	0.0053	0.1177	0.0018	0.1097	−0.0017	0.0917	−0.0013	0.0668
$\widehat{R}_{KL2calo}$	0.0053	0.1177	0.0018	0.1097	−0.0017	0.0917	−0.0013	0.0668
$\widehat{R}_{Dcali}$	0.0056	0.1196	0.0014	0.1132	−0.0022	0.0945	−0.0014	0.0676
$\widehat{R}_{Dcalr}$	0.0051	0.1195	0.0013	0.1133	−0.0022	0.0945	−0.0016	0.0676
$\widehat{R}_{Dcalo}$	0.0051	0.1195	0.0013	0.1133	−0.0022	0.0945	−0.0016	0.0676

Table 11.2 RB and RE for several sample sizes of the estimators of $R(0.8, 0.2)$. SRSWOR from the 2016 SPANISH LIVING CONDITIONS SURVEY

	RB	RE	RB	RE	RB	RE	RB	RE
Estimator	$n = 100$		$n = 125$		$n = 150$		$n = 200$	
$1/\exp(0.5 + b_1/3)$								
$\widehat{R}_{HT}$	1.0708	1	1.2001	1	1.3566	1	1.5867	1
$\widehat{R}^{(1)}_{cal}$	0.3746	0.1456	0.3339	0.0729	0.3264	0.055	0.328	0.0384
$\widehat{R}^{(2)}_{cal}$	0.036	0.0174	0.0212	0.0087	0.0183	0.0063	0.0166	0.0035
$\widehat{R}_{KL1cali}$	0.0375	0.0169	0.0225	0.0084	0.0189	0.0061	0.0176	0.0035
$\widehat{R}_{KL1calr}$	0.0372	0.0167	0.0229	0.0085	0.019	0.0061	0.0172	0.0034
$\widehat{R}_{KL1calo}$	0.0372	0.0167	0.0224	0.0085	0.019	0.0061	0.0174	0.0034
$\widehat{R}_{KL2cali}$	0.0374	0.0168	0.0224	0.0084	0.0189	0.0061	0.0177	0.0035
$\widehat{R}_{KL2calr}$	0.0374	0.0169	0.0224	0.0085	0.0189	0.0061	0.0176	0.0034
$\widehat{R}_{KL2calo}$	0.0374	0.0169	0.0224	0.0085	0.0189	0.0061	0.0176	0.0034
$\widehat{R}_{Dcali}$	0.0371	0.0172	0.0232	0.0088	0.0182	0.0063	0.0179	0.0035
$\widehat{R}_{Dcalr}$	0.0372	0.0173	0.0233	0.0088	0.0182	0.0063	0.0178	0.0035
$\widehat{R}_{Dcalo}$	0.0372	0.0173	0.0233	0.0088	0.0182	0.0063	0.0178	0.0035
$1/\exp(0.25 + b_1/2)$								
$\widehat{R}_{HT}$	1.3582	1	1.5555	1	1.7488	1	1.935	1
$\widehat{R}^{(1)}_{cal}$	0.3003	0.0607	0.2905	0.036	0.2966	0.0289	0.2842	0.0209
$\widehat{R}^{(2)}_{cal}$	0.0192	0.0083	0.0252	0.0055	0.021	0.0034	0.0155	0.0021
$\widehat{R}_{KL1cali}$	0.0204	0.0084	0.0243	0.0052	0.0217	0.0032	0.0155	0.0021
$\widehat{R}_{KL1calr}$	0.0201	0.0082	0.0243	0.0052	0.0213	0.0032	0.0156	0.0021
$\widehat{R}_{KL1calo}$	0.0206	0.0083	0.0239	0.0052	0.0215	0.0032	0.0156	0.0021
$\widehat{R}_{KL2cali}$	0.0205	0.0084	0.0241	0.0052	0.0216	0.0032	0.0155	0.0021
$\widehat{R}_{KL2calr}$	0.0201	0.0083	0.0243	0.0052	0.0216	0.0032	0.0156	0.0021
$\widehat{R}_{KL2calo}$	0.0201	0.0083	0.0243	0.0052	0.0216	0.0032	0.0156	0.0021
$\widehat{R}_{Dcali}$	0.0197	0.0088	0.0232	0.0053	0.0222	0.0033	0.0155	0.0021
$\widehat{R}_{Dcalr}$	0.0195	0.0088	0.0231	0.0053	0.0222	0.0033	0.0156	0.0021
$\widehat{R}_{Dcalo}$	0.0195	0.0088	0.0231	0.0053	0.0222	0.0033	0.0156	0.0021

(continued)

Table 11.2 (continued)

Estimator	RB	RE	RB	RE	RB	RE	RB	RE
	$n = 100$		$n = 125$		$n = 150$		$n = 200$	
$1/\exp(0.75 + b_1/8)$								
$\widehat{R}_{HT}$	0.848	1	0.848	1	1.1232	1	1.3591	1
$\widehat{R}^{(1)}_{cal}$	0.4107	0.3017	0.4107	0.3017	0.3705	0.0972	0.3527	0.0585
$\widehat{R}^{(2)}_{cal}$	0.0342	0.0236	0.0342	0.0236	0.0255	0.009	0.0156	0.0051
$\widehat{R}_{KL1cali}$	0.0316	0.0212	0.0316	0.0212	0.0243	0.0087	0.0149	0.005
$\widehat{R}_{KL1calr}$	0.0316	0.0212	0.0316	0.0212	0.0241	0.0087	0.0148	0.0049
$\widehat{R}_{KL1calo}$	0.0318	0.0214	0.0318	0.0214	0.0243	0.0087	0.0149	0.005
$\widehat{R}_{KL2cali}$	0.0316	0.0212	0.0316	0.0212	0.0243	0.0087	0.015	0.005
$\widehat{R}_{KL2calr}$	0.0316	0.0213	0.0316	0.0213	0.0243	0.0087	0.0149	0.005
$\widehat{R}_{KL2calo}$	0.0316	0.0213	0.0316	0.0213	0.0243	0.0087	0.0149	0.005
$\widehat{R}_{Dcali}$	0.0334	0.0231	0.0334	0.0231	0.0226	0.0085	0.014	0.0049
$\widehat{R}_{Dcalr}$	0.0332	0.023	0.0332	0.023	0.0226	0.0085	0.0139	0.0049
$\widehat{R}_{Dcalo}$	0.0332	0.023	0.0332	0.023	0.0226	0.0085	0.0139	0.0049
$1/\exp(0.125 + b_1/2)$								
$\widehat{R}_{HT}$	1.6643	1	1.8632	1	2.0019	1	2.2985	1
$\widehat{R}^{(1)}_{cal}$	0.2462	0.0296	0.2345	0.0194	0.228	0.0149	0.2266	0.0105
$\widehat{R}^{(2)}_{cal}$	0.0203	0.0056	0.0184	0.0033	0.0137	0.0024	0.0114	0.0014
$\widehat{R}_{KL1cali}$	0.0199	0.0053	0.0196	0.0032	0.0134	0.0023	0.0119	0.0013
$\widehat{R}_{KL1calr}$	0.0199	0.0053	0.0195	0.0032	0.0136	0.0023	0.0119	0.0013
$\widehat{R}_{KL1calo}$	0.0198	0.0052	0.0196	0.0032	0.0136	0.0023	0.012	0.0013
$\widehat{R}_{KL2cali}$	0.0198	0.0053	0.0196	0.0032	0.0134	0.0023	0.0119	0.0013
$\widehat{R}_{KL2calr}$	0.0198	0.0053	0.0196	0.0032	0.0134	0.0023	0.0119	0.0013
$\widehat{R}_{KL2calo}$	0.0198	0.0053	0.0196	0.0032	0.0134	0.0023	0.0119	0.0013
$\widehat{R}_{Dcali}$	0.0195	0.0054	0.0189	0.0033	0.0134	0.0024	0.0118	0.0013
$\widehat{R}_{Dcalr}$	0.0195	0.0054	0.0189	0.0033	0.0134	0.0024	0.0118	0.0013
$\widehat{R}_{Dcalo}$	0.0195	0.0054	0.0189	0.0033	0.0134	0.0024	0.0118	0.0013

Table 11.3 RB and RE for several sample sizes of the estimators of $R(0.9, 0.1)$. SRSWOR from the 2016 SPANISH LIVING CONDITIONS SURVEY

	RB	RE	RB	RE	RB	RE	RB	RE
Estimator	$n = 100$		$n = 125$		$n = 150$		$n = 200$	
$1/\exp(0.5 + b_1/3)$								
$\widehat{R}_{HT}$	0.6336	1	0.6867	1	0.8431	1	0.9662	1
$\widehat{R}^{(1)}_{cal}$	1.0228	2.4065	1.0892	2.0116	1.238	1.8797	1.4022	1.9122
$\widehat{R}^{(2)}_{cal}$	0.1069	0.2128	0.0593	0.0704	0.052	0.0431	0.0357	0.0244
$\widehat{R}_{KL1cali}$	0.0908	0.1241	0.0556	0.0675	0.0476	0.0403	0.0334	0.0236
$\widehat{R}_{KL1calr}$	0.0913	0.1238	0.0552	0.0661	0.0477	0.0403	0.0334	0.0236
$\widehat{R}_{KL1calo}$	0.0913	0.1238	0.0551	0.066	0.0476	0.0403	0.0334	0.0237
$\widehat{R}_{KL2cali}$	0.0908	0.124	0.0557	0.0675	0.0477	0.0403	0.0334	0.0236
$\widehat{R}_{KL2calr}$	0.0912	0.1242	0.0558	0.0675	0.0477	0.0403	0.0335	0.0237
$\widehat{R}_{KL2calo}$	0.0912	0.1242	0.0558	0.0675	0.0477	0.0403	0.0335	0.0237
$\widehat{R}_{Dcali}$	0.0923	0.1271	0.0604	0.1281	0.0485	0.0416	0.0342	0.0244
$\widehat{R}_{Dcalr}$	0.0926	0.1273	0.0604	0.1279	0.0485	0.0416	0.0344	0.0244
$\widehat{R}_{Dcalo}$	0.0926	0.1273	0.0604	0.1279	0.0485	0.0416	0.0344	0.0244
$1/\exp(0.25 + b_1/2)$								
$\widehat{R}_{HT}$	0.8144	1	0.9677	1	1.0907	1	1.2854	1
$\widehat{R}^{(1)}_{cal}$	0.9912	1.4969	1.1483	1.4556	1.2233	1.31	1.4571	1.3506
$\widehat{R}^{(2)}_{cal}$	0.0773	0.0842	0.0534	0.0398	0.0453	0.0256	0.0315	0.0126
$\widehat{R}_{KL1cali}$	0.0704	0.075	0.0484	0.0352	0.0431	0.0236	0.0295	0.0121
$\widehat{R}_{KL1calr}$	0.07	0.0747	0.048	0.0351	0.0428	0.0236	0.0293	0.012
$\widehat{R}_{KL1calo}$	0.0708	0.0751	0.0484	0.0352	0.0432	0.0237	0.0294	0.0121
$\widehat{R}_{KL2cali}$	0.0705	0.075	0.0483	0.0352	0.0431	0.0236	0.0294	0.0121
$\widehat{R}_{KL2calr}$	0.0708	0.0753	0.0485	0.0353	0.0431	0.0236	0.0294	0.0121
$\widehat{R}_{KL2calo}$	0.0708	0.0753	0.0485	0.0353	0.0431	0.0236	0.0294	0.0121
$\widehat{R}_{Dcali}$	0.0719	0.0804	0.051	0.0369	0.0444	0.0253	0.0307	0.0125
$\widehat{R}_{Dcalr}$	0.0723	0.0806	0.0511	0.037	0.0444	0.0253	0.0306	0.0125
$\widehat{R}_{Dcalo}$	0.0723	0.0806	0.0511	0.037	0.0444	0.0253	0.0306	0.0125

(continued)

Table 11.3 (continued)

	RB	RE	RB	RE	RB	RE	RB	RE
Estimator	$n = 100$		$n = 125$		$n = 150$		$n = 200$	
$1/\exp(0.75 + b_1/8)$								
$\widehat{R}_{HT}$	0.4666	1	0.531	1	0.65	1	0.7915	1
$\widehat{R}^{(1)}_{cal}$	0.9788	2.7242	1.0596	2.6561	1.1641	2.3383	1.3844	2.5426
$\widehat{R}^{(2)}_{cal}$	0.1035	0.2375	0.0761	0.1307	0.0554	0.0734	0.0431	0.0447
$\widehat{R}_{KL1cali}$	0.0944	0.2169	0.0725	0.1268	0.0534	0.0717	0.0399	0.037
$\widehat{R}_{KL1calr}$	0.0936	0.2147	0.0722	0.1268	0.0538	0.072	0.04	0.0368
$\widehat{R}_{KL1calo}$	0.0924	0.2081	0.0726	0.1271	0.0535	0.0718	0.04	0.0371
$\widehat{R}_{KL2cali}$	0.0941	0.2168	0.0725	0.1268	0.0534	0.0717	0.0399	0.0371
$\widehat{R}_{KL2calr}$	0.0944	0.2169	0.0726	0.1269	0.0535	0.072	0.04	0.0371
$\widehat{R}_{KL2calo}$	0.0944	0.2169	0.0726	0.1269	0.0535	0.072	0.04	0.0371
$\widehat{R}_{Dcali}$	0.0989	0.2464	0.0728	0.1289	0.0551	0.0737	0.0408	0.0379
$\widehat{R}_{Dcalr}$	0.099	0.2465	0.0728	0.1289	0.0551	0.0737	0.041	0.038
$\widehat{R}_{Dcalo}$	0.099	0.2465	0.0728	0.1289	0.0551	0.0737	0.041	0.038
$1/\exp(0.125 + b_1/2)$								
$\widehat{R}_{HT}$	1.0535	1	1.1436	1	1.2894	1	1.5446	1
$\widehat{R}^{(1)}_{cal}$	0.9491	0.9409	1.0499	1.0116	1.1124	0.9087	1.3752	0.9261
$\widehat{R}^{(2)}_{cal}$	0.0468	0.0363	0.0508	0.0281	0.038	0.0159	0.0364	0.0087
$\widehat{R}_{KL1cali}$	0.0401	0.0319	0.0463	0.0254	0.0386	0.0157	0.0361	0.0082
$\widehat{R}_{KL1calr}$	0.0403	0.032	0.0463	0.0252	0.0385	0.0156	0.0362	0.0082
$\widehat{R}_{KL1calo}$	0.0397	0.0317	0.0465	0.0253	0.0385	0.0156	0.0362	0.0082
$\widehat{R}_{KL2cali}$	0.0402	0.032	0.0463	0.0254	0.0386	0.0157	0.0361	0.0082
$\widehat{R}_{KL2calr}$	0.0401	0.032	0.0464	0.0254	0.0386	0.0157	0.0361	0.0082
$\widehat{R}_{KL2calo}$	0.0401	0.032	0.0464	0.0254	0.0386	0.0157	0.0361	0.0082
$\widehat{R}_{Dcali}$	0.0434	0.0347	0.0481	0.0274	0.0396	0.0164	0.0362	0.0083
$\widehat{R}_{Dcalr}$	0.0434	0.0348	0.0482	0.0274	0.0396	0.0164	0.0362	0.0083
$\widehat{R}_{Dcalo}$	0.0434	0.0348	0.0482	0.0274	0.0396	0.0164	0.0362	0.0083

Table 11.4 RB and RE for several sample sizes of the estimators of $R(0.9, 0.2)$. SRSWOR from the 2016 SPANISH LIVING CONDITIONS SURVEY

	RB	RE	RB	RE	RB	RE	RB	RE
Estimator	$n = 100$		$n = 125$		$n = 150$		$n = 200$	
$1/\exp(0.5 + b_1/3)$								
$\widehat{R}_{HT}$	0.5727	1	0.683	1	0.7759	1	0.9313	1
$\widehat{R}^{(1)}_{cal}$	0.9516	1.9501	1.0774	1.987	1.2049	2.0104	1.4164	2.0432
$\widehat{R}^{(2)}_{cal}$	0.0293	0.0526	0.0277	0.034	0.0225	0.0214	0.0157	0.0117
$\widehat{R}_{KL1cali}$	0.0283	0.0452	0.027	0.0309	0.0212	0.0208	0.0151	0.0115
$\widehat{R}_{KL1calr}$	0.029	0.0451	0.0269	0.0309	0.0214	0.0209	0.0152	0.0115
$\widehat{R}_{KL1calo}$	0.0291	0.0453	0.0271	0.0309	0.0212	0.0207	0.0152	0.0115
$\widehat{R}_{KL2cali}$	0.0286	0.0453	0.0271	0.0308	0.0212	0.0208	0.0151	0.0115
$\widehat{R}_{KL2calr}$	0.0288	0.0454	0.0272	0.0309	0.0212	0.0208	0.0152	0.0115
$\widehat{R}_{KL2calo}$	0.0288	0.0454	0.0272	0.0309	0.0212	0.0208	0.0152	0.0115
$\widehat{R}_{Dcali}$	0.0309	0.0471	0.0268	0.0315	0.0222	0.0209	0.0156	0.0117
$\widehat{R}_{Dcalr}$	0.0311	0.0472	0.027	0.0316	0.0222	0.0209	0.0157	0.0117
$\widehat{R}_{Dcalo}$	0.0311	0.0472	0.027	0.0316	0.0222	0.0209	0.0157	0.0117
$1/\exp(0.25 + b_1/2)$								
$\widehat{R}_{HT}$	0.7775	1	0.8934	1	1.0167	1	1.2284	1
$\widehat{R}^{(1)}_{cal}$	0.9419	1.3871	1.0599	1.3864	1.2055	1.3996	1.4514	1.4305
$\widehat{R}^{(2)}_{cal}$	0.0105	0.0219	0.019	0.0189	0.0193	0.0122	0.014	0.0066
$\widehat{R}_{KL1cali}$	0.0067	0.0203	0.0184	0.0182	0.0191	0.0119	0.0136	0.0065
$\widehat{R}_{KL1calr}$	0.0069	0.0203	0.0181	0.0181	0.0191	0.0119	0.0135	0.0064
$\widehat{R}_{KL1calo}$	0.0067	0.0203	0.0182	0.0181	0.0191	0.0119	0.0136	0.0064
$\widehat{R}_{KL2cali}$	0.0066	0.0203	0.0184	0.0182	0.0191	0.0119	0.0136	0.0065
$\widehat{R}_{KL2calr}$	0.0067	0.0203	0.0183	0.0182	0.0191	0.0119	0.0136	0.0065
$\widehat{R}_{KL2calo}$	0.0067	0.0203	0.0183	0.0182	0.0191	0.0119	0.0136	0.0065
$\widehat{R}_{Dcali}$	0.0074	0.0219	0.0207	0.0193	0.0199	0.0124	0.014	0.0066
$\widehat{R}_{Dcalr}$	0.0074	0.0219	0.0207	0.0193	0.0199	0.0124	0.014	0.0066
$\widehat{R}_{Dcalo}$	0.0074	0.0219	0.0207	0.0193	0.0199	0.0124	0.014	0.0066

(continued)

Table 11.4 (continued)

	RB	RE	RB	RE	RB	RE	RB	RE
Estimator	$n = 100$		$n = 125$		$n = 150$		$n = 200$	
$1/\exp(0.75 + b_1/8)$								
$\widehat{R}_{HT}$	0.4117	1	0.4871	1	0.5832	1	0.7334	1
$\widehat{R}_{cal}^{(2)}$	0.8896	2.348	0.9978	2.4013	1.1243	2.4448	1.3552	2.5469
$\widehat{R}_{cal}^{(2)}$	0.0387	0.0783	0.0272	0.0491	0.0235	0.0338	0.0147	0.0178
$\widehat{R}_{KL1cali}$	0.035	0.0713	0.024	0.0459	0.0224	0.0322	0.0148	0.0174
$\widehat{R}_{KL1calr}$	0.0365	0.0719	0.0237	0.0458	0.0225	0.0323	0.0148	0.0173
$\widehat{R}_{KL1calo}$	0.0356	0.0715	0.024	0.0459	0.0223	0.0322	0.0148	0.0173
$\widehat{R}_{KL2cali}$	0.0353	0.0713	0.024	0.0459	0.0224	0.0322	0.0148	0.0174
$\widehat{R}_{KL2calr}$	0.0357	0.0716	0.0241	0.0459	0.0224	0.0322	0.0149	0.0174
$\widehat{R}_{KL2calo}$	0.0357	0.0716	0.0241	0.0459	0.0224	0.0322	0.0149	0.0174
$\widehat{R}_{Dcali}$	0.0359	0.0744	0.0251	0.0478	0.0231	0.0329	0.0145	0.0174
$\widehat{R}_{Dcalr}$	0.0366	0.0748	0.0251	0.0478	0.0231	0.0329	0.0145	0.0174
$\widehat{R}_{Dcalo}$	0.0366	0.0748	0.0251	0.0478	0.0231	0.0329	0.0145	0.0174
$1/\exp(0.125 + b_1/2)$								
$\widehat{R}_{HT}$	1.0141	1	1.119	1	1.2479	1	1.4519	1
$\widehat{R}_{cal}^{(1)}$	0.9229	0.9531	0.9997	0.9341	1.1202	0.9487	1.2918	0.9429
$\widehat{R}_{cal}^{(2)}$	0.0221	0.0158	0.016	0.0109	0.0149	0.0077	0.014	0.0045
$\widehat{R}_{KL1cali}$	0.0222	0.0156	0.0154	0.0104	0.0146	0.0076	0.0139	0.0044
$\widehat{R}_{KL1calr}$	0.0228	0.0156	0.0155	0.0104	0.0145	0.0076	0.0138	0.0044
$\widehat{R}_{KL1calo}$	0.0227	0.0156	0.0155	0.0104	0.0146	0.0076	0.0138	0.0044
$\widehat{R}_{KL2cali}$	0.0223	0.0156	0.0154	0.0104	0.0146	0.0076	0.0139	0.0044
$\widehat{R}_{KL2calr}$	0.0225	0.0156	0.0155	0.0104	0.0146	0.0076	0.014	0.0044
$\widehat{R}_{KL2calo}$	0.0225	0.0156	0.0155	0.0104	0.0146	0.0076	0.014	0.0044
$\widehat{R}_{Dcali}$	0.0237	0.0165	0.0163	0.0108	0.0153	0.0078	0.0146	0.0045
$\widehat{R}_{Dcalr}$	0.0239	0.0165	0.0163	0.0108	0.0153	0.0078	0.0147	0.0045
$\widehat{R}_{Dcalo}$	0.0239	0.0165	0.0163	0.0108	0.0153	0.0078	0.0147	0.0045

Table 11.5 RB and RE for several sample sizes of the estimators of $R(0.95, 0.2)$. SRSWOR from the 2016 SPANISH LIVING CONDITIONS SURVEY

Estimator	RB	RE	RB	RE	RB	RE	RB	RE
	$n = 100$		$n = 125$		$n = 150$		$n = 200$	
$1/\exp(0.5 + b_1/3)$								
$\widehat{R}_{HT}$	0.2525	1	0.3418	1	0.4087	1	0.5525	1
$\widehat{R}^{(1)}_{cal}$	0.6794	2.4489	0.8063	2.5222	0.8924	2.5047	1.0903	2.5143
$\widehat{R}^{(2)}_{cal}$	0.0519	0.1743	0.0374	0.0959	0.033	0.0638	0.023	0.0322
$\widehat{R}_{KL1cali}$	0.0418	0.1339	0.0339	0.0882	0.0312	0.0602	0.0222	0.0311
$\widehat{R}_{KL1calr}$	0.0414	0.1327	0.0343	0.0883	0.0311	0.0603	0.0219	0.0309
$\widehat{R}_{KL1calo}$	0.042	0.1334	0.0342	0.0882	0.0311	0.0602	0.0221	0.031
$\widehat{R}_{KL2cali}$	0.0418	0.1339	0.0338	0.0882	0.0312	0.0602	0.0222	0.0311
$\widehat{R}_{KL2calr}$	0.0421	0.134	0.034	0.0882	0.0312	0.0602	0.0223	0.0311
$\widehat{R}_{KL2calo}$	0.0421	0.134	0.034	0.0882	0.0312	0.0602	0.0223	0.0311
$\widehat{R}_{Dcali}$	0.0444	0.145	0.0368	0.0946	0.0346	0.0649	0.023	0.0325
$\widehat{R}_{Dcalr}$	0.0446	0.145	0.0369	0.0947	0.0346	0.0649	0.0231	0.0325
$\widehat{R}_{Dcalo}$	0.0446	0.145	0.0369	0.0947	0.0346	0.0649	0.0231	0.0325
$1/\exp(0.25 + b_1/2)$								
$\widehat{R}_{HT}$	0.4557	1	0.5132	1	0.5897	1	0.7267	1
$\widehat{R}^{(1)}_{cal}$	0.773	1.8254	0.8512	1.8864	0.9553	1.9073	1.1336	1.9188
$\widehat{R}^{(2)}_{cal}$	0.048	0.0928	0.0317	0.0538	0.0266	0.0411	0.0215	0.02
$\widehat{R}_{KL1cali}$	0.0443	0.0702	0.0297	0.0511	0.0232	0.0366	0.0231	0.0199
$\widehat{R}_{KL1calr}$	0.0441	0.0702	0.0293	0.0502	0.0232	0.0366	0.0229	0.02
$\widehat{R}_{KL1calo}$	0.0441	0.0697	0.0295	0.0503	0.0235	0.0367	0.0229	0.0199
$\widehat{R}_{KL2cali}$	0.0443	0.0702	0.0297	0.0512	0.0232	0.0366	0.0231	0.02
$\widehat{R}_{KL2calr}$	0.0446	0.0704	0.0298	0.0512	0.0232	0.0366	0.0231	0.02
$\widehat{R}_{KL2calo}$	0.0446	0.0704	0.0298	0.0512	0.0232	0.0366	0.0231	0.02
$\widehat{R}_{Dcali}$	0.0484	0.0829	0.0322	0.054	0.0247	0.038	0.0235	0.0201
$\widehat{R}_{Dcalr}$	0.0487	0.083	0.0323	0.0541	0.0247	0.038	0.0235	0.0201
$\widehat{R}_{Dcalo}$	0.0487	0.083	0.0323	0.0541	0.0247	0.038	0.0235	0.0201

(continued)

Table 11.5 (Continued)

	RB	RE	RB	RE	RB	RE	RB	RE
Estimator	$n = 100$		$n = 125$		$n = 150$		$n = 200$	
$1/\exp(0.75 + b_1/8)$								
$\widehat{R}_{HT}$	0.1243	1	0.1843	1	0.2609	1	0.389	1
$\widehat{R}^{(1)}_{cal}$	0.6477	3.2315	0.7304	3.1825	0.8417	3.2802	1.0211	3.2011
$\widehat{R}^{(2)}_{cal}$	0.0583	0.2974	0.0409	0.1495	0.0359	0.1049	0.0251	0.0512
$\widehat{R}_{KL1cali}$	0.0455	0.228	0.0414	0.1439	0.0311	0.092	0.0225	0.049
$\widehat{R}_{KL1calr}$	0.0464	0.2295	0.0415	0.1444	0.0308	0.0917	0.0227	0.0491
$\widehat{R}_{KL1calo}$	0.0462	0.2279	0.0416	0.1439	0.031	0.0921	0.0227	0.0491
$\widehat{R}_{KL2cali}$	0.0457	0.2278	0.0415	0.144	0.0311	0.092	0.0224	0.049
$\widehat{R}_{KL2calr}$	0.0462	0.2285	0.0416	0.144	0.0311	0.092	0.0226	0.0491
$\widehat{R}_{KL2calo}$	0.0462	0.2285	0.0416	0.144	0.0311	0.092	0.0226	0.0491
$\widehat{R}_{Dcali}$	0.0451	0.2283	0.0459	0.1554	0.0337	0.0975	0.0238	0.0507
$\widehat{R}_{Dcalr}$	0.046	0.2289	0.0459	0.1554	0.0337	0.0975	0.024	0.0507
$\widehat{R}_{Dcalo}$	0.046	0.2289	0.0459	0.1554	0.0337	0.0975	0.024	0.0507
$1/\exp(0.125 + b_1/2)$								
$\widehat{R}_{HT}$	0.587	1	0.703	1	0.8148	1	0.9631	1
$\widehat{R}^{(1)}_{cal}$	0.7991	1.488	0.945	1.5212	1.0747	1.51	1.2532	1.5233
$\widehat{R}^{(2)}_{cal}$	0.039	0.0577	0.0293	0.0352	0.0272	0.0218	0.0192	0.0122
$\widehat{R}_{KL1cali}$	0.0312	0.0483	0.0277	0.0319	0.0265	0.0212	0.0189	0.012
$\widehat{R}_{KL1calr}$	0.0318	0.0486	0.0282	0.0321	0.0262	0.0211	0.019	0.012
$\widehat{R}_{KL1calo}$	0.0315	0.0485	0.0279	0.032	0.0262	0.0211	0.0191	0.012
$\widehat{R}_{KL2cali}$	0.0311	0.0483	0.0277	0.0319	0.0264	0.0212	0.0189	0.012
$\widehat{R}_{KL2calr}$	0.0313	0.0484	0.0278	0.0319	0.0265	0.0212	0.019	0.012
$\widehat{R}_{KL2calo}$	0.0313	0.0484	0.0278	0.0319	0.0265	0.0212	0.019	0.012
$\widehat{R}_{Dcali}$	0.0304	0.0496	0.0295	0.0329	0.0272	0.0218	0.0198	0.0123
$\widehat{R}_{Dcalr}$	0.0306	0.0497	0.0294	0.0329	0.0272	0.0218	0.0198	0.0124
$\widehat{R}_{Dcalo}$	0.0306	0.0497	0.0294	0.033	0.0272	0.0218	0.0198	0.0124

Table 11.6 RB and RE for several sample sizes of the estimators of $R(0.95, 0.5)$. SRSWOR from the 2016 SPANISH LIVING CONDITIONS SURVEY

	RB	RE	RB	RE	RB	RE	RB	RE
Estimator	$n = 100$		$n = 125$		$n = 150$		$n = 200$	
$1/\exp(0.5 + b_1/3)$								
$\widehat{R}_{HT}$	−0.4625	1	−0.4695	1	−0.4355	1	−0.4268	1
$\widehat{R}^{(1)}_{cal}$	0.5605	2.9425	0.6607	3.5969	0.7903	4.6128	0.9173	5.4401
$\widehat{R}^{(2)}_{cal}$	0.0305	0.2341	0.0106	0.1129	0.0129	0.1031	0.0115	0.0672
$\widehat{R}_{KL1cali}$	0.0221	0.1788	0.0101	0.1072	0.0113	0.0919	0.0109	0.0655
$\widehat{R}_{KL1calr}$	0.0223	0.1756	0.0109	0.1079	0.0114	0.0914	0.0114	0.0654
$\widehat{R}_{KL1calo}$	0.0229	0.1786	0.0108	0.1076	0.0114	0.0918	0.0114	0.0656
$\widehat{R}_{KL2cali}$	0.022	0.1786	0.0102	0.1075	0.0113	0.0919	0.0109	0.0655
$\widehat{R}_{KL2calr}$	0.0232	0.1788	0.0107	0.1075	0.0114	0.0919	0.0115	0.0656
$\widehat{R}_{KL2calo}$	0.0232	0.1788	0.0107	0.1075	0.0114	0.0919	0.0115	0.0656
$\widehat{R}_{Dcali}$	0.0247	0.1849	0.0124	0.1181	0.0142	0.0971	0.0109	0.0665
$\widehat{R}_{Dcalr}$	0.0258	0.1848	0.0129	0.1183	0.0142	0.0971	0.0115	0.0666
$\widehat{R}_{Dcalo}$	0.0258	0.1848	0.0129	0.1183	0.0142	0.0971	0.0115	0.0666
$1/\exp(0.25 + b_1/2)$								
$\widehat{R}_{HT}$	−0.0577	1	0.0351	1	0.1056	1	0.2102	1
$\widehat{R}^{(1)}_{cal}$	0.6081	3.736	0.8029	4.5287	0.8816	4.7993	1.0674	5.0535
$\widehat{R}^{2}_{cal}$	0.0235	0.208	0.0201	0.1103	0.0111	0.0791	0.0104	0.0421
$\widehat{R}_{KL1cali}$	0.0195	0.1675	0.0211	0.1072	0.0103	0.0777	0.0093	0.0413
$\widehat{R}_{KL1calr}$	0.0207	0.1683	0.0218	0.1076	0.0101	0.0759	0.0094	0.0413
$\widehat{R}_{KL1calo}$	0.0205	0.1679	0.0215	0.1072	0.0101	0.076	0.0094	0.0412
$\widehat{R}_{KL2cali}$	0.0195	0.1674	0.0212	0.1073	0.0103	0.0776	0.0093	0.0413
$\widehat{R}_{KL2calr}$	0.0204	0.168	0.0219	0.1074	0.0104	0.0776	0.0096	0.0414
$\widehat{R}_{KL2calo}$	0.0204	0.168	0.0219	0.1074	0.0104	0.0776	0.0096	0.0414
$\widehat{R}_{Dcali}$	0.0205	0.1768	0.0224	0.1137	0.0131	0.0809	0.0104	0.043
$\widehat{R}_{Dcalr}$	0.0217	0.1774	0.023	0.114	0.0131	0.0809	0.0107	0.0431
$\widehat{R}_{Dcalo}$	0.0217	0.1774	0.023	0.114	0.0131	0.0809	0.0107	0.0431

(continued)

Table 11.6 (continued)

	RB	RE	RB	RE	RB	RE	RB	RE
Estimator	$n = 100$		$n = 125$		$n = 150$		$n = 200$	
$1/\exp(0.75 + b_1/8)$								
$\widehat{R}_{HT}$	−0.6162	1	−0.6236	1	−0.6225	1	−0.6319	1
$\widehat{R}_{cal1}$	0.5322	2.6084	0.6176	2.8892	0.7233	3.314	0.8914	4.2668
$\widehat{R}^2_{cal}$	0.027	0.2134	0.017	0.1204	0.0163	0.0954	0.0092	0.059
$\widehat{R}_{KL1cali}$	0.0126	0.1638	0.0187	0.113	0.0109	0.0818	0.007	0.0576
$\widehat{R}_{KL1calr}$	0.0138	0.1636	0.0189	0.113	0.0108	0.0812	0.0077	0.0576
$\widehat{R}_{KL1calo}$	0.014	0.1645	0.019	0.1131	0.011	0.0818	0.0077	0.0577
$\widehat{R}_{KL2cali}$	0.0125	0.1635	0.0187	0.1129	0.011	0.0818	0.007	0.0576
$\widehat{R}_{KL2calr}$	0.014	0.1644	0.0191	0.1131	0.011	0.0818	0.0077	0.0577
$\widehat{R}_{KL2calo}$	0.014	0.1644	0.0191	0.1131	0.011	0.0818	0.0077	0.0577
$\widehat{R}_{Dcali}$	0.015	0.1697	0.0231	0.1266	0.0145	0.0886	0.0087	0.0594
$\widehat{R}_{Dcalr}$	0.0165	0.1701	0.0234	0.1267	0.0145	0.0886	0.0094	0.0595
$\widehat{R}_{Dcalo}$	0.0165	0.1701	0.0234	0.1267	0.0145	0.0886	0.0094	0.0595
$1/\exp(0.125 + b_1/2)$								
$\widehat{R}_{HT}$	0.2315	1	0.3475	1	0.423	1	0.5538	1
$\widehat{R}_{cal1}$	0.6754	2.5757	0.8457	2.5934	0.9563	2.7198	1.1544	2.8213
$\widehat{R}^2_{cal}$	0.0171	0.1007	0.0112	0.0464	0.011	0.0319	0.0041	0.0181
$\widehat{R}_{KL1cali}$	0.0125	0.0877	0.013	0.046	0.0109	0.0308	0.0025	0.0178
$\widehat{R}_{KL1calr}$	0.0125	0.0871	0.0128	0.0458	0.0108	0.0305	0.0028	0.0178
$\widehat{R}_{KL1calo}$	0.0128	0.0872	0.013	0.0459	0.0111	0.0309	0.0028	0.0178
$\widehat{R}_{KL2cali}$	0.0125	0.0878	0.013	0.0459	0.011	0.0308	0.0025	0.0178
$\widehat{R}_{KL2calr}$	0.0133	0.0881	0.0131	0.046	0.011	0.0308	0.0028	0.0178
$\widehat{R}_{KL2calo}$	0.0133	0.0881	0.0131	0.046	0.011	0.0308	0.0028	0.0178
$\widehat{R}_{Dcali}$	0.0156	0.0983	0.0147	0.0477	0.0123	0.032	0.0029	0.0181
$\widehat{R}_{Dcalr}$	0.0165	0.0985	0.0148	0.0477	0.0123	0.032	0.0032	0.0181
$\widehat{R}_{Dcalo}$	0.0165	0.0985	0.0148	0.0477	0.0123	0.032	0.0032	0.0181

Regarding efficiency, in general, the proposed estimators show the best performance for all sample sizes. Finally, in terms of bias and efficiency, there are no differences between the three versions of the estimators (linear method, raking and logit (l; u)) for the estimators $\widehat{R}^{(3)}_{cal}$.

For the variance estimation and confidence intervals, we computed the coverage probability (CP) , the lower (L) and the upper (U) tail error rates of the 95% confidence intervals, in percentage and the average length (AL) of the confidence intervals for each estimator and each bootstrap method.

Concerning the variance estimation and confidence intervals, we used 1000 bootstrap replications from each initial sample with all bootstrap methods included in the study to compute CP, L, U and AL of the 95% confidence intervals for each percentile ratio estimator considered. Result from this simulation study for some percentile ratios are presented in Tables 11.7, 11.8, and 11.9.

From bootstrap estimates, it is observed that:

- Bootstrap methods produce intervals with high true coverage.
- None of the intervals constructed with each estimator have problems of lack of coverage.
- The intervals obtained from the proposed calibration estimators always provide intervals with less amplitude than the intervals obtained from $\widehat{R}_{HT}$ and $\widehat{R}^{(1)}_{cal}$.
- The last method [1] provides results very similar to the one in[2].

11.6 Conclusion

In this study we use calibration techniques to estimate poverty measures based on percentiles ratios in presence of missing data through a more efficient estimation of the distribution function. The simulation study included shows the improvement in bias and efficiency with the two proposed calibration techniques, $\widehat{R}^{(2)}_{cal}$ and $\widehat{R}^{(3)}_{cali}$. The first one is based in two-step calibration method [10]. In the first step, the weighting is designed to remove the non-response bias while in the second step the weighting is designed to decrease the sampling error in the estimation of the distribution function. The second method is based on calibration weighting with instrumental variables [11].

The results show a large decrease in bias and MSE for all ratio percentiles considered, for both calibration methods, and for the three versions of them based on linear, raking and logit response models, which shows the robustness of the adjustment method. Although the simulation results show that there is no uniformly better estimator than another among the proposed estimators (both with respect to bias and efficiency), the $\widehat{R}^{(2)}_{cal}$ and $\widehat{R}^{(3)}_{KL2cal}$ estimators are computationally simpler than the other alternatives which implies that they are a suitable option for the estimation of measures for wage inequality based on percentiles ratios.

In [10] is said that there are reasons for preferring the use of two calibration-weighting steps even when the sets of calibration variables used in both steps are the same or a subset of the calibration variables in a single step. These reasons, together with the good performance of the two-step estimator shown in the simulation study, suggest the choice of the estimator $\widehat{R}^{(2)}_{cal}$.

We used parametric methods to model the lack of response but we could use machine learning techniques as regression trees, spline regression, random forests etc. Other way to reduce the bias is to combine calibration technique with other techniques as the Propensity Score Adjustment [8]. Further research should 318 focus on extensions of those methods for general parameter estimation.

Table 11.7 AL, CP %, L % and U % for several resampling method of the estimators compared. SRSWOR from the 2016 SPANISH LIVING CONDITIONS SURVEY $R(0.9, 0.1)$

	AL	CP%	L%	U%	AL	CP%	L%	U%	AL	CP%	L%	U%	AL	CP%	L%	U%
Estimator	$n = 100$				$n = 125$				$n = 150$				$n = 200$			
$\widehat{R}_{HT}$	9.43	100	0	0	7.85	98.19	1.80	0	6.09	96.70	3.29	0	4.71	96.55	3.44	0
$\widehat{R}^1_{cal}$	7.23	100	0	0	5.80	98.19	0.90	0.90	5.15	96.70	2.74	0.54	3.73	96.55	2.58	0.86
$\widehat{R}^2_{cal}$	5.00	99.04	0	0.95	3.88	95.49	0	4.50	3.81	97.80	0	2.19	2.97	94.82	0.86	4.31
$\widehat{R}_{KL1cali}$	5.10	99.04	0	0.95	3.93	95.49	0	4.50	3.79	97.80	0.54	1.64	2.98	96.55	0.86	2.58
$\widehat{R}_{KL1calr}$	5.10	99.04	0	0.95	3.93	95.49	0	4.50	3.78	97.80	0.54	1.64	2.98	95.68	0.86	3.44
$\widehat{R}_{KL1calo}$	5.10	99.04	0	0.95	3.93	95.49	0	4.50	3.78	97.80	0.54	1.64	2.98	96.55	0.86	2.58
$\widehat{R}_{KL2cali}$	5.10	99.04	0	0.95	3.93	95.49	0	4.50	3.79	97.80	0.54	1.64	2.98	96.55	0.86	2.58
$\widehat{R}_{KL2calr}$	5.10	99.04	0	0.95	3.93	95.49	0	4.50	3.79	97.80	0.54	1.64	2.98	96.55	0.86	2.58
$\widehat{R}_{KL2calo}$	5.10	99.04	0	0.95	3.93	95.49	0	4.50	3.79	97.80	0.54	1.64	2.98	96.55	0.86	2.58
$\widehat{R}_{Dcali}$	5.14	99.04	0	0.95	3.95	96.39	0	3.60	3.85	97.25	0.54	2.19	3.00	96.55	0.86	2.58
$\widehat{R}_{Dcalr}$	5.14	99.04	0	0.95	3.9	96.39	0	3.60	3.85	97.25	0.54	2.19	3.00	96.55	0.86	2.58
$\widehat{R}_{Dcalo}$	5.14	99.04	0	0.95	3.95	96.39	0	3.60	3.85	97.25	0.549	2.19	3.00	96.55	0.86	2.58

	Antal and Tillé [2]															
$\widehat{R}_{HT}$	7.35	100	0	0	5.57	95.49	3.60	0.90	5.70	96.70	2.7473	0.54	4.19	95.68	4.31	0
$\widehat{R}^1_{cal}$	8.58	99.04	0	0.95	6.49	96.39	0	3.60	6.58	97.25	1.09	1.64	4.62	97.41	0.86	1.72
$\widehat{R}^2_{cal}$	5.16	98.09	0	1.90	3.91	94.59	0	5.40	3.96	96.70	0.54	2.74	3.07	94.82	1.72	3.44
$\widehat{R}_{KL1cali}$	4.91	98.09	0	1.90	3.80	94.59	0.90	4.50	3.82	96.70	1.09	2.19	3.03	95.68	1.72	2.58
$\widehat{R}_{KL1calr}$	4.88	98.09	0	1.90	3.79	94.59	0.90	4.50	3.82	96.70	1.09	2.19	3.02	95.68	1.72	2.58
$\widehat{R}_{KL1calo}$	4.91	98.09	0	1.90	3.79	94.59	0.90	4.50	3.82	96.70	1.09	2.19	3.03	95.68	1.72	2.58
$\widehat{R}_{KL2cali}$	4.91	98.09	0	1.90	3.80	94.59	0.90	4.50	3.82	96.70	1.09	2.19	3.03	95.68	1.72	2.58
$\widehat{R}_{KL2calr}$	4.91	98.09	0	1.90	3.80	94.59	0.90	4.50	3.82	96.70	1.09	2.19	3.03	95.68	1.72	2.58
$\widehat{R}_{KL2calo}$	4.91	98.09	0	1.90	3.80	94.59	0.90	4.50	3.82	96.70	1.09	2.19	3.03	95.68	1.72	2.58
$\widehat{R}_{Dcali}$	4.95	98.09	0	1.90	3.85	93.69	0.90	5.40	3.88	97.25	0.54	2.19	3.06	96.55	1.72	1.72
$\widehat{R}_{Dcalr}$	4.95	98.09	0	1.90	3.85	93.69	0.90	5.40	3.89	97.25	0.54	2.19	3.06	96.55	1.72	1.72
$\widehat{R}_{Dcalo}$	4.95	98.09	0	1.90	3.85	93.69	0.90	5.40	3.89	97.25	0.54	2.19	3.06	96.55	1.72	1.72
	Antal and Tillé [1]															
$\widehat{R}_{HT}$	7.37	99.04	0	0.95	6.46	99.09	0	0.90	6.06	96.70	3.29	0	4.39	95.68	4.31	0
$\widehat{R}^1_{cal}$	8.83	99.04	0	0.95	6.98	94.59	0	5.40	6.52	96.70	1.09	2.19	4.30	98.27	0.86	0.86
$\widehat{R}^2_{cal}$	5.15	98.09	0	1.90	4.07	93.69	0	6.30	3.92	96.15	1.09	2.74	3.08	96.55	0.86	2.58
$\widehat{R}_{KL1cali}$	5.01	98.09	0	1.90	3.93	93.69	0.90	5.40	3.90	95.60	1.09	3.29	3.06	96.55	0.86	2.58
$\widehat{R}_{KL1calr}$	4.95	98.09	0	1.90	3.94	93.69	0.90	5.40	3.89	95.60	1.09	3.29	3.06	96.55	0.86	2.58
$\widehat{R}_{KL1calo}$	4.95	98.09	0	1.90	3.94	93.69	0.90	5.40	3.89	95.60	1.09	3.29	3.06	96.55	0.86	2.58
$\widehat{R}_{KL2cali}$	5.01	98.09	0	1.90	3.93	93.69	0.90	5.40	3.90	95.60	1.09	3.29	3.06	96.55	0.86	2.58
$\widehat{R}_{KL2calr}$	5.01	98.09	0	1.90	3.93	93.69	0.90	5.40	3.90	95.60	1.09	3.29	3.06	96.55	0.86	2.58
$\widehat{R}_{KL2calo}$	5.01	98.09	0	1.90	3.93	93.69	0.90	5.40	3.90	95.60	1.09	3.29	3.06	96.55	0.86	2.58
$\widehat{R}_{Dcali}$	5.12	98.09	0	1.90	4.09	94.59	0.90	4.50	3.94	96.70	0.54	2.74	3.09	97.41	0.86	1.72
$\widehat{R}_{Dcalr}$	5.12	98.09	0	1.90	4.09	94.59	0.90	4.50	3.94	96.70	0.54	2.74	3.09	97.41	0.86	1.72
$\widehat{R}_{Dcalo}$	5.12	98.09	0	1.90	4.09	94.59	0.90	4.50	3.94	96.70	0.54	2.74	3.09	97.41	0.86	1.72

Table 11.8 AL, CP %,L % and U % for several resampling method of the estimators compared. SRSWOR from the 2016 SPANISH LIVING CONDITIONS SURVEY $R(0.9, 0.2)$

	AL	CP%	L%	U%	AL	CP%	L%	U%	AL	CP%	L%	U%	AL	CP%	L%	U%
	Booth et al. [3]															
Estimator	$n = 100$				$n = 125$				$n = 150$				$n = 200$			
$\widehat{R}_{HT}$	4.97	100	0	0	4.89	98.60	1.39	0	3.45	97.5	2.5	0	2.69	94.3	5.6	0
$\widehat{R}^1_{cal}$	4.79	97.29	2.70	0	3.81	96.50	1.39	2.09	2.82	97.5	1.5	1	2.1	96.41	2.56	1.02
$\widehat{R}^2_{cal}$	2.75	98.19	1.80	0	2.27	96.50	0.69	2.79	1.86	93.5	3	3.5	1.50	95.38	0.51	4.10
$\widehat{R}_{KL1cali}$	2.82	98.19	1.80	0	2.18	95.80	0	4.19	1.86	94	2	4	1.50	94.87	0.51	4.61
$\widehat{R}_{KL1calr}$	2.81	98.19	1.80	0	2.18	95.80	0	4.19	1.85	94	2	4	1.50	94.87	0.51	4.61
$\widehat{R}_{KL1calo}$	2.81	98.19	1.80	0	2.18	95.80	0	4.19	1.85	94	2	4	1.50	94.87	0.51	4.61
$\widehat{R}_{KL2cali}$	2.82	98.19	1.80	0	2.18	95.80	0	4.19	1.86	94	2	4	1.50	94.87	0.51	4.61
$\widehat{R}_{KL2calr}$	2.82	98.19	1.80	0	2.18	95.80	0	4.19	1.86	94	2	4	1.50	94.87	0.51	4.61
$\widehat{R}_{KL2calo}$	2.82	98.19	1.80	0	2.18	95.80	0	4.19	1.86	94	2	4	1.50	94.87	0.51	4.61
$\widehat{R}_{Dcali}$	2.75	98.19	1.80	0	2.19	95.10	0.69	4.19	1.87	94	2	4	1.52	95.38	0.51	4.10
$\widehat{R}_{Dcalr}$	2.75	98.19	1.80	0	2.19	95.10	0.69	4.19	1.87	94	2	4	1.52	95.38	0.51	4.10
$\widehat{R}_{Dcalo}$	2.7	98.19	1.80	0	2.19	95.10	0.69	4.19	1.87	94	2	4	1.52	95.38	0.51	4.10

	Antal and Tillé [2]															
$\widehat{R}_{HT}$	3.20	99.09	0.90	0	3.84	97.20	2.79	0	2.82	94	5.5	0.5	2.26	93.84	6.15	0
$\widehat{R}^1_{cal}$	5.1	100	0	0	4.31	95.80	0	4.19	3.40	97	0	3	2.53	98.46	0	1.53
$\widehat{R}^2_{cal}$	2.36	100	0	0	1.94	95.10	1.39	3.49	1.75	95.5	1.5	3	1.46	93.84	1.53	4.61
$\widehat{R}_{KL1cali}$	2.32	100	0	0	1.93	95.80	0.69	3.49	1.72	94	1.5	4.5	1.45	94.3	1.02	4.61
$\widehat{R}_{KL1calr}$	2.32	100	0	0	1.93	95.80	0.69	3.49	1.72	94	1.5	4.5	1.45	94.3	1.02	4.61
$\widehat{R}_{KL1calo}$	2.32	100	0	0	1.93	95.80	0.69	3.49	1.72	94	1.5	4.5	1.45	94.3	1.02	4.61
$\widehat{R}_{KL2cali}$	2.32	100	0	0	1.93	95.80	0.69	3.49	1.72	94	1.5	4.5	1.45	94.35	1.02	4.61
$\widehat{R}_{KL2calr}$	2.32	100	0	0	1.93	95.80	0.69	3.49	1.72	94	1.5	4.5	1.45	94.35	1.02	4.61
$\widehat{R}_{KL2calo}$	2.32	100	0	0	1.93	95.80	0.69	3.49	1.72	94	1.5	4.5	1.45	94.3	1.02	4.61
$\widehat{R}_{Dcali}$	2.33	100	0	0	1.96	95.80	0.69	3.49	1.74	93.5	1.5	5	1.47	94.87	0.51	4.61
$\widehat{R}_{Dcalr}$	2.33	100	0	0	1.96	95.80	0.69	3.49	1.74	93.5	1.5	5	1.47	94.87	0.51	4.61
$\widehat{R}_{Dcalo}$	2.33	100	0	0	1.96	95.80	0.69	3.49	1.74	93.5	1.5	5	1.47	94.87	0.51	4.61
	Antal and Tillé [1]															
$\widehat{R}_{HT}$	3.62	99.09	0.90	0	4.15	97.20	2.79	0	2.98	96	3.5	0.5	2.45	94.87	5.12	0
$\widehat{R}^1_{cal}$	5.28	100	0	0	4.24	96.50	0	3.49	3.48	96.5	0	3.5	2.34	96.92	1.02	2.05
$\widehat{R}^2_{cal}$	2.36	100	0	0	1.930	95.80	0.69	3.49	1.75	94.5	1.5	4	1.44	94.35	1.53	4.10
$\widehat{R}_{KL1cali}$	2.40	100	0	0	1.95	95.10	0.69	4.19	1.75	93.5	3	3.5	1.44	94.87	1.02	4.10
$\widehat{R}_{KL1calr}$	2.39	100	0	0	1.95	95.10	0.69	4.19	1.75	93.5	3	3.5	1.44	95.38	0.51	4.10
$\widehat{R}_{KL1calo}$	2.39	100	0	0	1.95	95.10	0.69	4.19	1.75	93.5	3	3.5	1.44	95.38	0.51	4.10
$\widehat{R}_{KL2cali}$	2.40	100	0	0	1.95	95.10	0.69	4.19	1.75	93.5	3	3.5	1.44	94.87	1.02	4.10
$\widehat{R}_{KL2calr}$	2.40	100	0	0	1.95	95.10	0.69	4.19	1.75	93.5	3	3.5	1.44	94.87	1.02	4.10
$\widehat{R}_{KL2calo}$	2.40	100	0	0	1.95	95.10	0.69	4.19	1.75	93.5	3	3.5	1.44	94.87	1.02	4.10
$\widehat{R}_{Dcali}$	2.43	100	0	0	1.98	95.80	0.69	3.49	1.76	93.5	3	3.5	1.46	94.3	1.53	4.10
$\widehat{R}_{Dcalr}$	2.43	100	0	0	1.98	95.80	0.69	3.49	1.76	93.5	3	3.5	1.46	94.35	1.53	4.10
$\widehat{R}_{Dcalo}$	2.43	100	0	0	1.98	95.80	0.69	3.49	1.76	93.5	3	3.5	1.46	94.3	1.53	4.10

Table 11.9 AL, CP %,L % and U % for several resampling method of the estimators compared. SRSWOR from the 2016 SPANISH LIVING CONDITIONS SURVEY $R(0.8, 0.2)$

	AL	CP%	L%	U%	AL	CP%	L%	U%	AL	CP%	L%	U%	AL	CP%	L%	U%
Estimator	Booth et al. [3]															
	$n = 100$				$n = 125$				$n = 150$				$n = 200$			
$\widehat{R}_{HT}$	1.59	98.03	1.96	0	1.42	92.92	6.19	0.88	1.25	95.19	4.80	0	1.01	97.39	2.60	0
$\widehat{R}^1_{cal}$	1.56	98.03	0.98	0.98	1.37	93.80	4.42	1.76	1.20	96.15	0.96	2.88	0.99	97.39	0.86	1.73
$\widehat{R}^2_{cal}$	1.39	98.03	0.98	0.98	1.22	96.46	1.76	1.76	1.07	97.11	0.96	1.92	0.90	97.39	0.86	1.73
$\widehat{R}_{KL1cali}$	1.38	99.01	0	0.98	1.22	97.34	0.88	1.76	1.07	94.23	1.92	3.84	0.90	95.65	0.86	3.47
$\widehat{R}_{KL1calr}$	1.39	99.01	0	0.98	1.22	97.34	0.88	1.76	1.07	94.23	1.92	3.84	0.90	95.65	0.86	3.47
$\widehat{R}_{KL1calo}$	1.39	99.01	0	0.98	1.22	97.34	0.88	1.76	1.07	94.23	1.92	3.84	0.90	95.65	0.86	3.47
$\widehat{R}_{KL2cali}$	1.38	99.01	0	0.98	1.22	97.34	0.88	1.76	1.07	94.23	1.92	3.84	0.90	95.65	0.86	3.47
$\widehat{R}_{KL2calr}$	1.38	99.01	0	0.98	1.22	97.34	0.88	1.76	1.07	94.23	1.92	3.84	0.90	95.65	0.86	3.47
$\widehat{R}_{KL2calo}$	1.38	99.01	0	0.98	1.22	97.341	0.88	1.76	1.07	94.23	1.92	3.84	0.90	95.65	0.86	3.47
$\widehat{R}_{Dcali}$	1.42	98.03	0.98	0.98	1.23	96.46	1.76	1.76	1.08	94.23	1.92	3.84	0.90	98.26	0	1.73
$\widehat{R}_{Dcalr}$	1.42	98.03	0.98	0.98	1.23	96.46	1.76	1.76	1.08	94.23	1.92	3.84	0.90	98.26	0	1.73
$\widehat{R}_{Dcalo}$	1.42	98.03	0.98	0.98	1.23	96.46	1.76	1.76	1.08	94.23	1.92	3.84	0.90	98.26	0	1.73

	Antal and Tillé [2]															
$\widehat{R}_{HT}$	1.55	98.03	1.96	0	1.40	92.92	5.30	1.76	1.21	94.23	5.76	0	0.99	96.52	2.60	0.86
$\widehat{R}^1_{cal}$	1.62	98.03	0.98	0.98	1.45	97.34	0.88	1.76	1.23	95.19	0.96	3.84	1.00	96.52	0	3.47
$\widehat{R}^2_{cal}$	1.40	98.03	0.98	0.98	1.22	97.34	0.88	1.76	1.09	97.11	0	2.88	0.89	96.52	0.86	2.60
$\widehat{R}_{KL1cali}$	1.37	99.01	0	0.98	1.22	97.34	0.88	1.76	1.07	94.23	0	5.76	0.89	96.52	0.86	2.60
$\widehat{R}_{KL1calr}$	1.37	99.01	0	0.98	1.21	97.34	0.88	1.76	1.07	94.23	0	5.76	0.89	96.52	0.86	2.60
$\widehat{R}_{KL1calo}$	1.37	99.01	0	0.98	1.21	97.34	0.88	1.76	1.07	94.23	0	5.76	0.89	96.52	0.86	2.60
$\widehat{R}_{KL2cali}$	1.37	99.01	0	0.98	1.22	97.34	0.88	1.76	1.07	94.23	0	5.76	0.89	96.52	0.86	2.60
$\widehat{R}_{KL2calr}$	1.37	99.01	0	0.98	1.22	97.34	0.88	1.76	1.07	94.23	0	5.76	0.89	96.52	0.86	2.60
$\widehat{R}_{KL2calo}$	1.37	99.01	0	0.98	1.22	97.34	0.88	1.76	1.07	94.23	0	5.76	0.89	96.52	0.86	2.60
$\widehat{R}_{Dcali}$	1.41	98.03	0.98	0.98	1.23	96.46	1.76	1.76	1.09	96.15	0	3.84	0.89	95.65	0.86	3.47
$\widehat{R}_{Dcalr}$	1.41	98.03	0.98	0.98	1.23	96.46	1.76	1.76	1.09	96.15	0	3.84	0.89	95.65	0.86	3.47
$\widehat{R}_{Dcalo}$	1.41	98.03	0.98	0.98	1.23	96.46	1.76	1.76	1.09	96.15	0	3.84	0.89	95.65	0.86	3.47
	Antal and Tillé [1]															
$\widehat{R}_{HT}$	1.61	96.07	2.94	0.98	1.39	93.80	4.42	1.76	1.23	96.15	3.84	0	1.01	97.39	2.60	0
$\widehat{R}^1_{cal}$	1.58	99.01	0.98	0	1.39	92.03	4.42	3.53	1.22	97.11	0.96	1.92	0.98	97.39	0	2.60
$\widehat{R}^2_{cal}$	1.39	98.03	0.98	0.98	1.19	97.34	0.88	1.76	1.08	97.11	0	2.88	0.89	94.78	1.73	3.47
$\widehat{R}_{KL1cali}$	1.40	99.01	0	0.98	1.20	98.23	0	1.76	1.08	95.19	0.96	3.84	0.90	94.78	1.73	3.47
$\widehat{R}_{KL1calr}$	1.40	99.01	0	0.98	1.20	98.23	0	1.76	1.08	94.23	0.96	4.80	0.89	94.78	1.73	3.47
$\widehat{R}_{KL1calo}$	1.40	99.01	0	0.98	1.20	98.23	0	1.76	1.08	94.23	0.96	4.80	0.89	94.78	1.73	3.47
$\widehat{R}_{KL2cali}$	1.40	99.01	0	0.98	1.20	98.23	0	1.76	1.08	95.19	0.96	3.84	0.90	94.78	1.73	3.47
$\widehat{R}_{KL2calr}$	1.40	99.01	0	0.98	1.20	98.23	0	1.76	1.08	95.19	0.96	3.84	0.90	94.78	1.73	3.47
$\widehat{R}_{KL2calo}$	1.40	99.01	0	0.98	1.20	98.23	0	1.76	1.08	95.19	0.96	3.84	0.90	94.78	1.73	3.47
$\widehat{R}_{Dcali}$	1.44	98.03	0.98	0.98	1.21	96.46	1.76	1.76	1.09	97.11	0	2.88	0.90	94.78	1.73	3.47
$\widehat{R}_{Dcalr}$	1.45	98.03	0.98	0.98	1.21	96.46	1.76	1.76	1.09	97.11	0	2.88	0.90	94.78	1.73	3.47
$\widehat{R}_{Dcalo}$	1.45	98.03	0.98	0.98	1.21	96.46	1.76	1.76	1.09	97.11	0	2.88	0.90	94.78	1.73	3.47

Acknowledgments The work was supported by the Ministerio de Ciencia e Innovación, Spain, under grant PID2019-106861RB-I00./ 10.13039/501100011033 and IMAG-Maria de Maeztu CEX2020-001105-M/AEI/10.13039/501100011033.

References

1. Antal, E., Tillé, Y.: A direct bootstrap method for complex sampling designs from a finite population. J. Am. Stat. Assoc. **106**(494), 534–543 (2011)
2. Antal, E., Tillé, Y.: A new resampling method for sampling designs without replacement: The doubled half bootstrap. Comput. Stat. **29**(5), 1345–1363 (2014)
3. Booth, J.G., Butler, R.W., Hall, P.: Bootstrap methods for finite populations. J. Am. Stat. Assoc. **89**(428), 1282–1289 (1994)
4. Burtless, G.: Effects of growing wage disparities and changing family composition on the US income distribution. Eur. Econ. Rev. **43**(4–6), 853–865 (1999)
5. Chauvet, G.: Bootstrap pour un tirage à plusieurs degrés avec échantillonnage à forte entropie à chaque degré, Working Papers 2007-39, Center for Research in Economics and Statistics (2007)
6. Deville, J.C.: Generalized calibration and application to weighting for non-response COMP-STAT. In: Bethlehem, J.G., van der Heijden, P.G.M. (eds.), Proceedings in Computational Statistics, 14th Symposium, Utrecht, pp. 65–76. Springer, New York (2000)
7. Dickens, R., Manning, A.: Has the national minimum wage reduced UK wage inequality? J. R. Stat. Soc. Ser. A **167**(4), 613–626 (2004)
8. Ferri-García, R., Rueda, M.D.M.: Efficiency of propensity score adjustment and calibration on the estimation from non-probabilistic online surveys. SORT **1**(2), 159–162 (2018)
9. Jones, A.F., Jr., Weinberg, D.H.: The changing shape of the nation's income distribution. Curr. Popul. Rep. **60**, 1–11 (2000)
10. Kott, P.S., Liao, D.: One step or two? Calibration weihting form a complete list frame with nonresponse. Surv. Methodol. **41**(1), 165–181 (2015)
11. Kott, P.S., Liao, D.: Calibration weighting for nonresponse that is not missing at random: allowing more calibration than response-model variables. J. Surv. Stat. Methodol. **5**, 159–174 (2017)
12. Machin, S., Manning, A., Rahman, L.: Where the minimum wage bites hard: introduction of minimum wages to a low wage sector. J. Eur. Econ. Assoc. **1**(1), 154–180 (2003)
13. Rao, J.N.K., Kovar, J.G., Mantel, H.J.: On estimating distribution function and quantiles from survey data using auxiliary information. Biometrika **77**(2), 365–375 (1990)
14. Rueda, M., Martínez, S., Martínez, H., Arcos, A.: Estimation of the distribution function with calibration methods. J. Stat. Plann. Infer. **137**(2), 435–448 (2007)
15. Rueda, M., Martínez-Puertas, S., Illescas, M.: Treating nonresponse in the estimation of the distribution function. Math. Comput. Simul. **186**, 136–144 (2020)

Chapter 12
Numerical Methods Based on Spline Quasi-Interpolating Operators for Hammerstein Integral Equations

Domingo Barrera, Abdelmonaim Saou, and Mohamed Tahrichi

Abstract In this paper, we propose collocation type method, its iterated version and Nyström method based on discrete spline quasi-interpolating operators to solve Hammerstein integral equation. We present an error analysis of the approximate solutions and we show that the iterated solution of collocation type exhibits a superconvergence as in the case of the Galerkin method. Finally, we provide numerical tests, that confirm the theoretical results.

Keywords Quasi-interpolants · Spline functions · Collocation method · Nyström method · Hammerstein integral equation

12.1 Introduction

The issue of integral equations is one of the most useful mathematical tools in pure and applied mathematics. It has huge applications in many physical problems. Many initial and boundary value problems associated with ordinary differential equations (ODEs) and partial differential equations (PDEs) can be transformed into resolution problems of some approximate integral equations (Ref. [16]).

In this paper we are interested in Hammerstein nonlinear integral equation given by

$$u(x) = f(x) + \int_a^b k(x,t)\psi(t,u(t))\,dt, \quad x \in [a,b] \tag{12.1}$$

D. Barrera
Department of Applied Mathematics, University of Granada, Granada, Spain
e-mail: dbarrera@ugr.es

A. Saou · M. Tahrichi (✉)
Team ANAA, ANO Laboratory, Faculty of Sciences, University Mohammed First, Oujda, Morocco

D. Barrera et al. (eds.), *Mathematical and Computational Methods for Modelling, Approximation and Simulation*, SEMA SIMAI Springer Series 29,
https://doi.org/10.1007/978-3-030-94339-4_12

where f, k and ψ are continuous functions, with ψ nonlinear with respect to the second variable and u is the function to be determined. This type of equations appears in nonlinear physical phenomena such as electromagnetic fluid dynamics and reformulation of boundary value problems with a nonlinear boundary condition, see [7, 15]. Classical methods of solving (12.1) are projection methods. Within the commonly used projection methods, there are Galerkin and collocation methods based respectively on a series of orthogonal and interpolating projectors in finite dimensional approximation spaces. Both methods have been studied by many authors, see for example [6, 11, 14]. The idea of improving Galerkin and collocation solutions by an iteration technique was introduced by Sloan [23]. Then several authors applied this idea for different types of equations, see for example [8]. In [13], the authors introduced a new collocation-type method for the numerical solution of (12.1) and its superconvergence properties were studied in [12]. In [4], they used superconvergent Nyström and degenerate kernel methods to solve (12.1). More recently, for smooth kernel or less smooth along the diagonal, the authors in [5] introduced superconvergent product integration method to approximate the solution of (12.1). For Hammerstein integral equation with singular kernel, Gelerkin-type/modified Galerkin-type and Kantorovich methods are studied in [3].

Recently, many authors have been interested in using spline quasi-interpolating operators for the approximation of solution of integral equations. In particular, in [21] Fredholm integral equation is solved by using degenerate kernel methods based on quasi-interpolating operators. A new quadrature rule based on integrating spline quasi-interpolant is derived in [20] and used to solve Fredholm integral equation by Nyström method. A modified Kulkarni's method based on spline quasi-interpolating operators is investigated in [1]. The authors in [9], used spline quasi-interpolants that are projectors for the numerical solution of Fredholm integral equation by Galerkin, Kantorovich, Sloan and Kulkarni's schemes. The same operators are used in [10] to solve Urysohn nonlinear integral equations by using collocation and Kulkarni's schemes.

In this paper we investigate collocation and Nyström type methods based on spline quasi-interpolants that are not necessary projectors to solve Hammerstein integral equation. Here is an outline of the paper. In Sects. 12.2 and 12.3 we recall the definitions and main properties of the spline quasi-interpolants and a quadrature formulas associated, and present their convergence properties. In Sect. 12.4 we introduce the collocation and Nyström methods based in spline quasi-interpolants. In Sect. 12.5 we give a general framework for the error analysis of the approximate and the iterated solutions. Finally, in Sect. 12.6 we provide some numerical results, illustrating the approximation properties of the proposed methods.

12.2 A Family of Discrete Spline Quasi-Interpolants

Let $\mathcal{X}_n := \{x_k, 0 \leq k \leq n\}$ be the uniform partition of the interval $I = [a, b]$ into n equal subintervals, i.e. $x_k := a + kh$, with $h = (b - a)/n$ and $0 \leq k \leq n$. We consider the space $\mathcal{S}_d(I, \mathcal{X}_n)$ of splines of degree d and class C^{d-1} on this partition.

Its canonical basis is formed by the $n + d$ normalized B-splines $\{B_k, k \in \mathcal{J}\}$, with $\mathcal{J} := \{1, 2, \dots, n + d\}$. The support of each B_k is the interval $[x_{k-d-1}, x_k]$ if we add multiple knots at the endpoints.

The discrete quasi-interpolants of degree $d > 1$ is a spline operator of the form

$$Q_d f := \sum_{k \in \mathcal{J}} \mu_k(f) B_k, \tag{12.2}$$

where the coefficients $\mu_k(f)$ are linear combinations of values of f on either the set T_n (for d even) or on the set $\mathcal{X}_n$ (for d odd), where

$$T_n := \{t_0 = x_0,\ t_k = \frac{1}{2}\left(x_{k-1} + x_k\right), k = 1, \dots, n,\ t_{n+1} = b\},$$
$$\mathcal{X}_n := \{x_k,\ k = 0, \dots, n\}.$$

Therefore, for d even, we set $f(T_n) = \left\{f_k = f(t_k), 0 \le k \le n + 1\right\}$, and for d odd, we set $f\left(\mathcal{X}_n\right) = \left\{f_k = f(x_k), 0 \le k \le n\right\}$. The coefficients $\mu_k(f)$ for $d + 1 \le k \le n$, have the following form

$$\mu_k(f) := \begin{cases} \sum_{i=0}^{d} \alpha_{i+1,k} f_{k-d+i+1}, & \text{if } d \text{ is even,} \\ \sum_{i=1}^{d} \alpha_{i,k} f_{k-d+i-1}, & \text{if } d \text{ is odd,} \end{cases}$$

where α_{ik} are calculated such that the quasi-interpolants Q_d reproduce the space $\mathbb{P}_d$ of all polynomials of total degree at most d, i.e.

$$Q_d p = p, \quad \forall p \in \mathbb{P}_d.$$

The extremal coefficients $\mu_k(f)$ for $1 \le k \le d$ and $n + 1 \le k \le n + d$ have particular expressions.

The quasi-interpolants Q_d can be written under the following quasi Lagrange form

$$Q_d f = \sum_{j=0}^{\mathrm{n_d}} f_j Lj,$$

where $\mathrm{n_d} := n + 1$ if d is even, $\mathrm{n_d} := n$ if d is odd and L_j are linear combinations of finite number of B-splines B_j.

Since μ_k are continuous linear functionals, the operator Q_d is uniformly bounded on $C([a, b])$ and classical results in approximation theory provide that for any subinterval $I_k = [x_{k-1}, x_k]$, $1 \le k \le n$ and for any function f, we have

$$\|f - Q_d f\|_{\infty, I_k} \le (1 + \|Q_d\|) dist_{\infty, I_k}(f, \mathbb{P}_d),$$

where

$$dist_{\infty,I_k}(f, \mathbb{P}_d) = \inf_{p\in\mathbb{P}_d} \|f - p\|_{\infty,I_k}.$$

Therefore, if $f \in C^{d+1}([a,b])$, we get

$$\|f - Q_d f\|_\infty \le C_1 h^{d+1} \|f^{(d+1)}\|_\infty, \tag{12.3}$$

for some constant C_1 independent of h. As usual $\|f - p\|_{\infty,I_k} = \max_{x\in I_k} |f(x) - p(x)|$ and $\|f - p\|_\infty = \max_{x\in[a,b]} |f(x) - p(x)|$.

In what follows, we give two examples of spline quasi-interpolants denoted by Q_2 and Q_3.

- Q_2 is the C^1 quadratic spline quasi-interpolant exact on $\mathbb{P}_2$ and defined by (see e.g. [17])

$$Q_2 f := \sum_{k=1}^{n+2} \mu_k(f) B_k, \tag{12.4}$$

where the coefficient functionals $\mu_k(f)$ are given by

$$\begin{aligned} &\mu_1(f) = f_0, \quad \mu_2(f) = -\frac{1}{3} f_0 + \frac{3}{2} f_1 - \frac{1}{3} f_2, \\ &\mu_k(f) = -\frac{1}{8} f_{k-2} + \frac{5}{4} f_{k-1} - \frac{1}{8} f_k, \quad 3 \le k \le n, \\ &\mu_{n+1}(f) = -\frac{1}{3} f_{n-1} + \frac{3}{2} f_n - \frac{1}{3} f_{n+1}, \quad \mu_{n+2}(f) = f_{n+1}. \end{aligned} \tag{12.5}$$

In [17] the author has proved that the quasi-interpolant Q_2 is uniformly bounded and its infinity norm is given by

$$\|Q_2\|_\infty = \frac{305}{207} \approx 1.4734.$$

- Q_3 is the C^2 cubic spline quasi-interpolant exact on $\mathbb{P}_3$ and defined by (see e.g. [18])

$$Q_3 f := \sum_{k=1}^{n+3} \mu_k(f) B_k, \tag{12.6}$$

where the coefficient functionals $\mu_k(f)$ are given by

$$\begin{aligned}
&\mu_1(f) = f_0, \ \mu_2(f) = \frac{1}{18}\big(7f_0 + 18f_1 - 9f_2 + 2f_3\big), \\
&\mu_k(f) = \frac{1}{6}\big(- f_{k-3} + 8f_{k-2} - f_{k-1}\big), \quad 3 \le k \le n+1, \qquad (12.7) \\
&\mu_{n+2}(f) = \frac{1}{18}\big(2f_{n-3} - 9f_{n-2} + 18f_{n-1} + 7f_n\big), \ \mu_{n+3}(f) = f_n.
\end{aligned}$$

The infinity norm of the quasi-interpolant Q_3 is uniformly bounded and it is given by

$$\left\| Q_3 \right\|_\infty = 1.631.$$

In the case of even degree, the quasi-interpolant Q_d present some interesting properties. The first one concern superconvergence at some evaluation points. These superconvergence points depends upon the degree d. The following theorem provide them explicitly for the case of Q_2.

Theorem 12.1 *If $f \in C^4([a, b])$, then*

$$\begin{aligned}
|Q_2 f(x_i) - f(x_i)| &= O(h^4), \quad 0 \le i \le n, \ i \ne 1, n-1. \\
|Q_2 f(t_i) - f(t_i)| &= O(h^4), \quad 0 \le i \le n+1, \ i \ne 1, 2, n-1, n.
\end{aligned} \qquad (12.8)$$

Proof Using the exact values of Bsplines on x_i, t_i and the definition of coefficient functionals, we can show that $e_3(x_i) = e_3(t_i) = 0$ for all points included in (12.8), where e_3 represents the error of Q_2 on the monomial x^3. For the excluded points in (12.8), these values are not zero. Next, following the same logical scheme as in the proof of Lemma 4.1 in [9], we can get (12.8). □

12.3 Quadrature Rules Based on Q_d Defined on a Uniform Partition

Let f be a continuous function on the interval $[a, b]$, we consider the numerical evaluation of the integral

$$\mathcal{I}(f) := \int_a^b f(x)dx$$

by quadrature rules based on Q_d defined on a uniform partition. These rules are defined by

$$\mathcal{I}_{Q_d}(f) := \int_a^b Q_d f(x)dx$$
$$= \begin{cases} h\sum_{j=0}^{n+1} \omega_j f_j \text{ if } d \text{ even} \\ h\sum_{j=0}^{n} \omega_j f_j \text{ if } d \text{ odd} \end{cases}$$

where $\omega_j = \frac{1}{h}\int_a^b L_j(x)dx$ and L_j are quasi-Lagrange functions associated with Q_d.

In particular, for $d = 2$ and $d = 3$ we obtain the following quadrature rules

$$\mathcal{I}_{Q_2}(f) := h\sum_{j=0}^{n+1} \omega_j f_j, \quad \text{et} \quad \mathcal{I}_{Q_3}(f) := h\sum_{j=0}^{n} \omega_j f_j.$$

Where the ω_j weights are given explicitly in the following table.

j	0	1	2	3	4	...	$n-4$	$n-3$	$n-2$	$n-1$	n	$n+1$
ω_j $(d=2)$	$\frac{1}{9}$	$\frac{7}{8}$	$\frac{73}{72}$	1	1	...	1	1	1	$\frac{73}{72}$	$\frac{7}{8}$	$\frac{1}{9}$
ω_j $(d=3)$	$\frac{23}{72}$	$\frac{4}{3}$	$\frac{19}{24}$	$\frac{19}{18}$	1	...	1	$\frac{19}{18}$	$\frac{19}{24}$	$\frac{4}{3}$	$\frac{23}{72}$	–

As Q_d is exact on $\mathbb{P}_d$, we deduce that the associated quadrature formulas $\mathcal{I}_{Q_d}$ are also exact on $\mathbb{P}_d$. Therefore, the error $\mathcal{E}_{Q_d}(f) := \mathcal{I}(f) - \mathcal{I}_{Q_d}(f) = O(h^{d+1})$. However, in the case of an even degree d, this last error is more accurate. Indeed, the following theorem holds true (for the proof see [2]).

Theorem 12.2 *Let d be an even number and let Q_d be the quasi-interpolant defined by (12.2). For any function $f \in C^{d+2}([a, b])$, we have*

$$\int_a^b (f(t) - Q_d f(t))dt = O(h^{d+2}). \tag{12.9}$$

Moreover, for any weight function $g \in \mathcal{W}^{1,1}$ (i.e. $\|g'\|_1$ bounded), we have

$$\int_a^b g(t)(f(t) - Q_d f(t))dt = O(h^{d+2}). \tag{12.10}$$

Particular cases of the rule $\mathcal{I}_{Q_d}$ for $d = 2$ and $d = 4$ are studied in depth in [19] and [20] respectively. The authors in these papers have provided explicit error estimations and they have made comparisons with similar rules of interpolatory type.

12.4 Methods Based on Q_d

Let us consider the Hammerstein integral equation (12.1) given in operator form as

$$u - \mathcal{K}u = f, \tag{12.11}$$

where $\mathcal{K}$ in the Hammerstein integral operator defined on $\mathcal{L}^\infty[a, b]$ by

$$\mathcal{K}u(x) = \int_a^b k(x, t)\psi(t, u(t))dt.$$

Since the kernel k is assumed to be continuous, the operator $\mathcal{K}$ is compact from $\mathcal{L}^\infty[a, b]$ to $C[a, b]$. In what follows, we propose two methods based on the quasi-interpolant operator Q_d to solve (12.11).

12.4.1 Collocation Type Method and Its Iterated Version

Recall that a spline quasi-interpolant of degree d is an operator defined on $C[a, b]$ by:

$$\begin{aligned} Q_d : C[a, b] &\longrightarrow \mathcal{S}_d(I, X_n) \\ f &\longrightarrow \textstyle\sum_{j=1}^{n+d} \mu_j(f)B_j. \end{aligned}$$

We propose to approximate the integral operator $\mathcal{K}$ in (12.11) by $\mathcal{K}_n^c := Q_d\mathcal{K}$ and the second member f by $Q_d f$. The approximate equation is then given by

$$u_n^c - Q_d\mathcal{K}u_n^c = Q_d f, \tag{12.12}$$

where

$$Q_d\mathcal{K}u_n^c = \sum_{i=1}^{n+d} \mu_i\left(\mathcal{K}u_n^c\right)B_i.$$

The approximate solution u_n^c is a spline function, then we can write

$$u_n^c = \sum_{i=1}^{n+d} \alpha_i B_i.$$

Replacing u_n^c in Eq. (12.12), we obtain

$$\sum_{i=1}^{n+d} \alpha_i B_i - \sum_{i=1}^{n+d} \mu_i \Big(\mathcal{K}\big(\sum_{j=1}^{n+d} \alpha_j B_j \big) \Big) B_i = \sum_{i=1}^{n+d} \mu_i(f) B_i, \tag{12.13}$$

since the family $\{B_i,\ 1 \leq i \leq n+d\}$ is a basis of $\mathcal{S}_d(I, \mathcal{X}_n)$, we can identify the coefficients and we obtain the following nonlinear system:

$$\alpha_i - \mu_i \Big(\mathcal{K}\big(\sum_{j=1}^{n+d} \alpha_j B_j \big) \Big) = \mu_i(f), \qquad i = 1, 2, \ldots, n+d. \tag{12.14}$$

Another interesting solution to consider is the following iterated one

$$\hat{u}_n^c := \mathcal{K}\left(u_n^c\right) + f. \tag{12.15}$$

Applying Q_d on both sides, we find

$$Q_d \hat{u}_n^c = Q_d \mathcal{K}\left(u_n^c\right) + Q_d f = u_n^c. \tag{12.16}$$

Replacing in (12.15), we find that $\hat{u}_n^c$ satisfy the following equation

$$\hat{u}_n^c = \mathcal{K} Q_d \hat{u}_n^c + f. \tag{12.17}$$

We show later that the iterated solution $\hat{u}_n^c$ is more accurate than u_n^c.

Remark 12.1 It is important to note the presence of integrals in system (12.14) and also in the expression of iterated solution (12.15). When implementing the method these integrals were calculated numerically using high accuracy quadrature rules, like those defined on [22], to imitate exact integration .

12.4.2 Nyström Method

In the Nyström method the operator $\mathcal{K}$ in (12.11) is approximated by

$$\mathcal{K}_n^N u := \sum_{j=0}^{\mathrm{n_d}} \omega_j k(., \xi_j) \psi(\xi_j, u(\xi_j)),$$

where ξ_j and ω_j are respectively the nodes and the weights of the quadrature rule $\mathcal{I}_{Q_d}$ based on Q_d. More precisely, ξ_j are given by t_i for d even and by x_i for d odd.

Hence, the corresponding approximate equation is given by

$$u_n^N - \sum_{j=0}^{\mathrm{n_d}} \omega_j k(., \xi_j) \psi(\xi_j, u_n^N(\xi_j)) = f. \tag{12.18}$$

Taking this last equation at the points ξ_i, we get the following non linear system

$$u_n^N(\xi_i) - \sum_{j=0}^{\mathrm{n_d}} \omega_j k(\xi_i, \xi_j) \psi(\xi_j, u_n^N(\xi_j)) = f(\xi_i), \quad 0 \leq i \leq \mathrm{n_d}. \tag{12.19}$$

By solving this system, we obtain the approximate solution u_n^N at points ξ_i. Over all the domain, u_n^N is given by the following interpolation formula

$$u_n^N = \sum_{j=0}^{\mathrm{n_d}} \omega_j k(, \xi_j) \psi(\xi_j, u_n^N(\xi_j)) + f.$$

Remark 12.2 It should be noted that the Nyström method is completely discrete because the system (12.19) does not contain integrals to be evaluated numerically. Which makes this method one of the easiest methods to implement.

12.5 Error Analysis

Let u^* be the unique solution of (12.1), and let $\mathfrak{a}$ and $\mathfrak{b}$ be real numbers such that

$$[\min_{x\in[a,b]} u^*(x), \max_{x\in[a,b]} u^*(x)] \subset [\mathfrak{a}, \mathfrak{b}].$$

Define $\Omega = [a, b] \times [\mathfrak{a}, \mathfrak{b}]$. We assume throughout this paper unless stated otherwise, the following conditions on f, k and ψ:

(C.1) $k \in C(\Omega)$.

(C.2) $f \in C([a, b])$.

(C.3) $\psi(t, x)$ is continuous in $t \in [a, b]$ and Lipschitz continuous in $x \in [\mathfrak{a}, \mathfrak{b}]$, i.e., there exists a constant $q_1 > 0$ such that

$$\left|\psi(t, x_1) - \psi(t, x_2)\right| \leq q_1 |x_1 - x_2|, \quad \text{for all } x_1, x_2 \in [\mathfrak{a}, \mathfrak{b}].$$

(C.4) The partial derivative $\psi^{(0,1)}$ of ψ with respect to the second variable exists and is Lipschitz continuous, i.e., there exists a constant $q_2 > 0$ for which

$$\left|\psi(t, x_1)^{(0,1)} - \psi(t, x_2)^{(0,1)}\right| \leq q_2 |x_1 - x_2|, \quad \text{for all } x_1, x_2 \in [\mathfrak{a}, \mathfrak{b}].$$

Condition (C.4) implies that the operator $\mathcal{K}$ is Fréchet-differentiable and that $\mathcal{K}'(u^*)$ is Mq_2-Lipschitz, where

$$\mathcal{K}'(u^*)v(s) = \int_a^b k(s,t)\frac{\partial \psi}{\partial u}\big(t, u^*(t)\big)v(t)dt,$$

and

$$M = \sup_{s\in[a,b]} \int_a^b |k(s,t)|dt.$$

Furthermore, the operator $\mathcal{K}'(u^*)$ is compact.

12.5.1 Collocation and Nyström Solutions

It is easy to see that the operators $\mathcal{K}_n^c$ and $\mathcal{K}_n^N$ are Fréchet differentiables and

$$(\mathcal{K}_n^c)'(u^*)v(s) = \sum_{i=1}^{n+d} \mu_i\Big(\mathcal{K}'(u^*)v\Big)B_i(s),$$

$$(\mathcal{K}_n^N)'(u^*)v(s) = \sum_{j=1}^{n+d} \omega_j k(s,\xi_j)\frac{\partial \psi}{\partial u}(\xi_j, u^*(\xi_j))v(\xi_j).$$

Throughout the rest of this paper, we denote by L the operator $\mathcal{K}'(u^*)$ and by L_n either the operator $(\mathcal{K}_n^c)'(u^*)$ or the operator $(\mathcal{K}_n^N)'(u^*)$.

The following lemmas state some properties needed to prove the existence and the convergence of the approximate solutions. Their proofs are consequences of conditions (C.1)–(C.4), the fact that L_n is linear operator and that Q_d converges to the identity operator pointwise.

Lemma 12.1 *Assume that 1 is not an eigenvalue of L. Then for n large enough, 1 is not in the spectrum of L_n and $(I - L_n)^{-1}$ exists as a bounded linear operator, i.e.,*

$$\left\|(I - L_n)^{-1}\right\|_\infty \le C_1,$$

for a suitable constant C_1 independent of n.

Lemma 12.2 *Assume that 1 is not an eigenvalue of L. Then for n large enough, L_n is Lipschitz continuous on $B(u^*, \delta)$ for $\delta > 0$.*

Put $Tu = \mathcal{K}u + f$. Equation (12.11) becomes

$$u = Tu, \tag{12.20}$$

Generally, the previous equation is approximated by

$$u_n = T_n u_n, \tag{12.21}$$

where T_n is a sequence of approximating operators. We quote the following theorem from [24] which gives conditions on T_n to ensure the convergence of u_n to the exact solution.

Theorem 12.3 *Suppose that the Eq. (12.20) has a unique solution u^* and the following conditions are satisfied*

(i) T_n is Fréchet-differentiable and $(I - T_n'(u^*))^{-1}$ exists and is uniformly bounded.
(ii) For certain values of $\delta > 0$ and $0 < q < 1$, the inequalities

$$\sup_{\|u-u^*\|\leq\delta} \left\| \left[I - T_n'\left(u^*\right)\right]^{-1} \left(T_n'(u) - T_n'\left(u^*\right)\right) \right\|_\infty \leq q,$$

$$\alpha := \left\| \left[I - T_n'\left(u^*\right)\right]^{-1} \left(T_n u^* - T u^*\right) \right\|_\infty \leq \delta(1-q),$$

are valid.

Then the approximate equation (12.21) has a unique solution u_n in $B(u^*, \delta)$ such that

$$\frac{\alpha}{1+q} \leq \left\|u_n - u^*\right\|_\infty \leq \frac{\alpha}{1-q}. \tag{12.22}$$

We now give our result about the existence and uniqueness of the approximate solution for collocation and Nyström methods studied in this paper. Recall that for collocation method, $T_n u = \mathcal{K}_n^c u + Q_d f$, and for Nyström method $T_n u = \mathcal{K}_n^N u + f$.

Theorem 12.4 *Let u^* be the unique solution of Eq. (12.20). Under the assumptions (C.1)–(C.4), there exists a real number $\delta_0 > 0$ such that the approximate equation (12.21) has a unique solution u_n in $B(u^*, \delta_0)$ for a sufficiently large n. Moreover, the error estimate (12.22) holds.*

Proof We give the proof in the case of collocation method. The proof is similar for Nyström case. From Theorem 12.3, it suffices to prove that (i) and (ii) are satisfied. Lemma 12.1 ensures that (i) is valid for sufficiently large n say for all $(n > N_1)$. Moreover from Lemma 12.2, for $\left\|u - u^*\right\|_\infty \leq \delta$ and $n > N_1$, we have

$\left\| T_n'(u) - T_n'\left(u^*\right)\right\|_\infty \leq m\delta$, where m is the Lipschitz constant of L_n. Hence

$$\begin{aligned}
&\left\| \left[I - T_n'\left(u^*\right)\right]^{-1} \left(T_n'(u) - T_n'\left(u^*\right)\right)\right\|_\infty \\
&\leq \left\| \left[I - T_n'\left(u^*\right)\right]^{-1}\right\|_\infty \left\| T_n'(u) - T_n'\left(u^*\right)\right\|_\infty \\
&\leq m\delta \left\| \left[I - T_n'\left(u^*\right)\right]^{-1}\right\|_\infty .
\end{aligned}$$

Therefore,

$$\sup_{\|u-u^*\|\leq\delta} \left\| \left[I - T_n'\left(u^*\right)\right]^{-1} \left(T_n'(u) - T_n'\left(u^*\right)\right)\right\|_\infty \leq q,$$

with $q = m\delta \left\| \left(I - T_n'\left(u^*\right)\right)^{-1}\right\|_\infty$. Here we take $\delta = \delta_0$ so small that $0 < q < 1$. Now we have

$$\begin{aligned}
\left\| Tu^* - T_n u^*\right\|_\infty &= \left\| \left(\mathcal{K}u^* - Q_d\mathcal{K}u^*\right) + \left(f - Q_d f\right)\right\|_\infty \\
&= \left\| (I - Q_d)\mathcal{K}u^* + (I - Q_d) f\right\|_\infty \\
&= \left\| (I - Q_d)(\mathcal{K}u^* + f)\right\|_\infty \\
&= \left\| (I - Q_d)u^*\right\|_\infty \xrightarrow[n\to\infty]{} 0,
\end{aligned}$$

i.e. there exists N_2 such that for $n > N_2$

$$\alpha = \left\| \left[I - T_n'\left(u^*\right)\right]^{-1} \left(Tu^* - T_n u^*\right)\right\| \leq \delta_0(1 - q).$$

Consequently, the condition (ii) is also valid. Hence, for $n > max\{N_1, N_2\}$, using Theorem 12.2, one can conclude that (12.21) has a unique solution in $B(u^*, \delta_0)$ and the inequalities (12.22) hold. □

Using the results obtained in Theorem 12.4 and error estimates (12.3)–(12.10), we give in the following theorem the error explicit estimates of the collocation and Nyström methods based on quasi-interpolants Q_d.

Theorem 12.5 *Let u_n be a unique solution of the approximate equation (12.21) in $B(u^*, \delta_0)$ for a sufficiently large n. Assume that*

(i) $k \in C^{0,d+1}([a, b] \times [a, b])$,
(ii) $\psi \in C^{d+1}([a, b] \times [a, b])$,

(iii) $f \in C^{d+1}([a,b])$.

Then

$$\|u_* - u_n\|_\infty = O(h^{d+1}).$$

Moreover, if d is even and if

(i) $k \in C^{0,d+2}([a,b] \times [a,b])$,
(ii) $\psi \in C^{d+2}([a,b] \times [a,b])$,
(iii) $f \in C^{d+2}([a,b])$,

then, in the case of Nyström method it holds

$$\|u_* - u_n^N\|_\infty = O(h^{d+2}).$$

Proof It is an immediate consequence of preceding result and estimates (12.3)–(12.10). □

12.5.2 Iterated Collocation Solution

Recall that the iterated solution satisfies the following equation:

$$\hat{u}_n^c - \mathcal{K}Q_d\hat{u}_n^c = f. \tag{12.23}$$

Define r_n by

$$r_n = \frac{\left\|\mathcal{K}(u^*) - \mathcal{K}(u_n^c) - L(u^* - u_n^c)\right\|_\infty}{\left\|u^* - u_n^c\right\|_\infty},$$

where $L = \mathcal{K}'(u^*)$. From Theorem 12.4 and the definition of L, we conclude that

$$r_n \xrightarrow[n\to\infty]{} 0.$$

Moreover, it is possible to see that

$$r_n \leq \frac{\xi}{2}\left\|u^* - u_n^c\right\|_\infty,$$

where ξ is a positive constant. We also use the following notation

$$a = \left\|(I-L)^{-1}\right\|_\infty.$$

The error estimate for the iterated solution is given in the following theorem.

Theorem 12.6 *Let u^* be a unique solution of Eq. (12.20). Assume that the assumptions (C.1)–(C.4) are satisfied. Then for a sufficiently large n, we have*

$$\left\|u^* - \hat{u}_n^c\right\|_\infty \leq \xi \left\|u^* - u_n^c\right\|_\infty^2 + a \left\|L(I - Q_d)\mathcal{K}u^*\right\|_\infty + a \left\|L(I - Q_d)L(u^* - u_n^c)\right\|_\infty$$

Proof We have

$$\begin{aligned}(I - L)(u^* - \hat{u}_n^c) &= \mathcal{K}u^* - \mathcal{K}u_n^c - L\left(u^* - u_n^c\right) + L\left(\hat{u}_n^c - u_n^c\right)\\ &= \mathcal{K}u^* - \mathcal{K}u_n^c - L\left(u^* - u_n^c\right) + L\left(I - Q_d\right)\mathcal{K}u_n^c\\ &= \mathcal{K}u^* - \mathcal{K}u_n^c - L\left(u^* - u_n^c\right)\\ &\quad - L\left(I - Q_d\right)\left(\mathcal{K}u^* - \mathcal{K}u_n^c - L\left(u^* - u_n^c\right)\right)\\ &\quad + L\left(I - Q_d\right)\left(\mathcal{K}u^* - L\left(u^* - u_n^c\right)\right)\\ &= \Big(I - L\left(I - Q_d\right)\Big)\left(\mathcal{K}u^* - \mathcal{K}u_n^c - L\left(u^* - u_n^c\right)\right)\\ &\quad + L\left(I - Q_d\right)\mathcal{K}u^*\\ &\quad - L\left(I - Q_d\right)L\left(u^* - u_n^c\right).\end{aligned}$$

Multiplying by $(I - L)^{-1}$, we find

$$\begin{aligned}u^* - \hat{u}_n^c &= \left(I - (I - L)^{-1}LQ_d\right)\left(\mathcal{K}u^* - \mathcal{K}u_n^c - L\left(u^* - u_n^c\right)\right)\\ &\quad + (I - L)^{-1}L\left(I - Q_d\right)\mathcal{K}u^*\\ &\quad - (I - L)^{-1}L\left(I - Q_d\right)L\left(u^* - u_n^c\right).\end{aligned}$$

Therefore,

$$\left\|u^* - \hat{u}_n^c\right\|_\infty \leq \tilde{\xi} r_n \left\|u^* - u_n^c\right\|_\infty + a \left\|L(I - Q_d)\mathcal{K}u^*\right\|_\infty + a \left\|L(I - Q_d)L(u^* - u_n^c)\right\|_\infty$$

Which completes the proof of the theorem. □

Now, we show a preliminary result before stating the main theorem of this subsection.

Lemma 12.3 *Let Q_d be a quasi-interpolant of even degree d defined on the uniform partition of the interval $[a, b]$ of meshlength h. Assume that $f, g, k \in C[a, b]$ and $\frac{\partial \psi}{\partial u} \in C^{d+2}\big([a, b] \times [a, b]\big)$. Then we have*

$$\left\| L\big(I - Q_d\big)g \right\|_\infty = O\big(h^{d+2}\big).$$

Proof By definition of L we have:

$$\begin{aligned}\Big(L\left(I - Q_d\right)g\Big)(s) &= \int_a^b k(s,t)\frac{\partial \psi}{\partial u}\big(t, u^*(t)\big)\left(I - Q_d\right)g(t)dt \\ &= \int_a^b q(s,t)\left(I - Q_d\right)g(t)dt.\end{aligned}$$

where $q(s,t) = k(s,t)\frac{\partial \psi}{\partial u}\big(t, u^*(t)\big)$.
This error corresponds to the error of the quadrature formula $\mathcal{I}_{Q_d}$ based on a quasi-interpolant Q_d with a certain weight function $q(s,t)$, when from the Sect. 12.3, if d is even, the order of convergence h^{d+2} is obtained. □

Theorem 12.7 *Assume that the assumptions of Theorem 12.6 are satisfied. Then the iterated solution of the collocation type method based on quasi-interpolant Q_d of even degree d satisfies*

$$\left\| u^* - \hat{u}_n^c \right\|_\infty = O\big(h^{d+2}\big).$$

Proof From Theorem 12.6, we have

$$\begin{aligned}\left\| u^* - \hat{u}_n^c \right\|_\infty \le \xi \left\| u^* - u_n^c \right\|_\infty^2 &+ a\left\| L\big(I - Q_d\big)\mathcal{K}u^* \right\|_\infty \\ &+ a\left\| L\big(I - Q_d\big)L\big(u^* - u_n^c\big) \right\|_\infty.\end{aligned}$$

On the one hand, using the error of the approximate solution and the previous lemma, it holds

$$\left\| u^* - u_n^c \right\|_\infty^2 = O\big(h^{2d+2}\big), \tag{12.24}$$

$$\left\| L\big(I - Q_d\big)\mathcal{K}u^* \right\|_\infty = O\big(h^{d+2}\big). \tag{12.25}$$

On the other hand, we have

$$\left\| L\big(I - Q_d\big)L\big(u^* - u_n^c\big) \right\|_\infty \le \xi \left\| \big(I - Q_d\big)L \right\|_\infty \left\| u^* - u_n^c \right\|_\infty.$$

Furthermore, it is easy to see that

$$\left\|(I - Q_d)L\right\|_\infty = O(h^{d+1}).$$

Then

$$\left\|L(I - Q_d)L(u^* - u_n^c)\right\|_\infty = O(h^{2d+2}). \tag{12.26}$$

Using (12.24), (12.25) and (12.26), we deduce that

$$\|u^* - \hat{u}_n^c\|_\infty = O(h^{d+2}).$$

which completes the proof of theorem. □

We recall that Q_2 is superconvergent on the set of evaluation points $\mathcal{X}_n$ (from Theorem 12.1). Therefore the following corollary holds.

Corollary 12.1 *Let u_n^c be collocation approximate solution obtained by using the spline quasi-interpolant Q_2. Then, we have the following superconvergence properties*

$$\begin{aligned} |u^*(x_i) - u_n^c(x_i)| &= O(h^4), \quad 0 \le i \le n,\ i \ne 1, n-1. \\ |u^*(t_i) - u_n^c(t_i)| &= O(h^4), \quad 0 \le i \le n+1,\ i \ne 1, 2, n-1, n. \end{aligned} \tag{12.27}$$

Proof From the Eq. (12.16) and Theorem 12.1 we obtain

$$\begin{aligned} |u^*(\xi_i) - u_n^c(\xi_i)| &= |u^*(\xi_i) - Q_2\hat{u}_n^c(\xi_i)| \\ &\le |u^*(\xi_i) - \hat{u}_n^c(\xi_i)| + |\hat{u}_n^c(\xi_i) - Q_2\hat{u}_n^c(\xi_i)| \\ &= O(h^4), \end{aligned}$$

where ξ_i are either x_i or t_i given in (12.27), hence the result. □

12.6 Numerical Results

In this section, we consider three examples of Hammerstein integral equations to illustrate the theory established in previous sections for collocation-type method, its iterated solution and Nyström method. As quasi- interpolating operators we use those given by (12.4) and (12.5) for the quadratic case, and by (12.6) and (12.7) for the cubic case. Note that the different nonlinear systems were solved using a *Newton-Raphson* algorithm.

For successively doubled values of n, we compute the maximum absolute errors

$$E^c_\infty := \|u^* - u^c_n\|_\infty, \; \hat{E}^c_\infty := \|u^* - \hat{u}^c_n\|_\infty, \; E^N_\infty := \|u^* - u^N_n\|_\infty,$$

and the maximum absolute error at the superconvergent points given by

$$ES^c = \max_i |u^*(\xi_i) - u^c_n(\xi_i)|.$$

where ξ_i are either x_i or t_i given in (12.27). Moreover, we present the corresponding numerical convergence orders *NCO*, obtained by the logarithm to base 2 of the ratio between two consecutive errors.

The following table gives the data of the three examples of equations considered. For all theses examples, we note that $[a, b] = [0, 1]$.

	Kernel k	Function ψ	Second member f	Exact solution u^*
Example 1	$\pi x \sin(\pi t)$	$\dfrac{t}{1+(u(t))^2}$	$\sin\left(\frac{\pi}{2}x\right) - 2x\ln(24-16\sqrt{2})$	$\sin\left(\frac{\pi}{2}x\right)$
Example 2	$x + 2\sin\left(\frac{\pi t}{4}\right)$	$-(u(t))^2$	$\dfrac{4(4-\sqrt{2})+3(2+\pi)x}{6\pi} + \cos\left(\frac{\pi}{4}x\right)$	$\cos\left(\frac{\pi}{4}x\right)$
Example 3	xt	$(u(t))^3$	$\exp(-2x) - \frac{1}{36}\left(1-\frac{7}{\exp(6)}\right)$	$\exp(-2x)$

- **Case of quadratic quasi-interpolants Q_2**
  The obtained results are reported in Tables 12.1, 12.2 and 12.3, which confirm the theoretical convergence orders predicted theoretically for each method. Moreover, we notice that the approximate collocation solution is superconvergent at x_i and t_i as stated in Corollary 12.1.
- **Case of cubic quasi-interpolants Q_3**
  The obtained results are reported in Tables 12.4, 12.5 and 12.6. In this case the degree of the quasi-interpolant is odd and the theoretical results obtained previously are well confirmed.

Table 12.1 E^c_∞, ES^c, $\hat{E}^c_\infty$, E^N_∞ and corresponding *NCO*

Example 1								
n	E^c_∞	*NCO*	ES^c	*NCO*	$\hat{E}^c_\infty$	*NCO*	E^N_∞	*NCO*
8	1.25(−04)	–	2.53(−05)	–	9.09(−06)	–	5.83(−05)	–
16	1.57(−05)	2.99	1.62(−06)	3.96	5.84(−07)	3.96	2.23(−06)	4.71
32	1.97(−06)	3.00	1.02(−07)	3.99	3.61(−08)	4.02	8.82(−08)	4.66
64	2.29(−07)	3.10	6.73(−09)	3.92	1.86(−09)	4.28	3.85(−09)	4.52
Theoretical value	–	03	–	04	–	04	–	04

Table 12.2 E^c_∞, ES^c, $\hat{E}^c_\infty$, E^N_∞ and corresponding *NCO*

Example 2								
n	E^c_∞	*NCO*	ES^c	*NCO*	$\hat{E}^c_\infty$	*NCO*	E^N_∞	*NCO*
8	1.08(−05)	–	8.10(−06)	–	6.71(−07)	–	6.17(−06)	–
16	1.37(−06)	2.97	1.06(−06)	2.94	4.36(−08)	3.95	4.08(−07)	3.92
32	1.74(−07)	2.98	1.35(−07)	2.97	2.65(−09)	4.04	2.61(−08)	3.97
64	2.01(−08)	3.10	1.69(−08)	3.00	1.62(−10)	4.03	1.64(−09)	3.98
Theoretical value	–	03	–	04	–	04	–	04

Table 12.3 E^c_∞, ES^c, $\hat{E}^c_\infty$, E^N_∞ and corresponding *NCO*

Example 3								
n	E^c_∞	*NCO*	ES^c	*NCO*	$\hat{E}^c_\infty$	*NCO*	E^N_∞	*NCO*
8	2.18(−04)	–	5.63(−05)	–	1.83(−06)	–	5.88(−05)	–
16	2.96(−05)	2.88	4.49(−06)	3.65	2.33(−07)	2.98	5.12(−06)	3.52
32	3.88(−06)	2.93	4.83(−07)	3.22	2.16(−08)	3.43	3.77(−07)	3.76
64	4.63(−07)	3.07	5.68(−08)	3.09	1.15(−09)	4.23	2.56(−08)	3.88
Theoretical value	–	03	–	04	–	04	–	04

Table 12.4 E^c_∞, $\hat{E}^c_\infty$, E^N_∞ and corresponding *NCO*

Example 1						
n	E^c_∞	*NCO*	$\hat{E}^c_\infty$	*NCO*	E^N_∞	*NCO*
8	7.07(−05)	–	1.15(−05)	–	6.40(−04)	–
16	4.66(−06)	3.92	8.43(−07)	3.77	3.76(−05)	4.09
32	2.93(−07)	3.99	5.49(−08)	3.94	1.48(−06	4.66
64	1.85(−08)	3.98	3.60(−09)	3.93	5.58(−08)	4.73
Theoretical value	–	04	–	04	–	04

Table 12.5 E^c_∞, $\hat{E}^c_\infty$, E^N_∞ and corresponding *NCO*

Example 2						
n	E^c_∞	*NCO*	$\hat{E}^c_\infty$	*NCO*	E^N_∞	*NCO*
8	4.38(−06)	–	1.18(−06)	–	2.91(−05)	–
16	2.93(−07)	3.90	1.13(−07)	3.38	2.59(−06)	3.49
32	1.91(−08)	3.94	8.29(−09)	3.77	1.80(−07)	3.84
64	1.25(−09)	3.94	5.77(−10)	3.84	1.17(−08)	3.94
Theoretical value	–	04	–	04	–	04

12.7 Conclusions

In this paper we have proposed the Nyström and collocation type methods based on the quasi-interpolants Q_d, and the iterated solution in order to numerically solve the Hammerstein equation, and we also have studied their order of convergence.

Table 12.6 E^c_∞, $\hat{E}^c_\infty$, E^N_∞ and corresponding $\mathcal{NCO}$

Example 3						
n	E^c_∞	$\mathcal{NCO}$	$\hat{E}^c_\infty$	$\mathcal{NCO}$	E^N_∞	$\mathcal{NCO}$
8	1.19(−04)	–	5.98(−06)	–	2.18(−05)	–
16	8.68(−06)	3.76	5.41(−07)	3.46	1.55(−05)	0.49
32	5.81(−07)	3.90	3.77(−08)	3.84	1.85(−06)	3.07
64	3.78(−08)	3.94	2.52(−09)	3.90	1.54(−07)	3.58
Theoretical value	–	04	–	04	–	04

Finally, we have presented some numerical examples, illustrating the approximation properties of the proposed methods.

References

1. Allouch, C., Sablonnière, P., Sbibih, D.: A modified Kulkarni's method based on a discrete spline quasi-interpolant. Math. Comput. Simul. **81**, 1991–2000 (2010)
2. Allouch, C., Sablonnière, P., Sbibih, D., Tahrichi, M.: Product integration methods based on discrete spline quasi-interpolants and application to weakly singular integral equations. J. Comput. Appl. Math. **233**(11), 2855–2866 (2010)
3. Allouch, C., Sbibih, D., Tahrichi, M.: Numerical solutions of weakly singular Hammerstein integral equations. Appl. Math. Comp. **329**, 118–128 (2018)
4. Allouch, C., Sbibih, D., Tahrichi, M.: Superconvergent Nyström and degenerate kernel methods for Hammerstein integral equations, J. Comput. Appl. Math. **258**, 30–41 (2014)
5. Allouch, C., Sbibih, D., Tahrichi, M.: Superconvergent product integration method for Hammerstein integral equations. J. integr. Eq. Appl. **30**(3), 1–28 (2019)
6. Atkinson, K.E.: A survey of numerical methods for solving nonlinear integral equations. J. Integr. Eq. Appl. **4**(1), 15–46 (1992)
7. Atkinson, K.E.: The Numerical Solution of Integral Equations of the Second Kind. Cambridge University Press, Cambridge (1997)
8. Atkinson, K.E., Potra, F.: Projection and iterated projection methods for nonlinear integral equations. SIAM J. Numer. Anal. 14, 1352–1373 (1987)
9. Dagnino, C., Remogna, S., Sablonnière, P.: On the solution of Fredholm equations based on spline quasi-interpolating projectors. BIT Numer. Math. **54**, 979–1008 (2014)
10. Dagnino, C., Dallefrate, A., Remogna, S.: Spline quasi-interpolating projectors for the solution of nonlinear integral equations. J. Comput. Appl. Math. **354**, 360–372 (2019)
11. Krasnoselskij, M.A., Vainikko, G.M., Zabreiko, P.P., Rutitskii, Y.B., Stetsenko, V.Y.: Approximate Solution of Operator Equations. Springer Science & Business Media, Berlin (1972)
12. Kumar, S.: Superconvergence of a collocation-type method for Hammerstein equations. IMA J. Numer. Anal. **7**, 313–325 (1987)
13. Kumar, S., Sloan, I.H.: A new collocation-type method for Hammerstein euqations. Math. Comput. **178**, 585–593 (1987)
14. Müller, P.H., Krasnosel'skii, M.A.: Topological methods in the theory of nonlinear integral equations. Zeitschrift Angewandte Mathematik und Mechanik **44**, 521–521 (1964)
15. O'Regan, D., Meehan, M.: Existence Theory for Nonlinear Integral and Integro-Differential Equations. Kluwer Academic, Dordrecht (1998)
16. Rahman, M.: Integral Equations and Their Applications. WIT Press, Ashurst (2007)

17. Sablonniére, P.: Quadratic spline quasi-interpolants on bounded domains of $\mathbb{R}^d$, d = 1, 2, 3. Rend. Sem. Mat. Univ. Pol. Torino **61**(3), 229–246 (2003)
18. Sablonniére, P.: Univariate spline quasi-interpolants and applications to numerical analysis (2005). arXiv Preprint math/0504022
19. Sablonniére, P.: A quadrature formula associated with a univariate quadratic spline quasi-interpolant. BIT Numer. Math. **47**, 825–837 (2007)
20. Sablonniére, P., Sbibih, D., Tahrichi, M.: Error estimate and extrapolation of a quadrature formula derived from a quartic spline quasi-interpolant. BIT Numer. Math. **50**(4), 843–862 (2010)
21. Sablonniére, P., Sbibih, D., Allouch, C.: Solving fredholm integral equations by approximating kernels by spline quasi-interpolants. Num. Algorithms **56**(3), 437–453 (2011)
22. Sablonniére, P., Sbibih, D., Tahrichi, M.: High-order quadrature rules based on spline quasi-interpolants and application to integral equations. Appl. Num. Math. **62**(5), 507–520 (2012)
23. Sloan, I.H.: Improvement by iteration for compact operator equations. Math. Comput. **30**(136), 758–764 (1976)
24. Vainikko, G.M.: Galerkin's perturbation method and the general theory of approximate methods for non-linear equation. USSR Comput. Math. Math. Phys. **7**(4), 1–41 (1967)